Macromolecules Incorporating Transition Metals
Tackling Global Challenges

Polymer Chemistry Series

Editor-in-chief:
Ben Zhong Tang, *The Hong Kong University of Science and Technology, Hong Kong, China*

Series editors:
Alaa S. Abd-El-Aziz, *University of Prince Edward Island, Canada*
Jianhua Dong, *National Natural Science Foundation of China, China*
Jeremiah A. Johnson, *Massachusetts Institute of Technology, USA*
Toshio Masuda, *Shanghai University, China*
Christoph Weder, *University of Fribourg, Switzerland*

Titles in the series:
1: Renewable Resources for Functional Polymers and Biomaterials
2: Molecular Design and Applications of Photofunctional Polymers and Materials
3: Functional Polymers for Nanomedicine
4: Fundamentals of Controlled/Living Radical Polymerization
5: Healable Polymer Systems
6: Thiol-X Chemistries in Polymer and Materials Science
7: Natural Rubber Materials: Volume 1: Blends and IPNs
8: Natural Rubber Materials: Volume 2: Composites and Nanocomposites
9: Conjugated Polymers: A Practical Guide to Synthesis
10: Polymeric Materials with Antimicrobial Activity: From Synthesis to Applications
11: Phosphorus-Based Polymers: From Synthesis to Applications
12: Poly(lactic acid) Science and Technology: Processing, Properties, Additives and Applications
13: Cationic Polymers in Regenerative Medicine
14: Electrospinning: Principles, Practice and Possibilities
15: Glycopolymer Code: Synthesis of Glycopolymers and their Applications

16: Hyperbranched Polymers: Macromolecules in-between Deterministic Linear Chains and Dendrimer Structures
17: Polymer Photovoltaics: Materials, Physics, and Device Engineering
18: Electrical Memory Materials and Devices
19: Nitroxide Mediated Polymerization: From Fundamentals to Applications in Materials Science
20: Polymers for Personal Care Products and Cosmetics
21: Semiconducting Polymers: Controlled Synthesis and Microstructure
22: Bio-inspired Polymers
23: Fluorinated Polymers: Volume 1: Synthesis, Properties, Processing and Simulation
24: Fluorinated Polymers: Volume 2: Applications
25: Miktoarm Star Polymers: From Basics of Branched Architecture to Synthesis, Self-assembly and Applications
26: Mechanochemistry in Materials
27: Macromolecules Incorporating Transition Metals: Tackling Global Challenges

How to obtain future titles on publication:
A standing order plan is available for this series. A standing order will bring delivery of each new volume immediately on publication.

For further information please contact:
Book Sales Department, Royal Society of Chemistry, Thomas Graham House, Science Park, Milton Road, Cambridge, CB4 0WF, UK
Telephone: +44 (0)1223 420066, Fax: +44 (0)1223 420247
Email: booksales@rsc.org
Visit our website at www.rsc.org/books

Macromolecules Incorporating Transition Metals
Tackling Global Challenges

Alaa S. Abd-El-Aziz
University of Prince Edward Island, Canada
Email: abdelaziz@upei.ca

Christian Agatemor
Harvard University, USA
Email: christmuno@seas.harvard.edu

and

Wai-Yeung Wong
Hong Kong Polytechnic University, Hong Kong
Email: wai-yeung.wong@polyu.edu.hk

ROYAL SOCIETY OF CHEMISTRY

THE QUEEN'S AWARDS FOR ENTERPRISE: INTERNATIONAL TRADE 2013

Polymer Chemistry Series No. 27

Print ISBN: 978-1-78262-899-6
PDF ISBN: 978-1-78801-036-8
EPUB ISBN: 978-1-78801-412-0
ISSN: 2044-0790

A catalogue record for this book is available from the British Library

The Royal Society of Chemistry is a charity, registered in England and Wales, Number 207890, and a company incorporated in England by Royal Charter (Registered No. RC000524), registered office: Burlington House, Piccadilly, London W1J 0BA, UK, Telephone: +44 (0) 207 4378 6556.

For further information see our web site at www.rsc.org

Printed in the United Kingdom by CPI Group (UK) Ltd, Croydon, CR0 4YY, UK

Preface

Incorporating transition metals into macromolecules is a proven strategy to access functional materials and is a vibrant research field in materials science. In this area, fundamental concepts of inorganic, synthetic, physical, analytical, polymer, and materials chemistry merge and are integrated with engineering models to construct materials with implications for various global challenges. The reactivity of transition metals permits their installation into macromolecular scaffolds to create platforms in which the functionality of the metal center is pivotal to the function of these macromolecules as catalysts, sensors, imaging probes, therapeutic agents, and intelligent drug delivery systems. Demonstrated successes of this strategy are emerging and are close to being translated from the laboratory to the market as exemplified by the phase II clinical trial of ProLindac™, a platinum-containing polymer-based anticancer prodrug.

The first chapter of this monograph outlines some of the global challenges and how materials, specifically transition-metal-containing macromolecules, offer innovative solutions to these problems. Chapter 2 focuses on the properties of transition metal complexes that influence their exploration in the design of functional macromolecules and Chapter 3 presents synthetic strategies used to construct these entities. Subsequent chapters discuss the applications of these macromolecules in energy, medicine, the environment, catalysis, and data storage. We hope that this monograph provides readers with an understanding of the fundamentals and applications of transition-metal-containing macromolecules as well as inspires more research to deepen our understanding and broaden the scope of applications.

Alaa S. Abd-El-Aziz, Christian Agatemor and Wai-Yeung Wong

Polymer Chemistry Series No. 27
Macromolecules Incorporating Transition Metals: Tackling Global Challenges
By Alaa S. Abd-El-Aziz, Christian Agatemor and Wai-Yeung Wong

Published by the Royal Society of Chemistry, www.rsc.org

Contents

Polymer Chemistry Series No. 27
Macromolecules Incorporating Transition Metals: Tackling Global Challenges
By Alaa S. Abd-El-Aziz, Christian Agatemor and Wai-Yeung Wong

Published by the Royal Society of Chemistry, www.rsc.org

CHAPTER 1

Introduction

1.1 Introduction

The contemporary positive transitions in globalization, economic prosperity, and quality of life mark an epoch in human evolution, as do the aftermaths of these changes, which ironically threaten the ecosystem, the planet and human survival.[1,2] Exponential growth on all fronts of human activities has, indeed, benefited our social, psychological, and biological well-being but has led to enormous pressures on the Earth's resources and the biosphere, pushing the natural system beyond sustainable limits, and challenging our achievements and survival. Climate change, food scarcity, diseases, dependence on unsustainable energy resources, and a widening gap between the rich and the poor are notable examples of these challenges (Figure 1.1).[2–16] A paradigm shift in our development strategies is crucial to a sustainable future, and indeed, proactive initiatives have been taken by several stakeholders. The United Nations' Sustainable Development Goals agreed upon by the 193 member states in September 2015 exemplify initiatives that seek to address many of the challenges that face contemporary humanity.[11] Materials scientists have also envisaged these problems for decades and have taken concrete steps to help realize a sustainable future.[1,5,7,17–27] This monograph presents selected contributions from materials scientists, specifically the research and development that uses transition-metal-containing macromolecules to address some of these challenges. Here, the focus is on three challenges: energy, climate change and the burden of diseases, which are central and interconnected to the overall problems facing humanity. For instance, climate change affects food security and biodiversity, energy supply is a major driver of global security, and diseases are a burden to the economy and adversely affect the holistic well-being of the individual.

Polymer Chemistry Series No. 27
Macromolecules Incorporating Transition Metals: Tackling Global Challenges
By Alaa S. Abd-El-Aziz, Christian Agatemor and Wai-Yeung Wong

Published by the Royal Society of Chemistry, www.rsc.org

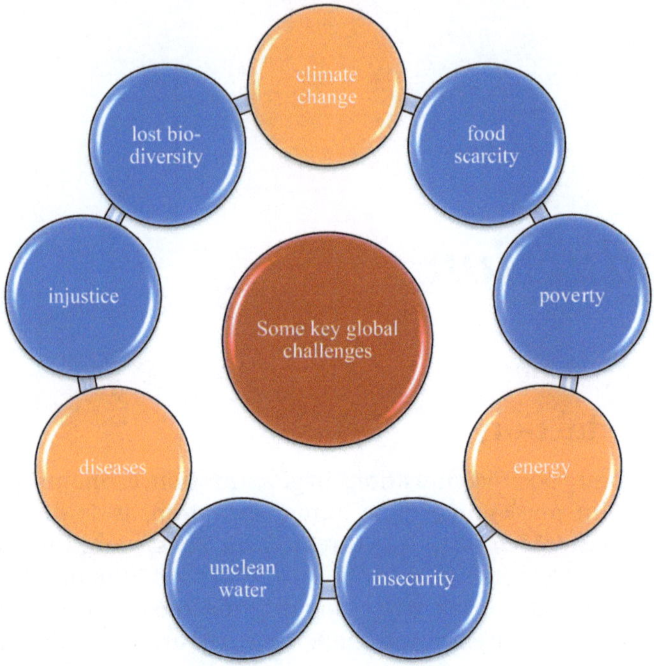

Figure 1.1 Some of the problems facing humanity in the 21st century with those discussed in this monograph being shown in the orange boxes.

1.2 Key Global Challenges

1.2.1 Diseases

Science has made unprecedented advancement in disease prevention, diagnosis, and treatment, which has resulted in a significant reduction in mortality and an increase in life expectancy. These achievements notwithstanding, a significant proportion of the world's population are yet to enjoy good health and well-being as exemplified by the 14.1 million new cancer cases diagnosed each year, 387 million people living with diabetes, and 1.1 billion people having high blood pressure.[28] Also, the progress in medicine is challenged by the substantial number of premature deaths, for instance, the over 15.8 million premature deaths from noncommunicable diseases such as cancer in 2015 (Figure 1.2).[29] The majority of these premature deaths are preventable if a healthier lifestyle is adopted or quality health care is provided. The scenario is bleak in developing countries where 82% of the deaths from noncommunicable diseases occur, and a projected $7 trillion USD will be lost over the next 13 years if the *status quo* remains.[30] While the situation is better in developed countries whose governments implement policies that reduce risk factors and enable viable health care systems, it is still imperative that the positive epidemiological transition in these countries be sustained and copied in developing countries.

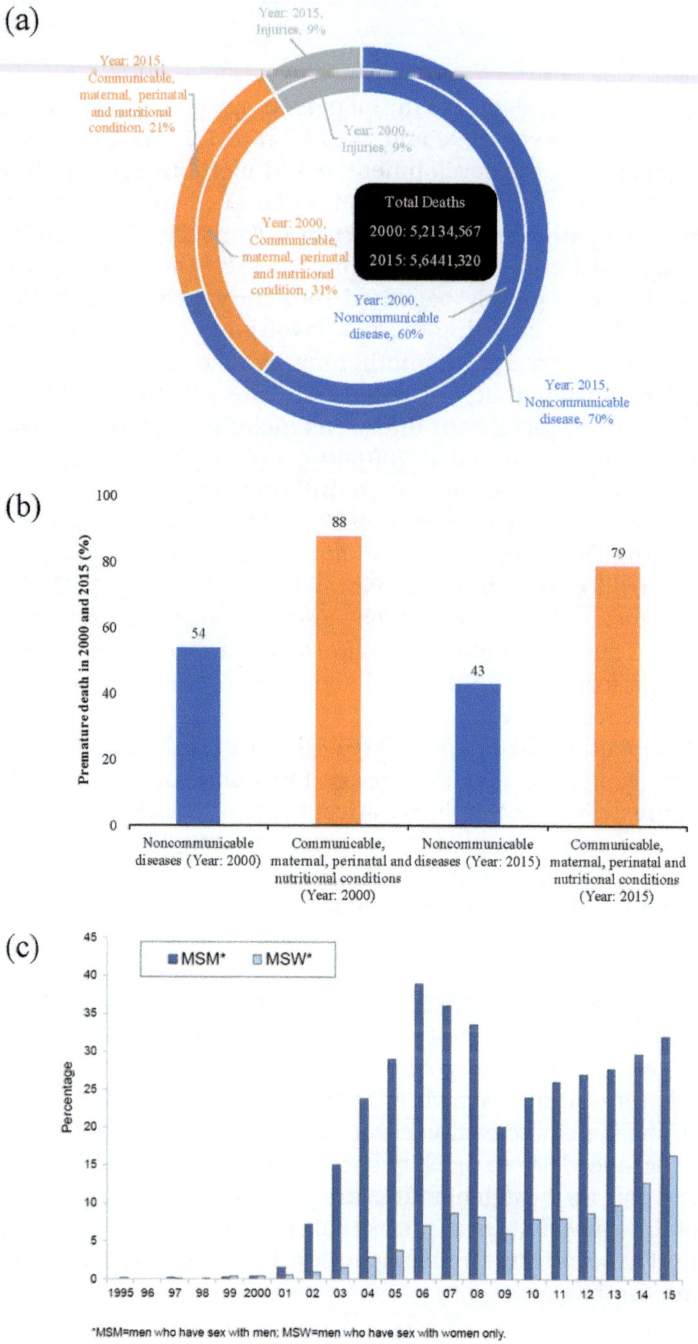

Figure 1.2 (a) Estimated global deaths and (b) global premature deaths from communicable and noncommunicable diseases in 2000 and 2015; (c) ciprofloxacin-resistant *Neisseria gonorrhoeae* isolates in the United States from 1995 to 2015. (a) and (b) Adapted from World Health Organization, Health Statistics and Information System. Estimates for 2000–2015, 2017,[29] © WHO 2017. (c) Reproduced from *Sexually Transmitted Disease Surveillance 2015: Gonococcal Isolate Surveillance Project (GISP) Supplement and Profiles*, Centers for Disease Control and Prevention, Atlanta: U.S. Department of Health and Human Services; 2017.

The commendable decrease in the percentage of deaths from communication diseases between 2000 and 2015 (Figure 1.2)[29] is due to advances in medicine, including the development of antimicrobial agents. However, the evolutionary process of natural selection coupled with the selective pressure exerted on pathogens by the overuse and misuse of drugs trigger a new threat: drug-resistant pathogens (Figure 1.2).[8,9,12] Undoubtedly, drugs such as antibiotics have greatly benefited humanity, protecting patients from fatal illnesses and mitigating the risk involved in complex medical procedures such as surgery and chemotherapy, but the emergence of resistance threatens the sustainability of these erstwhile effective medical interventions. Globally, drug-resistant infections including malaria, HIV/AIDS, and tuberculosis kill an estimated 700 000 people yearly, and the number of deaths is projected to increase to 10 million yearly by 2050 if no pre-emptive action is taken.[12] The burden of resistance extends beyond mortality, encompassing the economic, social and security landscapes with an estimated 100 trillion USD of the economy being at risk by 2050 if no action is taken to tackle the problem.[12] The efforts of materials scientists on the research and development of new, affordable and accessible antimicrobial agents/drugs form part of a global action plan to mitigate the burden of antimicrobial resistance.

The therapeutic efficacy and safety of drugs directly depend on the mode of delivery, which includes the dosage form and route of administration. Indeed, drug delivery technologies affect the pharmacokinetics and pharmacodynamics of drugs as well as convenience and compliance of patients.[18] The discovery of a new drug requires the development of a delivery technology that enables the therapeutic function and ensures patients' safety and compliance. Current research on the development of drug delivery technologies seeks to localize therapeutic function at the target cells because this approach maximizes the effect and minimizes the toxicity. The administration of a new generation of highly potent and useful drugs such as peptides, proteins, genes and polymer-based drugs remains challenging due to the inherent problems such as enzymatic degradation, renal clearance, and immune system activation associated with state-of-the-art delivery systems.[18,31–36] Indeed, these delivery barriers limit the clinical impact of many drugs and call for next-generation drug delivery technologies that ensure drug stability, specificity, and precise control over drug release, with the ultimate effect of localizing the therapeutic effects and guaranteeing patients' safety and compliance.

1.2.2 Energy and Climate Change

The prosperity and comfort inherent in contemporary lifestyles are strongly dependent on affordable and accessible energy. With a fast-growing population and economic prosperity, especially in developing countries, energy demand is expected to increase. The United States Energy Information Administration expects a 48% increase in 28 years, from the 549 quadrillion

British thermal units (Btu) of energy consumed globally in 2012 to 815 quadrillion Btu in 2040 (Figure 1.3).[6] Most likely, an extent of the projected increase could be offset by energy-efficient technologies as governments promote awareness about sustainability but a major energy crisis will persist if we maintain the existing situation. Currently, fossil fuels provide more than 80% of the global energy demand,[7] but the non-renewability of fossil fuel within a realistic time frame poses a sustainability issue. Though global fossil fuel resources are of a sufficient quantity to meet increasing demand in the short term, converting these resources into reserves in an economical and eco-friendly approach requires advanced innovative solutions that could hike fuel prices and eventually challenge sustainable development. Another key issue with fossil fuel technologies is the comparatively high carbon dioxide footprint, which results from the combustion process and contributes significantly to greenhouse gas (GHG) concentrations in the atmosphere.[1,5,7,22,37] The burning of fossil fuel also releases other GHGs including methane and nitrous oxide into the atmosphere, further contributing to the total concentration of GHGs in the atmosphere, which has steadily increased to an unprecedented level over the last 800 000 years.[7,14] A growing amount of empirical evidence implicates these anthropogenic GHGs, mainly carbon dioxide from fossil fuel, as unequivocal drivers of global warming, which causes climate change.[1,5,7,14,17,20,22,38,39] The threats posed by climate change to the planet, ecosystem and human survival are wide-ranging, affecting the physical, biological, social, and economic landscapes (Figure 1.3).[14]

1.3 Meeting the Global Challenges with Materials Science

Since humans have walked the planet, the fundamental knowledge of materials (ceramics, metals, and polymers) has contributed immensely to their survival. Evolutionists connect the emergence of the genus *Homo* with the development of sharp-edged stone tools among other factors such as climate change.[40,41] Archaeological evidence suggests that the dexterous exploitation of the macroscopic properties of materials by hominins predates the emergence of *Homo sapiens*, our species. Exemplarily, fossil artifacts revealed the use of mussel shells for tool making and engraving by *Homo erectus* about 0.43 ± 0.05 million years ago,[42] and the use of stones as anvils, hammers, and cutters by hominins assumed to be *Kenyanthropus platyops* or *Australopithecus afarensis* 3.3 million years ago (Figure 1.4).[40] The use of materials as technological solutions to routine chores and extraordinary challenges is, probably, as old as the evolution of hominins, and has shaped our evolution and enabled the conquest of the harsh environment.

Materials science as a scientific discipline is comparatively new and has evolved beyond mere serendipitous discoveries within our environment to encompass, as a central feature, the deliberate manipulation of the

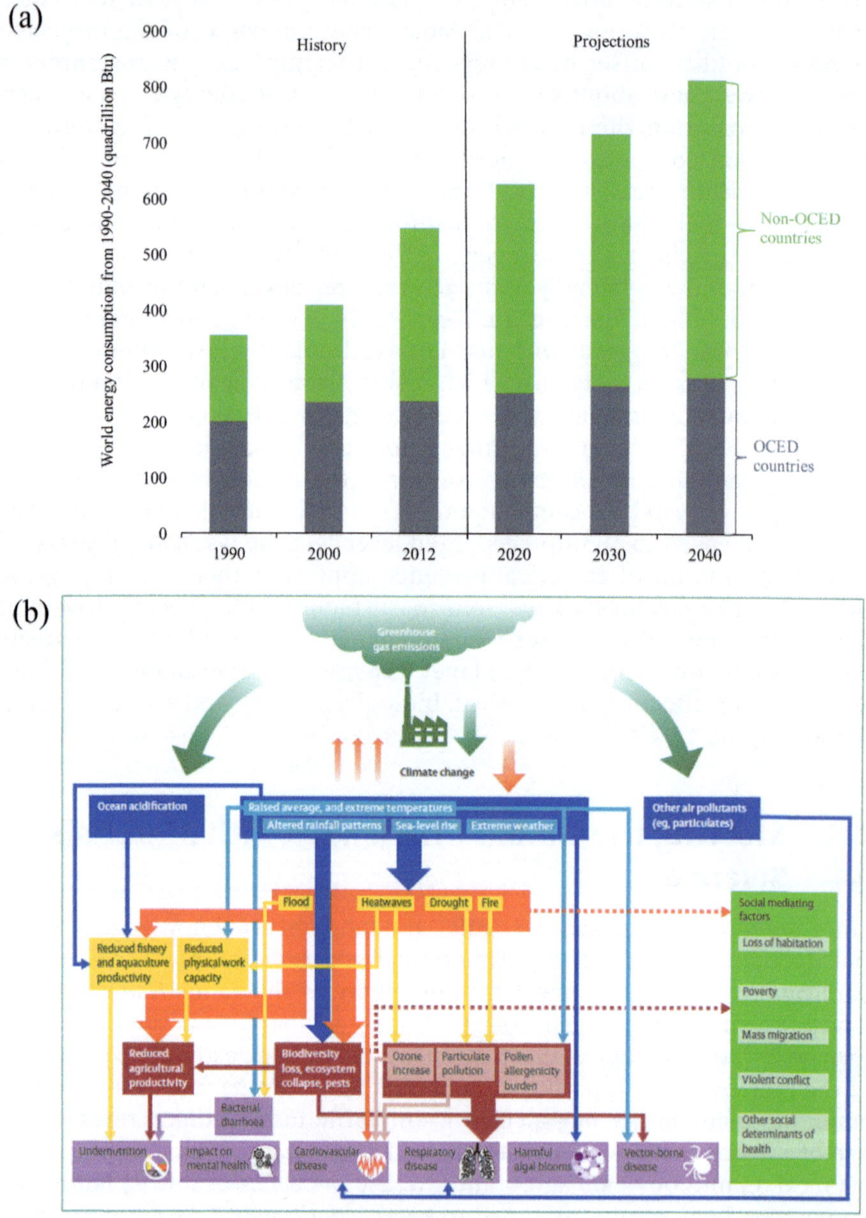

Figure 1.3 (a) Historical and projected global energy consumption. (b) Effect of climate change. (a) Adapted from International Energy Outlook 2016 with Projections to 2040, United States Energy Information Administration, 2016,[6] courtesy of the United States Energy Information Administration. (b) Reprinted from *The Lancet*, **386**, N. Watts, W. N. Adger, P. Agnolucci *et al.*, Health and climate change: policy responses to protect public health, 1861–1914,[14] with permission from Elsevier.

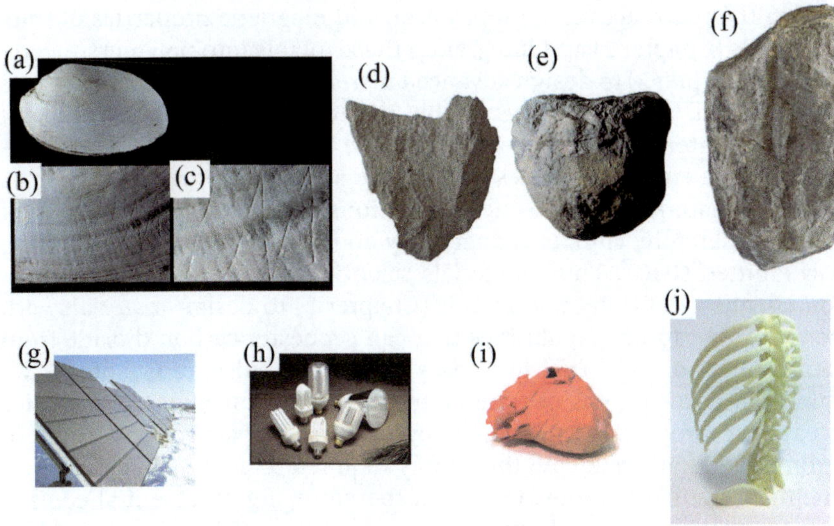

Figure 1.4 Some past, present, and future materials-based technologies that improve the quality of life. (a)–(c) Mussel shells used by *Homo erectus* about 0.43 ± 0.05 million years ago, (d)–(f) stones: anvil, hammer, and cutter supposedly used by *Kenyanthropus platyops* or *Australopithecus afarensis*, (g) photovoltaics solar energy panel, (h) energy efficient light bulbs, (i) 3D printed heart model, and (j) 3D printed ribs. (a)–(c) Reproduced by permission from Macmillan Publishers Ltd: *Nature*, (ref. 42), copyright 2015. (d)–(f) Reproduced by permission from Macmillan Publishers Ltd: Nature (ref. 40), copyright 2015. (g) and (h) reproduced from the National Renewable Energy Laboratory (NREL) Image Gallery (https://images.nrel.gov/bp/#/folder/207386/). (i) and (j) Reproduced with permission from the National Institute of Health (NIH) Image Gallery.

microstructure of matter to create custom-made or new-to-the-world materials for task-specific applications (Figure 1.4). Modern materials science is an interdisciplinary field that explores the fundamentals of physics, chemistry, and engineering to design, synthesize and fabricate ceramics, metals, and polymers as well as composites materials with wide-ranging implications for diverse facets of life including medicine, energy, and the environment. The ability to manipulate the microstructure of matter to create new materials enables innovative solutions to the global challenges of diseases, energy, and climate change. On energy, the United States Department of Energy's Basic Energy Sciences Advisory Committee asserted in a 2003 report that "Many of the current technological barriers related to energy hinge on improved materials. Thus, materials research is an area in which scientific advances could have a key impact on future energy security."[4] To keep the energy crisis in perspective, scientists must recognize that the goal is not just fossil fuel alternatives, rather affordable, accessible and clean energy that improves the quality of life of the 7.5 billion people on earth and sustain the earth's natural system for future generations. Towards this goal, materials scientists are exploiting the redox-active, photophysical, and magnetic properties of transition metals (Chapter 2) and integrating these metals into polymers/macromolecules (Chapter 3) to design advanced materials. The inherent properties of the metals in these materials enable the realization of energy-efficient and storage materials (Chapter 4). Also, these materials can efficiently harness renewable energy resources such as the sun and biomass (Chapter 4). While these materials provide energy solutions that reduce the emission of GHGs, and therefore, climate change, they do not mitigate the effect of previously emitted GHGs. Thus, materials scientists are exploring the coordination chemistry of transition metals (Chapter 2) to design materials with carbon dioxide capture capabilities that can sequester carbon dioxide from the atmosphere or industrial flue gases (Chapter 6). Also, in Chapter 6, we examine the development of transition metal-containing macromolecules with catalytic properties. Indeed, catalysts lower the activation energy of reactions, eventually reducing the energy required to achieve chemical and biochemical transformations. Catalysts, therefore, are an essential component of the sustainable development action plan since the chemical and biochemical industry are pivotal to economic prosperity.

Materials scientists are among the pioneers of innovative solutions to health care problems. By exploiting the redox-active, photophysical, and magnetic properties of transition metals, materials scientists are creating new materials with intelligent behavior, bioactivity, and sensing capability (Chapter 5). Materials with redox-active metals are rationally programmed to feature redox-responsive behavior that is integrated into drug delivery systems to realize precise control over drug release and to target specific diseased sites. The redox activity is strongly connected with bioactivity and is exploited to tackle drug-resistant pathogens and to treat cancer *via* redox-induced generation of reactive oxygen species. The photophysical and magnetic properties of transition metals are integrated into macromolecular frameworks to realize bioimaging technologies that enable deep tissue as well as 3D profiling of diseased tissues. Materials scientists are also at the frontline designing biosensors by incorporating redox-active and photoactive transition metals into macromolecular frameworks. Indeed, transition-metal-containing macromolecules enable the design of therapeutic and diagnostic technologies to combat challenging health care problems.

1.4 Conclusion

To summarize, humanity's achievements in the last decades are historically unprecedented, as are their side effects. Accelerated human activity, such as burning of fossil fuel to power industries, has afforded prosperity and comfort but has increased the level of GHGs to induce climate change with its mostly deleterious effects. Diseases are becoming increasingly difficult to treat due to misuse of antimicrobial agents, and the results are increased health care costs and premature deaths. Materials scientists are interested in solving these problems, and this monograph highlights their contributions towards a sustainable future.

References

1. D. Larcher and J. Tarascon, *Nat. Chem.*, 2015, **7**, 19.
2. A. J. McMichael, *N. Engl. J. Med.*, 2013, **368**, 1335.
3. World Health Organization, *Health in 2015 from Millennium Development Goals to Sustainable Development Goals*, World Health Organization, Geneva, Switzerland, 2015.
4. Oak Ridge National Laboratory of United States Department of Energy, *Basic Research Needs to Assure a Secure Energy Future, a Report from the Basic Energy Sciences Advisory Committee*, Oak Ridge National Laboratory, United States Department of Energy, Oak Ridge, Tennessee, United State, 2003.
5. D. Gielen, F. Boshell and D. Saygin, *Nat. Mater.*, 2016, **15**, 117.
6. United States Energy Information Administration, *International Energy Outlook 2016 with Projections to 2040*, United States Energy Information Administration, Washington, DC, United States, 2016.
7. S. Chu, Y. Cui and N. Liu, *Nat. Mater.*, 2017, **16**, 16.
8. World Health Organization, *Global Action Plan on Antimicrobial Resistance*, World Health Organization Document Production Service, Geneva, Switzerland, 2015.
9. L. J. Shallcross, S. J. Howard, T. Fowler and S. C. Davies, *Philos. Trans. R. Soc., B*, 2015, **370**, 20140082.
10. R. Sugden, R. Kelly and S. Davies, *Nat. Microbiol.*, 2016, **1**, 16187.
11. United Nations, *Transforming Our World: The 2030 Agenda for Sustainable Development*, United Nations, New York, United States, 2015.
12. Jim O'Neill (Chair), *Tackling Drug-resistant Infections Globally: Final Report and Recommendations, the Review on Antimicrobial Resistance*, 2016, https://amr-review.org/sites/default/files/160525_Final%20paper_with%20cover.pdf, accessed on August 16, 2017.
13. P. Vineis and C. P. Wild, *Lancet*, 2014, **383**, 549.
14. N. Watts, W. N. Adger, P. Agnolucci, J. Blackstock, P. Byass, W. Cai, S. Chaytor, T. Colbourn, M. Collins, A. Cooper, P. M. Cox, J. Depledge, P. Drummond, P. Ekins, V. Galaz, D. Grace, H. Graham, M. Grubb, A. Haines, I. Hamilton, A. Hunter, X. Jiang, M. Li, I. Kelman, L. Liang, M. Lott, R. Lowe, Y. Luo, G. Mace, M. Maslin, M. Nilsson, T. Oreszczyn, S. Pye, T. Quinn, M. Svensdotter, S. Venevsky, K. Warner, B. Xu, J. Yang, Y. Yin, C. Yu, Q. Zhang, P. Gong, H. Montgomery and A. Costello, *Lancet*, 2015, **386**, 1861.
15. C. L. Ventola, *P T.*, 2015, **40**, 277.
16. C. L. Ventola, *P T.*, 2015, **40**, 344.
17. S. A. Matlin, G. Mehta, H. Hopf and A. Krief, *Nat. Chem.*, 2015, **7**, 941.
18. M. W. Tibbitt, J. E. Dahlman and R. Langer, *J. Am. Chem. Soc.*, 2016, **138**, 704.
19. M. J. Webber, E. A. Appel, E. Meijer and R. Langer, *Nat. Mater.*, 2016, **15**, 13.
20. R. E. Smalley, *MRS Bull.*, 2005, **30**, 412.
21. B. C. Steele and A. Heinzel, *Nature*, 2001, **414**, 345.

22. M. Dresselhaus and I. Thomas, *Nature*, 2001, **414**, 332.
23. R. Langer and D. A. Tirrell, *Nature*, 2004, **428**, 487.
24. J. Wadsworth, *Mater. Sci. Forum*, 2007, **539**, 1.
25. P. Nugent, Y. Belmabkhout, S. D. Burd, A. J. Cairns, R. Luebke, K. Forrest, T. Pham, S. Ma, B. Space, L. Wojtas, M. Eddaoudi and M. J. Zaworotko, *Nature*, 2013, **495**, 80.
26. A. Polman, M. Knight, E. C. Garnett, B. Ehrler and W. C. Sinke, *Science*, 2016, **352**, 307.
27. A. S. Aricò, P. Bruce, B. Scrosati, J. Tarascon and W. Van Schalkwijk, *Nat. Mater.*, 2005, **4**, 366.
28. Centers for Disease Control and Prevention, *Global Noncommunicable Diseases: Infographics*, 2017, www.cdc.gov/globalhealth/healthprotection/ncd/resources/infographics.html, accessed on August 16, 2017.
29. World Health Organization, *Health Statistics and Information System. Estimates for 2000–2015*, 2017, www.who.int/healthinfo/global_burden_disease/estimates/en/index1.html, accessed on August 16, 2017.
30. World Health Organization, *Global Status Report on Communicable Diseases 2014*, World Health Organization Press, Geneva, Switzerland, 2014.
31. S. Mura, J. Nicolas and P. Couvreur, *Nat. Mater.*, 2013, **12**, 991.
32. E. Blanco, H. Shen and M. Ferrari, *Nat. Biotechnol.*, 2015, **33**, 941.
33. Z. Niu, I. Conejos-Sánchez, B. T. Griffin, C. M. O'Driscoll and M. J. Alonso, *Adv. Drug Delivery Rev.*, 2016, **106**, 337.
34. W. He, M. Parowatkin, V. Mailänder, M. Flechtner-Mors, R. Graf, A. Best, K. Koynov, K. Mohr, U. Ziener and K. Landfester, *Biomacromolecules*, 2015, **16**, 2282.
35. M. Karimi, A. Ghasemi, P. S. Zangabad, R. Rahighi, S. M. M. Basri, H. Mirshekari, M. Amiri, Z. S. Pishabad, A. Aslani, M. Bozorgomid, D. Ghosh, A. Beyzavi, A. Vaseghi, A. R. Aref, L. Haghani, S. Bahrami and M. R. Hamblin, *Chem. Soc. Rev.*, 2016, **45**, 1457.
36. X. J. Loh, T. Lee, Q. Dou and G. R. Deen, *Biomater. Sci.*, 2016, **4**, 70.
37. W. Moomaw, P. Burgherr, G. Heath, M. Lenzen, J. Nyboer and A. Verbruggen, Annex II: Methodology, in *IPCC Special Report on Renewable Energy Sources and Climate Change Mitigation*, ed. O. Edenhofer, R. Pichs-Madruga, Y. Sokona, K. Seyboth, P. Matschoss, S. Kadner, T. Zwickel, P. Eickemeier, G. Hansen, S. Schlömer and C. von Stechow, Cambridge University Press, Cambridge, United Kingdom and New York, USA, 2011.
38. *Intergovernmental Panel on Climate Change 2014, Climate Change 2014: Mitigation of Climate Change. Contribution of Working Group III to the Fifth Assessment Report of the Intergovernmental Panel on Climate Change*, ed. O. Edenhofer, R. Pichs-Madruga, Y. Sokona, E. Farahani, S. Kadner, K. Seyboth, A. Adler, I. Baum, S. Brunner, P. Eickemeier, B. Kriemann, J. Savolainen, S. Schlömer, C. von Stechow, T. Zwickel and J. C. Minx, Cambridge, United Kingdom and New York, USA, 2014.
39. D. Griggs, M. Stafford-Smith, O. Gaffney, J. Rockström, M. C. Öhman, P. Shyamsundar, W. Steffen, G. Glaser, N. Kanie and I. Noble, *Nature*, 2013, **495**, 305.

40. S. Harmand, J. E. Lewis, C. S. Feibel, C. J. Lepre, S. Prat, A. Lenoble, X. Boës, R. L. Quinn, M. Brenet, A. Arroyo, N. Taylor, S. Clément, G. Daver, J. Brugal, L. Leakey, R. A. Mortlock, J. D. Wright, S. Lokorodi, C. Kirwa, D. V. Kent and H. Roche, *Nature*, 2015, **521**, 310.
41. L. S. Leakey, P. V. Tobias and J. R. Napier, *Nature*, 1964, **202**, 7.
42. J. C. Joordens, F. d'Errico, F. P. Wesselingh, S. Munro, J. de Vos, J. Wallinga, C. Ankjærgaard, T. Reimann, J. R. Wijbrans, K. F. Kuiper, H. J. Mücher, H. Coqueugniot, V. Prié, I. Joosten, B. van der Haas, W. Lustenhouwer, J. J. G. Reijmer and W. Roebroeks, *Nature*, 2015, **518**, 228.

CHAPTER 2

Transition Metal Complexes as Attractive Motifs to Design Macromolecules

2.1 Introduction

Transition metal complexes, which include organometallic and coordination compounds, constitute a major class of compounds in chemistry and occupy a prominent niche in academic and industry research. Their significance is understandable given that transition metals represent more than half of the elements in the periodic table (Figure 2.1). Empirical data, as well as theoretical calculations, validate the unique redox activity, photoactivity, magnetic properties and coordination chemistry of these complexes. Undoubtedly, these properties inform the extensive exploration of these complexes as functional and structural motifs to design and construct macromolecules. Information about these properties is abundant in the literature,[1–5] but in this chapter, we focus on empirical data and fundamental concepts that lead chemists to explore these complexes as building blocks in macromolecules.

Whereas the use of molecular transition metal complexes is ideal in certain applications, the installation of these complexes into macromolecules extends the applications into the realms of materials science and engineering. As an example, redox-active transition metal complexes are incorporated into polymers to realize antimicrobial polymers, which can be fabricated into bioactive medical devices and implants to combat microbial infections.[6] A macromolecular scaffold provides a platform to amplify the functionality as several of the transition metal complexes can be grafted into a single scaffold. Also, the scaffold ensures chemical stability and prevents leaching, thereby

Polymer Chemistry Series No. 27
Macromolecules Incorporating Transition Metals: Tackling Global Challenges
By Alaa S. Abd-El-Aziz, Christian Agatemor and Wai-Yeung Wong

Published by the Royal Society of Chemistry, www.rsc.org

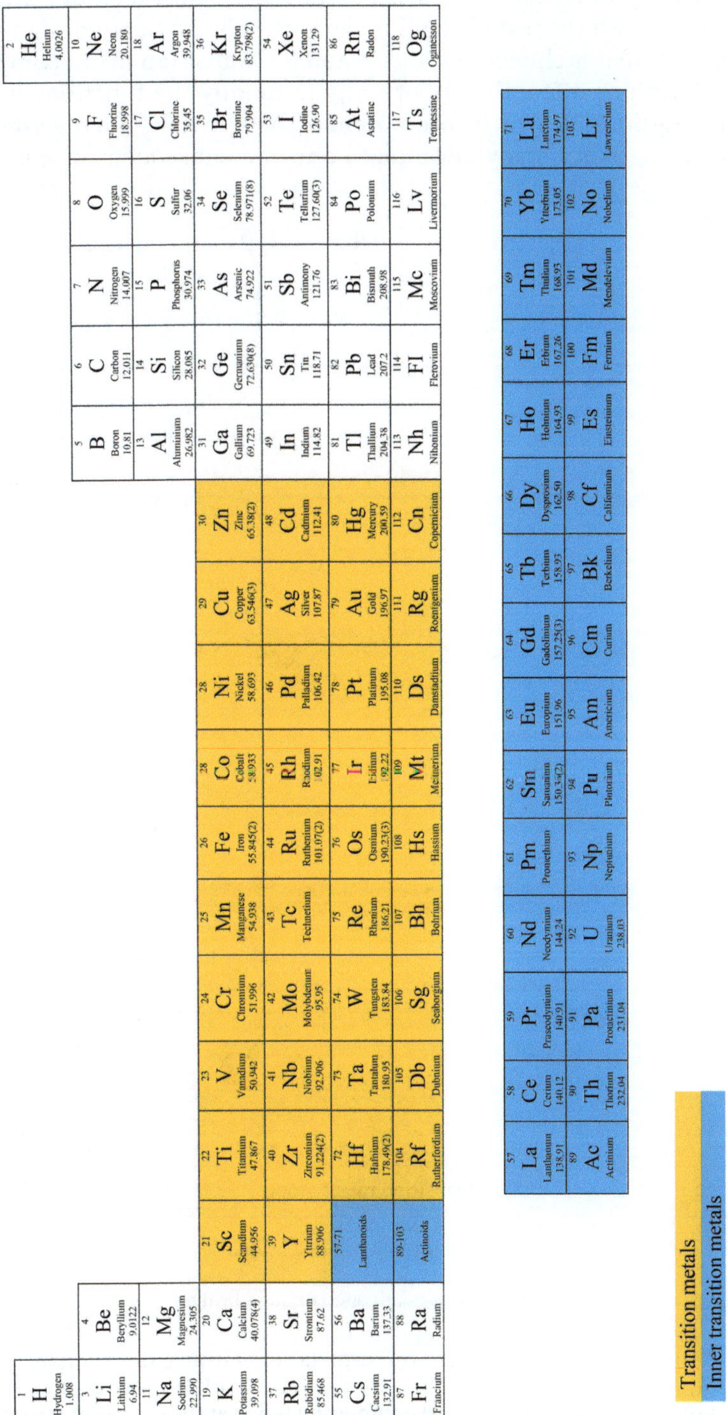

Figure 2.1 Periodic table of the elements. The transition metals, together with the inner transition metals (lanthanoids and actinoids) constitute more than half of the known elements in the periodic table.

guaranteeing longer lifetimes of activity, and better environmental compatibility compared with the use of small molecules.[6] The ability to design various macromolecular architectures and install the transition metal complex at different locations offers the opportunity to modify the functionality of the complex (Figure 2.2).[7] The incorporation of these metals into macromolecules, therefore, offers many advantages and the subsequent discussion

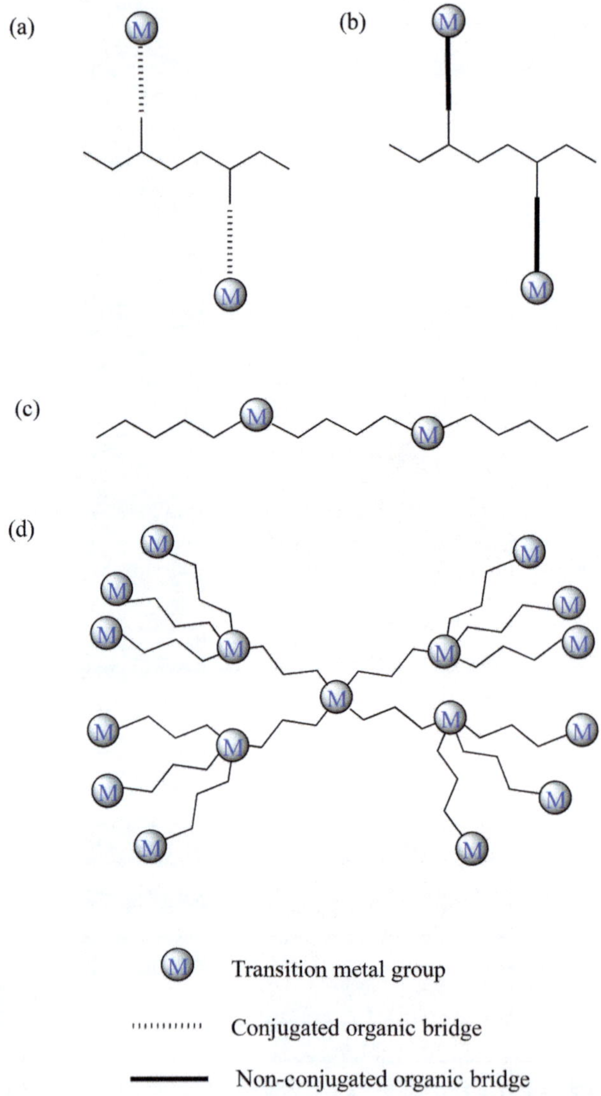

Figure 2.2 Architectural types of transition-metal-containing macromolecules. (a) and (b) The metal complex as a pendant; (c) metal complex in the polymer backbone; (d) metal complex installed at the core, branches and termini of a dendrimer or hyperbranched polymer.

focuses on the attractive properties of molecular transition metal complexes that are transferable into macromolecules.

2.2 Attractive Properties of Transition Metal Complexes

2.2.1 Chemistry of Transition Metals

The chemistry of transition metals primarily drives their continuous exploration as motifs to design macromolecules. Several inorganic chemistry textbooks[1–5] thoroughly cover the fundamentals of this chemistry, which is not the focus of this monograph; nonetheless, we outline some basic concepts and empirical evidence that inspire their use in the design of functional macromolecules. In principle, transition metals, typically, "have partially filled d sub-shell, or yield cations with incomplete d sub-shell",[8] but in practice, scientists include the f-block elements, lanthanides and actinides, regarding them as the inner transition metals (Figure 2.1). The chemistry of these elements depends on the number and electronic configuration of the d- or f-electrons, which also influence their redox-activity, photoactivity, and magnetism.[9,10] Of importance is the selective interaction of these metals with ligands to form cationic, anionic, or neutral complexes that retain the intrinsic functionalities of the ligand and the metal center as well as exhibit new properties.

Selectivity of a transition metal ion towards a ligand as well as the stability of the resulting metal–ligand complex depends on the chemistry of the metal and the ligand. The basic theory of hard and soft acids and bases (HSAB) in organometallic chemistry predicts the selectivity of the metal–ligand interactions.[11,12] As an illustration, a hard Lewis acid, Fe^{3+}, selectively coordinates to hard Lewis bases such as carboxylates and amines; a borderline acid, Fe^{2+}, favorably binds with borderline bases such as pyridine and azides; while a soft acid, Ag^{+}, preferentially interacts with soft bases such as thioethers and cyanide.[1–5] In practice, other factors, such as pH, interfere with the real affinity of these metals to ligands, lowering the specificity of the interactions and the overall stability of the metal complex, as predicted by the HSAB theory.[10] Nonetheless, the theory remains popular in chemistry and biology and is exploited to assemble coordination compounds and rationalize certain behaviors of transition metals. The construction of the 3D $[\{Ag_3Cu_3(\text{L-methioninato})_6(NO_3)_3(H_2O)_3\}7H_2O]_n$ coordination network from an L-methionine ligand and the borderline Lewis acid Cu^{2+} and soft acid Ag^{+} (Scheme 2.1) illustrates the concept.[13] The hard amino and carboxylate groups on the L-methionine ligand preferentially bind to the borderline Cu^{2+} ion while the soft thioether group appropriates the soft Ag^{+} ion to form a 3D heterobimetallic coordination network (Scheme 2.1).[13] In addition to the implication of the theory for synthesis, it explains the positive correlation between the softness of transition metal ions and their toxicity to planktonic cells and biofilms, since soft toxic metals such as Cd^{2+} and Ag^{+}

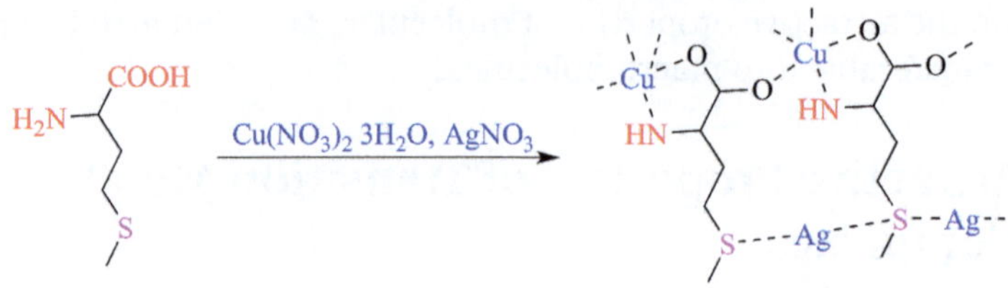

Scheme 2.1 A schematic of the preferential coordination of borderline Cu^{2+} to hard amino and carboxylic acid (red), and soft Ag^{+} to soft thioether (purple).

preferentially bind to the sulfhydryl (RSH) group in cellular proteins, causing damage.[9,14]

Transition metal complexes are known for the plethora of coordination geometries they form, depending on the number and electronic configuration of the d-electrons as well as the nature of the interacting ligand. The specificity of these geometries imposes a structural or functional bias on the transition metals. An illustrative example, though not with a transition metal, is the ability of periplasmic phosphate-binding proteins to differentiate and preferentially bind to the biologically indispensable phosphate in the presence of non-essential, toxic arsenates of similar pK_a, oxyanions, and radii based on a difference in the coordination geometry of the resulting complexes.[15] Also, a preferred geometry can favor the formation of an oxidation state to the exclusion of others. To illustrate this, Cu^{2+}, a d^9 ion, prefers a square planar, square pyramidal, or distorted octahedral geometry, but Cu^{+} with a d^{10} configuration is unlimited in the number of geometries it adopts.[10] A scenario that favors all geometries except the square planar, square pyramidal, or distorted octahedral will, therefore, drive a redox reaction towards the exclusive formation of Cu^{+} as the allowed geometries destabilize Cu^{2+} complexes.[10] Also, in the presence of a tetradentate ligand, Cu^{+} favors the formation of a tetrahedral dinuclear double-stranded complex, whereas Cu^{2+} assembles into a square pyramidal or distorted octahedron mononuclear complex (Figure 2.3).[16] The ability to access different geometries by changing the oxidation state is a useful strategy to control the structure and tune the properties.

2.2.2 Redox Properties of Transition Metals

The ability to acquire or lose discrete and variable numbers of electrons from the outermost atomic orbital is a principal characteristic of transition metals. This property enables these metals to engage in redox reactions, which are relevant in many chemical and biochemical processes. To an approximation, the redox chemistry of transition metal complexes fits the localized molecular orbital (MO) description, which views the redox process as metal (MC) or ligand centered (LC).[17] A simplified description of this approximation using a hypothetical octahedral transition metal complex portrays each MO as MC or LC, depending on the predominantly contributing orbital.[17]

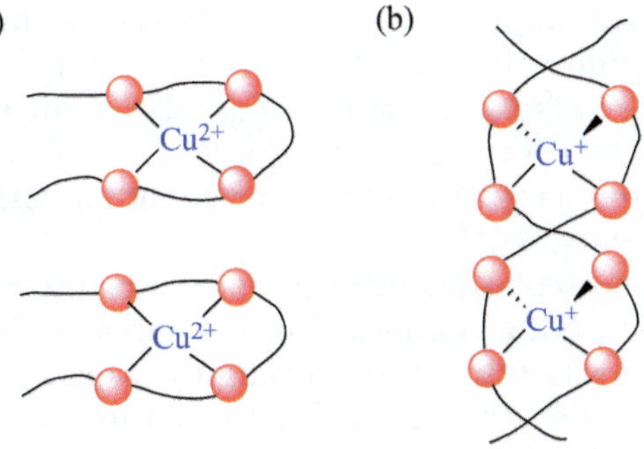

Figure 2.3 Redox-driven coordination geometry of copper. (a) Preferred square planar geometry of Cu^{2+}, (b) preferred tetrahedral geometry of Cu^{+}.

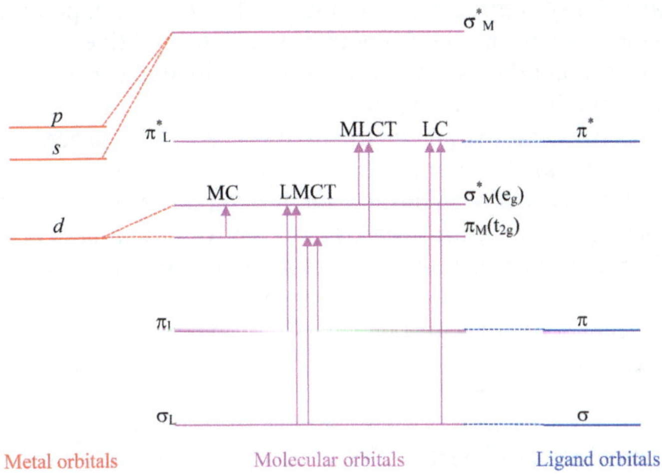

Figure 2.4 Localized molecular orbital approximation for an octahedral transition metal complex.

As illustrated in Figure 2.4, in the ground state, the bonding MOs, π_L, σ_L, π_M, are completely or partially occupied while the antibonding MOs, σ^*_M, π^*_L, σ^*_M, are empty.[17]

The low-energy bonding MOs, π_L, σ_L, are LC because the ligand orbitals contribute more to this orbital than the metal orbital, whereas the high-energy MO that corresponds to the highest occupied molecular orbital (HOMO) is mostly MC for the same reason.[17] Contribution to the lowest unoccupied molecular orbital (LUMO) depends on the relative energy ordering of metal and ligand.[17] In the excited, reduced or oxidized state, ground state electronic configuration is altered, for instance, oxidation removes electrons from the bonding MO while reduction adds electrons to the antibonding MO. Most often, the redox process results in transition metal complexes with an electronic configuration that violates the

18-electron rule, leading to a destabilized molecular integrity, and eventually, decomposition.[18] It is important to note that the 18-electron rule applies specifically to organometallic complexes, as the energy gap in classical transition metal coordination complexes is small, and is energetically favorable to electronic transition to the antibonding MOs without disruption of molecular integrity.[19]

A destabilized molecular integrity compromises the redox chemistry of redox-active species, but a complexed ligand can stabilize the species and restore redox activity as well as tune the affinity to acquire or lose electrons. As an example, ligands with delocalized electronic systems, appropriate symmetry, and low-lying antibonding orbitals ensure stability after a redox process perturbs the molecular integrity.[18] Furthermore, π-accepting ligands with extended delocalized systems, such as terpyridine, better stabilize transition metal ions in their low oxidation state than π-accepting ligands with lesser delocalized systems, such as pyridine.[20] The reduction potential of low oxidation state transition metals coordinated to terpyridine is accessible at a more positive potential compared with the analogous pyridine complex,[20] a reflection of the improved stability or better tendency of the metal center to acquire electrons during an electrochemical reduction process. A graphical example is provided by Long and coworkers, who observed a $Ni^{2+/+}$ redox couple at −1199 mV in a terpyridine complex compared with −1610 mV in a pyridine complex.[20] The shift towards the positive potential is also evident in the $Co^{2+/+}$ redox couple, which is accessible at −1197 mV in a terpyridine complex, in contrast with the less positive potential (−1466 mV) found in a pyridine complex.[20] Also, the respective reduction potential of the $Co^{2+/3+}$ and $Co^{2+/+}$ couples in a terpyridine complex shifts towards a positive potential by 75 and 61 mV when the terpyridine bears electron-withdrawing CF_3 groups.[20] Long and coworkers attributed the shift towards the positive potential to the stabilizing effect of the π-backbonding ability of the terpyridine on transition metal ions of lower oxidation state.[20]

The redox transformation of an 18-electron to a 17-electron or 19-electron transition metal complex invariably violates the 18-electron rule of organometallic chemistry and leads to a labile compound that instantaneously decomposes or reacts with air or moisture. From a synthesis perspective, such labile compounds enable the synthesis of model compounds of fundamental importance. As an example, the 18-electron η^6-arene-η^5-cyclopentadienyliron(II) complexes of the type $[\eta^6\text{-}C_6(R)_6\text{-}\eta^5\text{-}C_5(R)_5Fe]^+$ (R is an alkyl group) reduces to the 19-electron $[\eta^6\text{-}C_6(R)_6\text{-}\eta^5\text{-}C_5(R)_5Fe]$, which instantaneously reacts with O_2 or air to produce a superoxide radical, $O_2^{-\bullet}$, and regenerates the 18-electron precursor (Scheme 2.2).[21–25] The superoxide radical is sufficiently basic to abstract a hydrogen from a methyl group in the hexamethylbenzene ligand of the regenerated precursor to form a cyclohexadienyl ligand with an exocyclic double bond and a hydroperoxyl radical (Scheme 2.2).[21,25] The formation of radicals, superoxide and hydroperoxyl, is biologically significant, given their toxicity to eukaryotic and prokaryotic cells.

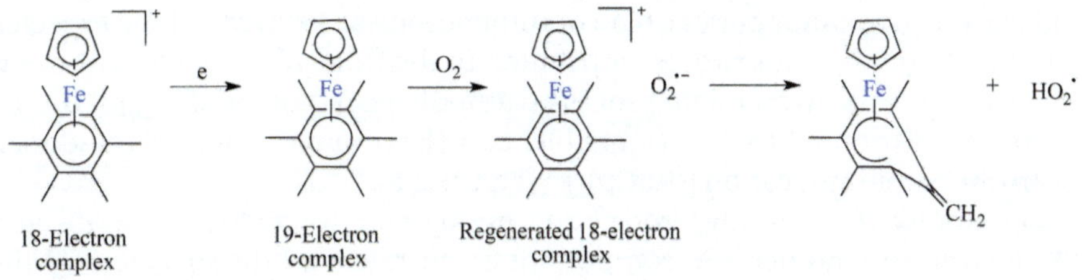

Scheme 2.2 Regeneration of an 18-electron complex from a 19-electron complex with the concomitant formation of radicals.

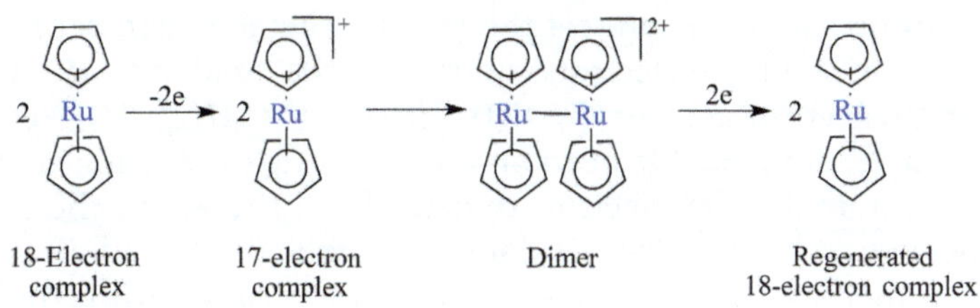

Scheme 2.3 Regeneration of an 18-electron complex from a 17-electron complex.

The tendency of redox-active organometallics to adhere to the 18-electron rule provides a synthetic access to many relevant chemical species, as demonstrated in the formation of the superoxide and hydroperoxyl radicals. The dimerization of 19-electron sandwich complexes of Rh^{2+}, Ir^{2+}, and Ru^{+} under ambient conditions to form 18-electron dimeric species with implications for n-doping in organic electronics, highlights the importance of the redox chemistry of 19-electron organometallic species.[26–28] The 17-electron organometallic compounds equally react to restore the 18-electron configuration, as observed in the dimerization of 17-electron complexes of Ru,[29–31] Os,[31] Rh,[31] and Co.[32] To illustrate this, the electrochemical oxidations of ruthenocene and osmocene proceed by a quasi-irreversible one-electron process in a CH_2Cl_2 solution containing 0.1 M tetrabutylammonium tetrakis(perfluoroaryl)borate at room temperature to produce the 17-electron ruthenocenium and osmocenium cations, respectively (Scheme 2.3).[29–31] These 17-electron cations are highly reactive species, dimerizing to regain the 18-electron configuration (Scheme 2.3), a chemical reaction that explains their quasi-irreversible redox behavior. The susceptibility of the highly reactive 17-electron ruthenocenium and osmocenium cations to dimerization contrasts with the stable 17-electron ferrocenium cation and is presumably due to the relatively longer bond length between the cyclopentadienyl carbons and the metal center in the ruthenocenium and osmocenium cations, compared with the bond length in the ferrocenium cation.[31] A longer carbon–metal bond length more likely opens the Ru or Os center to nucleophilic attack than the more sterically protected Fe center in the ferrocenium cation. The dimers are oxidizing agents, accepting two electrons to regenerate the neutral ruthenocene or osmocene (Scheme 2.3).[29–31]

Indeed, cyclic voltammograms (CV) of ruthenocene at 30 °C reveal the expected quasi-irreversible redox process attributed to the $[RuCp_2]^{0/+}$ couple and a new electrochemically irreversible process (cathodic peak potential (E_{pc}) = 0 V *vs.* ferrocene) attributed to an irreversible, two-electron reduction of the dimer, bis(ruthenocenium) cation $[(RuCp)_2]^{2+}$ (Figure 2.5).[29]

The ability to sterically protect the highly reactive redox-active 19- and 17-electron organometallic compounds from nucleophilic or electrophilic attack allows access to some of the strongest organometallic reducing and oxidizing agents. Astruc and coworkers accessed one of these reducing agents, the 19-electron $[\eta^6\text{-}C_6(CH_3)_6\text{-}\eta^5\text{-}C_5H_5Fe(\text{I})]$, by stabilizing the highly reactive 19-electron $[\eta^6\text{-}C_6H_6\text{-}\eta^5\text{-}C_5H_5Fe(\text{I})]$ *via* peralkylation of the arene ligand.[18,21–25] The alkyl groups sterically protect the reactive metal center, permitting the isolation of the stable 19-electron complex in the solid state (Scheme 2.4). The other 19-electron analogs, $[\eta^6\text{-}C_6H_{6-n}(CH_3)_n\text{-}\eta^5\text{-}C_5H_{5-m}(CH_3)_mFe(\text{I})]$ ($0 \le n \le 5$; $m = 0$ or 5), are unstable at room temperature, dimerizing at the arene ligand *via* intramolecular electron transfer from the Fe^+ to the arene carbon (Scheme 2.5).[18,21,23,25] The alkylation of the arene ligand electronically favors the dimerization, whereas peralkylation sterically thwarts the electron transfer to stabilize the 19-electron complex.[21,23,25] Also, peralkylation fine-tunes the half-wave potential ($E_{1/2}$) of the 18-electron $[\eta^6\text{-}C_6H_6\text{-}\eta^5\text{-}C_5H_5Fe(\text{II})]^+$, with permethylation of the cyclopentadienyl and arene ligands shifting the

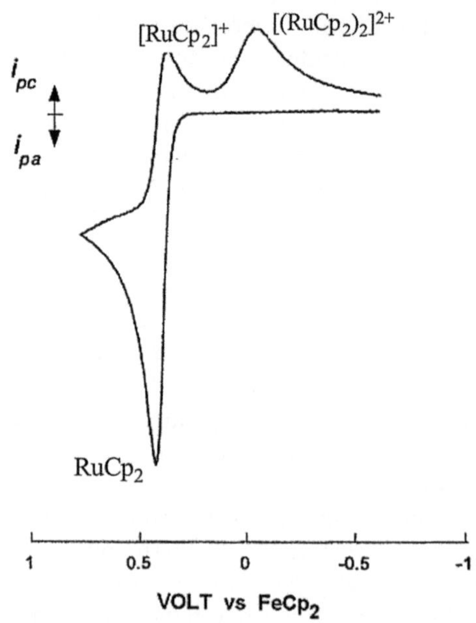

Figure 2.5 Cyclic voltammograms showing the one-electron quasi-irreversible oxidation of 18-electron ruthenocene ($RuCp_2$) and the two-electron irreversible reduction of bis(ruthenocenium) cation $[(RuCp)_2]^{2+}$. Reprinted with permission from ref. 29. Copyright (2003) American Chemical Society.

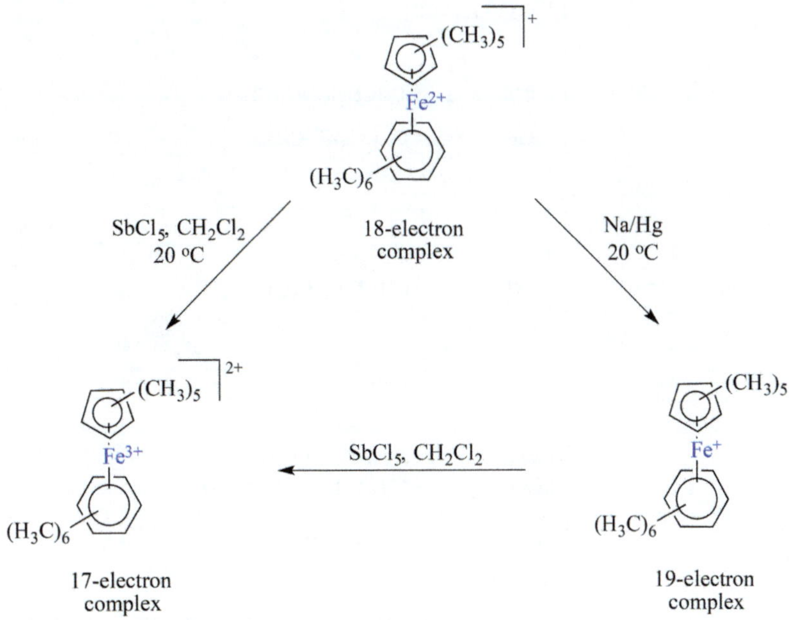

Scheme 2.4 Synthesis of stable 19- and 17-electron organometallic compounds. The peralkylation stabilizes the 19- and 17-electron compounds, inhibiting dimerization.

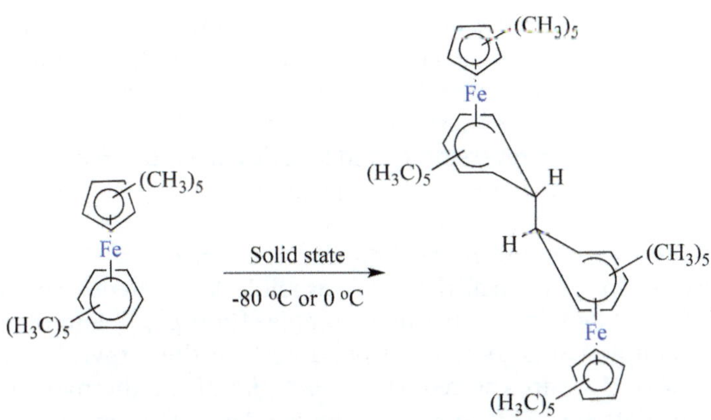

Scheme 2.5 Dimerization of an unstable 19-electron organometallic compound. The alkylation favors electron transfer from the 19-electron Fe center to the arene carbon, resulting in dimerization.

$E_{1/2}$ by 236 and 160 mV, respectively, relative to the non-methylated 18-electron $[\eta^6\text{-}C_6H_6\text{-}\eta^5\text{-}C_5H_5Fe(\text{II})]^+$ (Figure 2.6).[22] Furthermore, peralkylation stabilizes the 17-electron $[\eta^6$-arene-η^5-cyclopentadienyliron(III)$]^{2+}$, resulting in the synthesis of a robust organometallic oxidizing agent, 17-electron $[\eta^6\text{-}C_6(CH_3)_6\text{-}\eta^5\text{-}C_5H_5Fe(\text{III})]^{2+}$ (Scheme 2.4).[24]

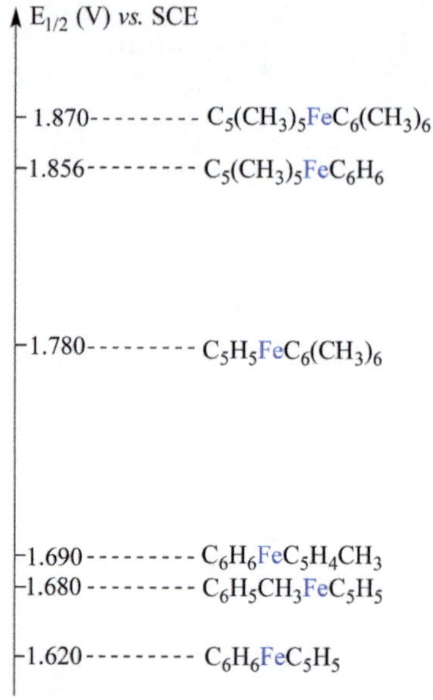

Figure 2.6 Effect of methylation on the electrochemical reduction of 18-electron η^6-arene-η^5-cyclopentadienyliron(II) to 19-electron η^6-arene-η^5-cyclopentadienyliron(I). Adapted with permission from ref. 23. Copyright (1981) American Chemical Society.

Another attractive feature of transition metal complexes is the possibility to fine-tune, *via* judicious selection of a ligand, the redox chemistry of the metal center. Indeed, the ligand influences the reduction potential, a thermodynamic parameter that describes the tendency of the complex to acquire electrons as well as the reversibility and kinetics of a redox reaction. Anchoring an electron-withdrawing perylene bisimide ligand to Ru^{2+} or Ir^{3+} terpyridine complexes, for instance, stabilizes the lower oxidation state of the metal centers, making oxidation more difficult.[33] Unequivocally, CV experiments with these complexes reveal that the reversible $Ru^{3+/2+}$ redox couple shifts to 1.02 V from 0.88 V in a terpyridine complex $[Ru(bpy)_3]^{2+}$ and the irreversible $Ir^{4+/3+}$ couple shifts to 1.46 V from 1.40 V in the terpyridine complex $[Cp^*Ir(bpy)Cl]^+$.[33] Again, the reduction potential of the thermodynamically favorable oxidation of Fe^{2+} to Fe^{3+} is strongly dependent on the complexed ligand, as evidenced by the 18-electron organometallic compounds ferrocene and η^6-arene-η^5-cyclopentadienyliron(II) complex. The $Fe^{2+/3+}$ redox couple in ferrocene is electrochemically accessible at 0.4 V *vs.* SHE, but at a higher positive potential (0.800–1.290 V *vs.* ferrocene) in the η^6-arene-η^5-cyclopentadienyliron(II) complex $[\eta^6\text{-}C_6(CH_3)_6\text{-}\eta^5\text{-}C_5H_5Fe(II)]^+$.[18,24] Also, the $Fe^{2+/+}$ redox couple is equally dependent on the ligand, as observed in the electrochemical reduction of 18-electron η^6-arene-η^5-cyclopentadienyliron(II) complexes to their 19-electron analogs (Table 2.1).[34] Changing the arene ligand from benzene to hexamethylbenzene, for example, shifts the cathodic peak

Table 2.1 Effect of arene substitution on the cathodic peak potential (E_{pc})[a] of 18-electron η^6-arene-η^5-cyclopentadienyliron(II) complex. Adapted from ref. 34 with permission from The Royal Society of Chemistry.

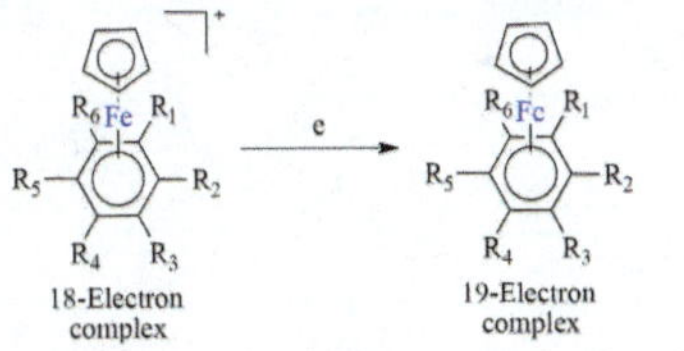

R_1	R_2	R_3	R_4	R_5	R_6	E_{pc} (V) *vs.* ferrocene
H	H	H	H	H	H	1.86
CH_3	H	H	H	H	H	1.91
CH_3	H	H	CH_3	H	H	1.97
CH_3	CH_3	H	CH_3	CH_3	H	1.99
CH_3	CH_3	CH_3	CH_3	CH_3	CH_3	2.06
H	H	H	H	H	CH_2CH_3	1.92
Cl	H	H	H	H	H	1.65
Cl	H	H	CH_3	H	H	1.71
Cl	H	H	Cl	H	H	1.48
NH_2	H	H	H	H	H	2.02
CHO	H	H	H	H	H	1.91

[a]Cyclic voltammograms were acquired at room temperature using a nitrogen-purged 6 mM solution of complex in propylene carbonate solution 0.1 M [n-Bu_4N][PF_6].

potential (E_{pc}) to a more negative potential, specifically from 1.86 V to 2.06 V *vs.* ferrocene (Table 2.1).[34] Theoretical calculations reveal that contribution of the metal center to the HOMO in 18-electron η^6-arene-η^5-cyclopentadienyliron(II) complexes changes as the structure of the arene ligand is modified.[24] For instance, calculations showed that the HOMOs in [η^6-C_6H_6-η^5-C_5H_5Fe(II)]$^+$ and [η^6-$C_6H_5NH_2$-η^5-C_5H_5Fe(II)]$^+$ are predominantly 76% and 69% localized on iron,[24] which, presumably, could contribute to the difference in the E_{pc} of these complexes (Table 2.1).

An important attribute of redox-active transition metals is their readily accessible multiple oxidation states, which has implications for biological and catalytic functions. Depending on the coordinated or covalently bound ligand, these oxidation states are stable and observable within the timescale of cyclic voltammetry. Figure 2.7 presents cyclic voltammograms showing the multiple reversible redox couples of some catalytic 3d transition metal complexes.[20] The terpyridine–Ni complex, for instance, is characterized by reversible one-electron redox waves at 1392 mV and 1199 mV attributed to $Ni^{2+/3+}$ and $Ni^{2+/+}$ redox couples, respectively (Figure 2.7).[20] The one-electron redox waves of $Fe^{2+/3+}$ and $Fe^{2+/+}$ couples appear at 686 mV and 2349 mV and those of $Co^{2+/3+}$ and $Co^{2+/+}$ at 235 mV and 1197 mV, respectively (Figure 2.7).[20] Multiple oxidation states are relevant in many applications as demonstrated by a molybdenum complex that drives the reduction of dinitrogen to ammonia *via* a catalytic pathway that involves a Mo^{3+} to Mo^{6+} cycle.[35] The redox

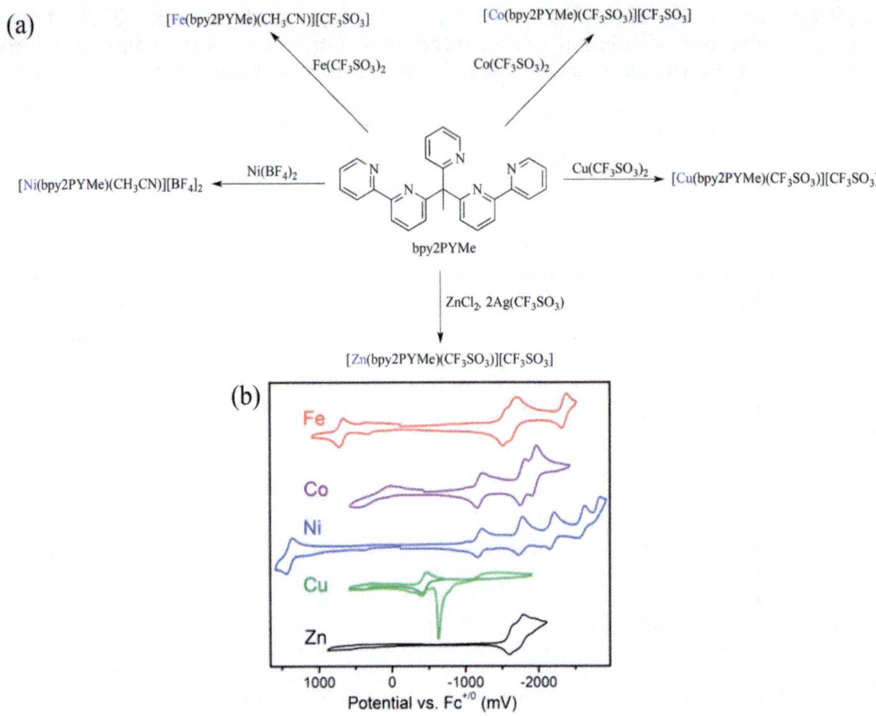

Figure 2.7 Some catalytic 3d transition metal complexes (a) with their cyclic voltammograms (b) showing the multiple oxidation states. The redox process of the terpyridine ligand is also evidenced by the CV. Reproduced from ref. 20 with permission from The Royal Society of Chemistry.

activity of the transition metals is retained when incorporated into a macromolecule, as demonstrated by Abd-El-Aziz and coworkers, who observed the redox waves of ferrocene and η^6-arene-η^5-cyclopentadienyliron(II) in a linear polymer (Figure 2.8).[36]

2.2.3 Photophysics and Photochemistry of Transition Metal Complexes

The realization that photoactive molecules are relevant in the fabrication of emitters, sensors, photocatalysts, and bioimaging probes inspires an accelerated interest in the investigation of the photoactivity of transition metal complexes. From a fundamental perspective, the photophysics and photochemistry of these complexes depend on the transition metal center, the ligand, and the microenvironment, and are distinct from those of organic molecules.[17] Whereas photophysical investigations focus on the physical processes that result from the absorption and emission of light by a material, photochemical studies examine the chemical changes. The fundamental principle in both processes involves the transition of an electron from

(a)

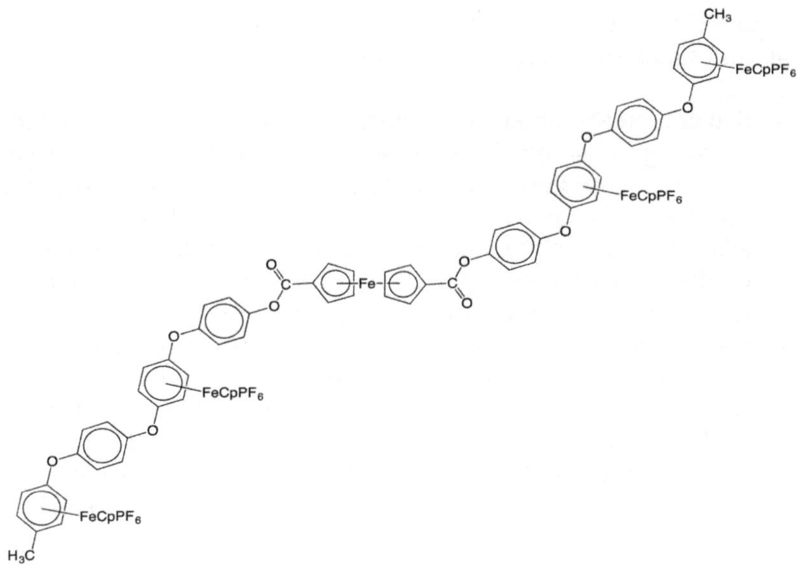

(b)

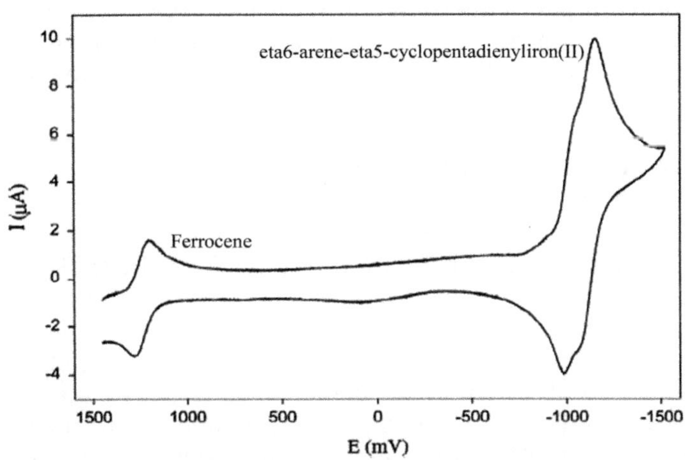

Figure 2.8 (a) Ferrocene and η^6-arene-η^5-cyclopentadienyliron(II) polymer and (b) redox wave of the redox-active moieties in the polymer. Reprinted with permission from ref. 36. Copyright (2005) American Chemical Society.

the ground state to excited state following the absorption of light, and the return of the excited state species to the ground state, with the simultaneous emission of light, transfer of energy, or an electron that can be engaged in a redox reaction.

Understanding the electronic transitions and the resulting excited states of transition metal complexes is complicated because of the interplay of the photophysics of the metal center with that of the ligand as well as that resulting from the metal–ligand interaction; nonetheless, three dominant transitions and states are recognized among others (Figure 2.3).[17]

1. d–d or metal-centered (MC) transitions that result in MC excited states following transitions between MOs, with predominant localization on the metal. Emission from this state is typically broad and structureless.
2. intraligand or ligand-centered (LC) transitions that yield LC excited states after transitions between MOs, with predominant localization on the ligand. The energy and band structure of the emission approximate those of the uncoordinated ligand.
3. charge transfer (CT) transitions, which are ligand-to-metal charge-transfer (LMCT) or metal-to-ligand charge transfer (MLCT) transitions and result from electronic transitions between MOs with different localizations. These transitions give rise to CT excited states and emission bands that are highly intense and structured. The LMCT and MLCT transitions reduce and oxidize the metal center, respectively.

Transition metal complexes, especially 4d and 5d metal complexes, possess strong spin–orbit coupling (respective spin–orbit coupling values (cm^{-1}) for C, Co, Ru, Ir, and Pt, are 30, 517, 878, 3600, 4052), which plays a dominant role in their CT states.[37,38] It is important to note that the interference of spin–orbit coupling in the CT states of transition metal complexes led some researchers to consider spin label and the term "intersystem crossing" rationally meaningless.[37,38] Nonetheless, current literature, including this monograph, continues to describe the photophysics of these complexes using these terms.[17,39–45]

Spin–orbit coupling efficiently facilitates excited state singlet to triplet intersystem crossing and radiative transition from the triplet state to the ground singlet state, leading to intense and long-lived phosphorescence. However, the presence of a transition metal with strong spin–orbit coupling does not guarantee phosphorescence, as established by Rybtchinski and coworkers, who observed strong fluorescence in palladium complexes of perylene diimide (PDI) (Figure 2.9).[46] Based on time-dependent density functional theory (TD-DFT) computation, Rybtchinski and coworkers rationalized that the absence of phosphorescence may be due to the weak electronic interaction between the σ-bond connecting the Pd to the frontier π-orbital of the PDI.[46] Obviously, a radiative pathway is crucial to observing phosphorescence after populating the excited triplet state *via* intersystem crossing. For instance, Castellano *et al.* found that the singlet-state fluorescence of PDI quenches without phosphorescence in the visible or near-IR region upon conjugation to Pt(II) *via* acetylene bridges.[47,48] Expectedly, the strong spin–orbit coupling in Pt(II) promotes efficient excited singlet PDI or ligand centered state (^{1}LC) intersystem crossing to populate the triplet PDI state (^{3}LC) with up to 55% quantum yield.[48] Indeed, an ultrafast transient absorption experiment proved that the ^{1}LC state undergoes intersystem crossing in

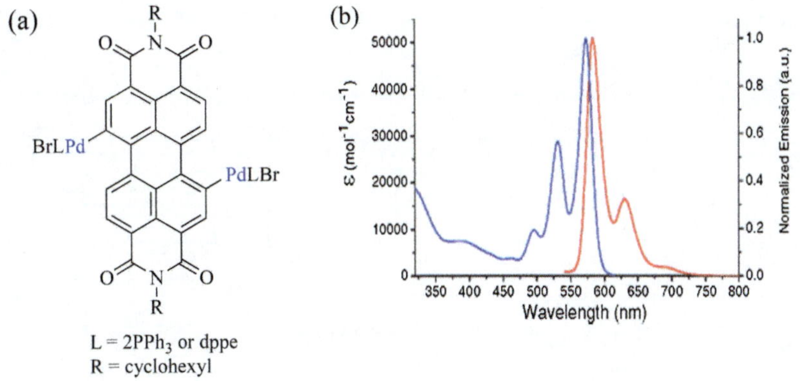

Figure 2.9 (a) Non-phosphorescent palladium perylene diimide complex, (b) photophysical properties of the complex. (Blue) UV-vis absorption and (red) fluorescent emission. Reprinted with permission from ref. 46. Copyright (2007) American Chemical Society.

2–4 ps to a ^{3}LC state that vibrationally relaxes in 12–19 ps.[48] Presumably, the absence of phosphorescence is due to vibrational relaxation, a non-radiative quenching pathway for the ^{3}LC state.[48] From an application perspective, the long-lived excited ^{3}LC state can be deployed in electron and energy transfer processes to facilitate relevant reactions, as demonstrated by the sensitization of singlet oxygen.[48] Using excited state lifetime measurements, Castellano and coworkers established that the stereochemistry as well as the chemistry of coordination ligands plays a fundamental role in the excited state dynamics.[48]

Phosphorescence is possible through a rational design that ensures substantial interactions between the transition metal orbitals and the PBI ligand to promote intersystem crossing and creates a radiative pathway during relaxation to the ground singlet state. Würthner *et al.* designed Ru^{2+} and Ir^{3+} complexes of azabenz-annulated PBI that emit near-IR phosphorescence in 11% and <1% quantum yield (Figure 2.10).[33] Using DFT and TD-DFT calculations, they reasoned that the phosphorescence of the Ru complex originates from a hybrid of ^{3}MLCT and ^{3}LC states and that of the Ir complex from the ^{3}LC state. Experimental and theoretical data proved that photophysics of the Ru and Ir complexes differ, with the Ru complex preferring a low-energy ^{3}MLCT transition in contrast with the Ir complex's preference for a purely ^{3}LC transition.[33] This difference explains their distinct excited states, as evidenced by their emission spectra, which is broad for the Ru complex and well-resolved for the Ir complex. The strong spin–orbit coupling in these complexes decreases the phosphorescence lifetime from 100 s for organic PBI to 4.2 s and 33 s for the Ru and Ir complexes, respectively.[33] The lesser effect of spin–orbit coupling on the heavier 5d Ir compared with the 4d Ru is counter-intuitive, but is attributed to the MLCT state in the Ru complex.[33]

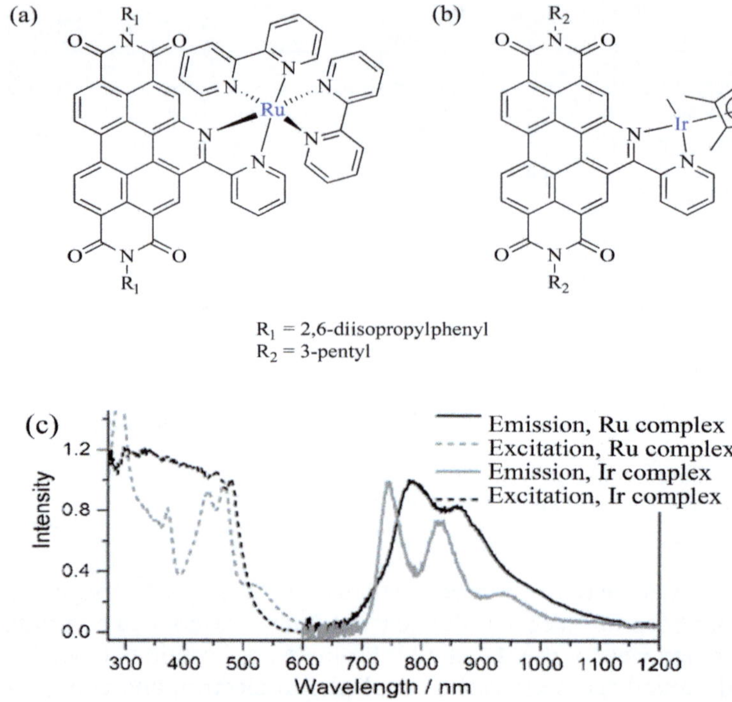

Figure 2.10 Phosphorescent Ru (a) and Ir (b) complexes. (c) Near-IR phosphorescence of complexes. Reproduced from M. Schulze, A. Steffen and F. Würthner, Near-IR Phosphorescent Ruthenium(II) and Iridium(III) Perylene Bisimide Metal Complexes, *Angewandte Chemie International Edition*,[33] © 2015 Wiley-VCH Verlag GmbH & Co. KGaA, Weinheim.

The nature of the coordinating ligand tunes the photophysics of transition metal complexes, a practicality that is explored in the design of sensing and imaging platforms. To illustrate this, the phosphorescence of the Ru and Ir complexes $[Ru(bpy)_2(DA\text{-phen})](PF_6)_2$ and $[Ir(ppy)_2(DA\text{-phen})](PF_6)$ (bpy: 2,2-bipyridine; DA-phen: 4,5-diamino-1,10-phenanthroline; and ppy: 2-phenylpyridine), is weak in aqueous media but improves after reaction with methylglyoxal (MGO).[49] The enhancement in phosphorescence implies sensitivity, which can be used to detect endogenous MGO, an ubiquitous metabolite.[50] The time-dependent increase in phosphorescence intensity of these complexes in a solution containing phosphate buffered saline (pH = 7.4) to an 800 M solution of MGO confirms the sensitivity.[49] Also, the emission intensity increases with the concentration of MGO, and these complexes successfully image intracellular MGO based on turn-on phosphorescence.[49] The lowest-lying triplet excited state (^{1}T) in $[Ru(bpy)_2DA\text{-phen})](PF_6)_2$ and $[Ir(ppy)_2(DA\text{-phen})](PF_6)$ originates from ^{3}LLCT/^{3}LMCT and ^{3}LLCT and ^{3}ILCT states and is quenched by electron transfer from the amino group on the 1,10-phenanthroline to the ligand

and metal center.[49] A reaction between the MGO and these amino groups eliminates this deleterious electron transfer process, creating an emissive low-lying excited state.[49]

An appropriately designed transition metal complex can function as a visible light antenna, harvesting light and funneling photogenerated energy or charges to drive chemical processes. A typical example is the photosensitization of photocatalysts, as found in binuclear Ru–Ru and Ru–Ir complexes that reduce water.[51] On visible light absorption, these complexes undergo electronic transition dominated by ^{1}MLCT, which passes to the ^{3}MLCT *via* intersystem crossing. Whereas in the Ru–Ru complex, transition from the ^{3}MLCT state to the ground state is direct; in the Ru–Ir complex, the high-energy ^{3}MLCT state of the Ir center transfers energy to the low-energy ^{3}MLCT state of the Ru center, which then passes to the ground state.[51] Thus, the emission of the Ru–Ir complex is inherently that of the Ru component. These complexes sensitize a platinum catalyst to generate hydrogen gas from water.[51]

The microenvironment of the transition metal complex influences its emissive behavior. A series of osmium complexes, for example, fails to emit in solution but emits in the solid state around the near-IR region with a 0.5–8.8% quantum yield.[52] Presumably, in solution, solvent collision or other non-radiative pathways quenches the emission, but these channels may be absent in the solid state. These emissions originate from a mixture of ^{3}MLCT and ^{3}LLCT states and are characterized by an atypically short phosphorescent lifetime of 26.4–431 ns.[52] The solid-state emissive property of these osmium complexes allows their use as dopants in the fabrication of organic light-emitting diodes.[52]

In the past decades, a surge in innovative applications of photons has led to many breakthrough developments, including two-photon probes for bioimaging. In the traditional one-photon probes, the excited states originate from the absorption of one photon in the UV–visible region (350–550 nm). In contrast, in a two-photon photophysical process, a material simultaneously absorbs two photons of the same or different energies to reach the excited state. Since Nobel laureate Maria Göppert-Mayer envisaged the concept in her Ph.D. thesis in 1931,[53] two-photon absorption is now a confirmed phenomenon that enables unprecedented deep and highly resolved bioimaging as well as 3D photochemistry, data storage, and microfabrication.[54–60] Practical application of two-photon absorption requires fluorophores with a large two-photon absorption cross section (δ) value, which typically is accessible in conjugated molecules with a high degree of intramolecular charge transfer.[54]

Transition metal complexes exhibit tailor-made specificity, including two-photon absorption as demonstrated by dinuclear ruthenium(II) polypyridyl[60] and zinc salen/salophen complexes (salen: *N*,*N*-ethylenebis-(salicylimine); salophen: *N*,*N*-phenylenebis(salicylimine)).[55–57] A well-thought-out functionalization of a salen/salophen ligand with an electron donor and electron acceptor imparts intramolecular charge transfer

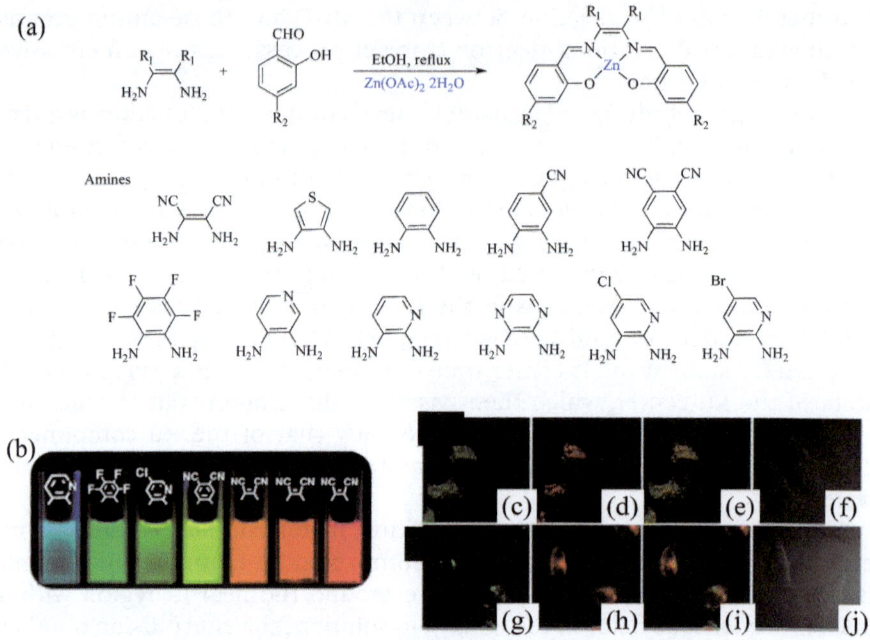

Figure 2.11 Schematics of some of the 48 zinc salen/salophen complexes obtained *via* changing the substituents at the diamine or the N-substituent at the 4-position. The diamines used are shown in (a). (b) Tuning the fluorescence emission by changing the diamine. (c)–(j) Two-photon imaging of HeLa cells. Reproduced from ref. 56 with permission of The Royal Society of Chemistry.

properties to achieve two-photon absorption.[55–57] Indeed, modification of the substituents on the diamine bridge, and to a lesser extent on the N- and O- substituents at the 4-position of the salicyaldehyde, of a zinc salen/salophen complex enables fine-tuning of the electronic properties of the complex to modulate the photophysical properties, including the δ values.[55–57] This is exemplified by a series of 48 zinc salen/salophen complexes obtained *via* changing the substituents at the diamine or the N-substituent at the 4-position (Figure 2.11).[56] Absorption bands at 370–430 nm and 440–510 nm, attributed to a π–π^* transition, are characteristic of salen/salophen ligands and an additional band at 550–700 nm, due to an intramolecular charge transfer transition, is evident in complexes designed from *cis*-2,3-diaminomaleonitrile.[56] Increasing the electron-withdrawing ability of the substituent on the amine resulted in a small bathochromic shift in the absorption and emission spectra, but modifying the N-substituent at the 4-position had no significant effect on the photophysical property.[56] In this study, the electron-acceptor, *cis*-2,3-diaminomaleonitrile, red-shifts the emission band more than other substituents and reduces the non-radiative decay process by up to a factor of 20. The δ values of the complexes are impressive, ranging from 3–215 GM[56] and are higher than those of commercially available dyes used in two-photon microscopy.[61] Also, the δ value is tuned by modifying the substituent on the diamine, as revealed by the almost linear relationship between δ values at 810 nm and the change in dipole moment upon transition to the excited state.[56] Whereas the N-substituent at the 4-position is insignificant in the photophysics, it tunes lipophilicity and subcellular localization during application of these complexes as a bioimaging platform. The practical application of these complexes is demonstrated by their use as one- and two-photon colour palettes for multicolor imaging of HeLa cells (Figure 2.11).[56]

2.2.4 Magnetic Property of Transition Metal Complexes

Magnetism is an attractive research theme in materials science due to its implications for diverse applications in many facets of life including medicine, electronics, and data processing. The presence of unpaired electrons impacts a magnetic behavior other than diamagnetism, which is a property of all materials. Therefore, transition metals with unpaired d- and f-electrons exhibit paramagnetism or a cooperative magnetic behavior depending on the relative alignments of the different spins.[62,63] Exchange interactions alongside other quantum mechanical effects enable coupling of spins to impact cooperative magnetic behavior such as ferromagnetism, antiferromagnetism, or ferrimagnetism.[62,63] Transition metal ions are the spin carrier in most magnetic materials including magnetic nanoparticles, single-molecule magnets, single-ion magnets, and single-chain magnets.[64] The presence of superparamagnetism in extremely small ferrimagnets and ferromagnets, such as single-molecule magnets and magnetic nanoparticles, has prompted a surge in the research and development of these materials.[64] Superparamagnetism exists where fluctuation in thermal energy overcomes the anisotropy energy to bring magnetization to zero on the removal of an external magnetic field.[65] Superparamagnetism is qualitatively similar to paramagnetism but is differentiated by its high magnetic susceptibility. To establish the existence of the superparamagnetic state in nanomagnets such as single-molecular magnets, researchers rely on AC susceptibility measurements, a technique that measures magnetic response as a function of a magnetic field at different times or temperatures. A frequency-dependent response of magnetic susceptibility suggests superparamagnetism.[66] Boča and coworkers, for example, found that the maxima of the out-of-phase component of the molar magnetic susceptibility of a superparamagnetic cobalt-based single-molecule magnet, $[Co(PPh_3)_2Br_2]$, are sensitive to the frequency at the applied (0.1 T) external field (Figure 2.12).[66]

The tunability of magnetic properties is an important attribute of magnetic systems derived from transition metal complexes and is realizable by structural modifications. To demonstrate this, a systematic distortion

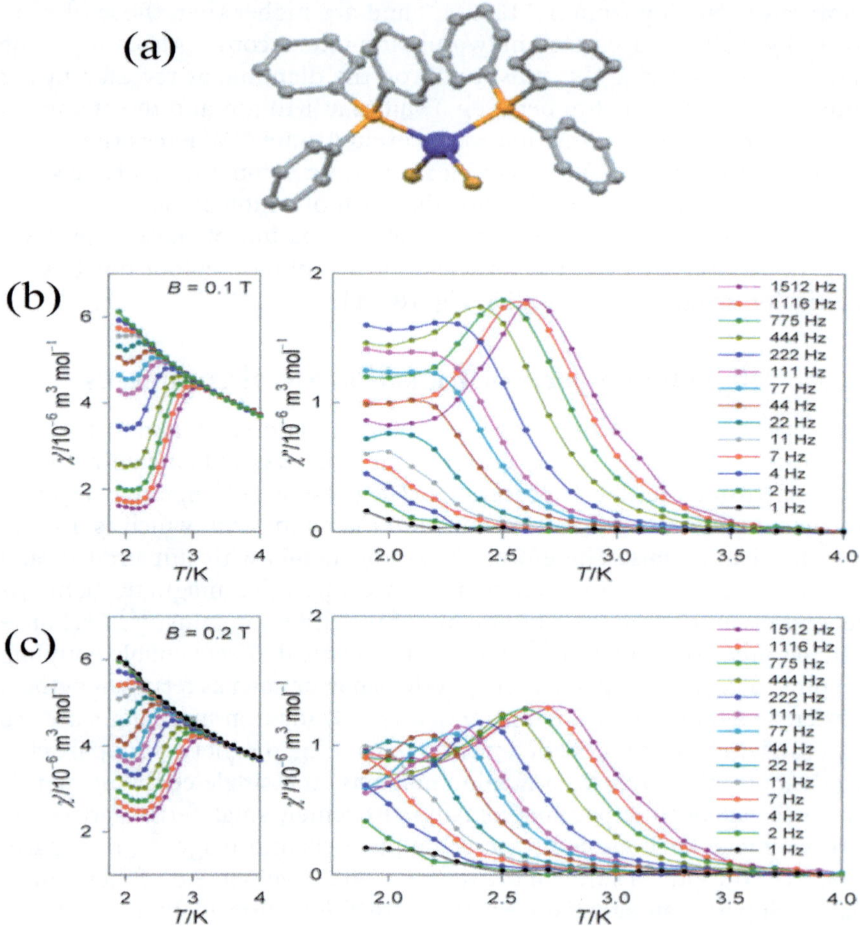

Figure 2.12 (a) Crystal structure of superparamagnetic $[Co(PPh_3)_2Br_2]$ (Co: purple; Br: yellow). (b) Left: in-phase; right: out-of-phase molar susceptibility of the complex at 0.1 T applied field. (c) Left: in-phase; right: out-of-phase molar susceptibility of the complex at 0.2 T applied field. (a) Reproduced from *Inorganic Chemistry Communications*, **35**, J. Titiš, J. Miklovič and R. Boča, Magnetostructural study of tetracoordinate cobalt(II) complexes, 72–75,[67] Copyright (2013), with permission from Elsevier. (b) and (c) Reprinted with permission from ref. 66. Copyright (2014) American Chemical Society.

of the tetrahedral geometry of a series of cobalt(II) complexes, $[CoX_2L_2]$, (X = Cl, Br, and NSC, and L = N- or P-donor ligand) tunes the D zero-field splitting parameter.[67] Specifically, the value of the D-parameter increases from −13 to +5 cm^{-1} as the angular distortion of the X–Co–X and L–Co–L angles from the ideal tetrahedron increases from −13.6 to +6.2°.[67] Even a subtle structural difference in transition-metal-based single-molecule magnets manifests as a noticeable difference in the various spin Hamiltonian parameters including the *g*-anisotropy (*g*-values) and the *D* and *E* zero-field splitting parameters. The magnetic properties of a series of five-coordinated cobalt(II) complexes with distorted trigonal bipyramidal geometry exemplifies a scenario where noticeable differences in the spin Hamiltonian manifest in complexes with similar geometry.[68] In this series of complexes, replacing a rigid [(1-phenyl-1*H*-1,2,3-triazol-4-yl)methyl] amine (tpta) ligand with a flexible [(1-benzyl-1*H*-1,2,3-triazol-4-yl)methyl] amine (tbta) ligand doubles the absolute value of the *D*- and *E*-parameters and changing from a Cl to NCS ancillary ligand increases the $|E/D|$ and the *g*-values.[68] In contrast, using a Cl ligand in tetrahedral scorpionate complexes, $Tp^{R,R'}Co(X)$ (X = Cl or NCS), decreases the $|E/D|$ value (0.012–0.061) and increases the *D*-parameter (+10.88 to +12.72 cm^{-1}) relative to using NCS ($|E/D|$ = 0.056–0.15; *D*-parameter = +2.39 to +3.34 cm^{-1}).[69]

A fundamental understanding of the spin relaxation process is crucial to tuning the energy barrier to trap the magnetic state in a favorable high-spin energy well. The relaxation process in mononuclear transition-metal-based single-molecule magnets is temperature-dependent and is typically realized predominantly *via* a direct or tunneling mechanism at low temperatures and Raman or Orbach mechanism at high temperatures.[70,71] The chemistry of the nanomagnets modulates these relaxation mechanisms. Changing the counterions, (ClO_4^- or BPh_4^-), in cobalt complexes, $[Co(tpm)_2][ClO_4]_2$ and $[Co(tpm)_2][BPh_4]_2·2MeCN$ (tpm = tris(pyrazol-1-yl)methane), for instance, tunes the Raman relaxation mechanism at higher temperatures.[71] It is impressive to note that the uniaxial anisotropies (92 cm^{-1} and 93 cm^{-1}) of these Co complexes are among the highest reported for transition metal complexes to date and that tunneling is the predominant relaxation mechanism at low temperature.[71]

Evidence of a temperature-dependent spin relaxation process exists in a series of cobalt complexes, $[Co(C_3S_5)_2]·2X$ (X = Bu_4N^+, Ph_4P^+, PPN^+, or $[K(18\text{-}c\text{-}6)]^+$).[72] Below 3 K, the spin predominantly relaxes by tunneling as indicated by the independence of the relaxation time on temperature. A transition to a predominantly Raman mechanism occurs as the temperature increases to 22 K, and then the Orbach mechanism is predominant, as indicated by the linear temperature logarithm of the relaxation time relationship.[72] Also, in this series of cobalt complexes $[Co(C_3S_5)_2]·2X$ (X = Bu_4N^+, Ph_4P^+, PPN^+, or $[K(18\text{-}c\text{-}6)]^+$), deviation of the dihedral angle from the 90°, the ideal D_{2d} symmetry expected of the complexes, tunes the tunneling frequency, with the most distorted $[Co(C_3S_5)_2]·2[Bu_4N]^+$ (dihedral angle = 76.5°) exhibiting the highest frequency (61 Hz) and the least distorted $[Co(C_3S_5)_2]_2·[K(18\text{-}c\text{-}6)]^+$ (dihedral angle = 83.1°) exhibiting the lowest frequency (0.26 Hz).[72] Deviation from the dihedral angle influences other magnetic properties, thereby extending the frontiers of the application of these nanomagnets beyond the traditional information processing to the realms of quantum computing. For instance, the least distorted complex exhibited a low *E*-parameter and a magnetic hysteresis, hallmarks of traditional data storage magnetic platforms, whereas the most distorted complex features a large *E*-parameter without a magnetic hysteresis, suggestive of quantum information processing.[72] A predominant relaxation mechanism implies the existence of several mechanisms, which are tunable by changing the applied external magnetic field. The $[Co(PPh_3)_2Br_2]$ single-molecule magnet, for instance, relaxes *via* the tunneling mechanism at an applied field of 0.1 T but *via* tunneling alongside an unidentified mechanism at 0.2 T.[66]

Thermal decomposition of organometallic or coordination compounds in an inert atmosphere provides a versatile synthetic route to magnetic nanoparticles with control over size, shape and magnetic properties (Figure 2.13).[65,73–78] Magnetic nanoparticles are an attractive class of materials, where each particle behaves as a single magnetic domain with the spin aligning in the direction of an applied field on actuation by neighboring particles and an external field.[65] Park and coworkers, for instance, synthesized Mn_3O_4 nanoparticles *via* the thermal decomposition of $Mn(acac)_2$ (acac = acetylacetonate) in oleylamine.[73] The size of monodisperse nanoparticles is tunable by control of the reaction temperature, with nanoparticles of 6 nm, 10 nm, and 15 nm in diameter obtained at 150 °C, 180 °C, and 250 °C, respectively.[73] Addition of water to the reaction mixture limits the oxidation of $Mn(acac)_2$ due to decomposition, leading to the formation of MnO nanoparticles.[73] The Mn_3O_4 nanoparticles are ferromagnetic at low temperature and paramagnetic at room temperature.[73] Size is an important property because it tunes the blocking temperatures (T_B). As an example, the 6 nm, 10 nm, and 15 nm Mn_3O_4 nanoparticles feature T_B of 36 K, 40 K, and 41 K (Figure 2.13).[73] In contrast with bulk MnO, which is antiferromagnetic, the MnO nanoparticles are weakly ferromagnetic at low temperature.[73] The inverse relationship between particle size and T_B of the MnO nanoparticles contrasts with the direct correlation found in Mn_3O_4 nanoparticles.[73]

Surface anisotropy plays a fundamental role in magnetism; therefore, the link between surface and shape suggests a possibility to tune magnetic properties by controlling the shape of magnetic nanoparticles. Zhang and coworkers found a shape-dependent magnetic property in $CoFe_2O_4$ nanoparticles.[77] Control of the rate of growth of theses nanoparticles reversibly switches between spherical and cubic shapes, enabling control of the coercivity (H_c), which is lower in the nanocubes compared with the nanospheres. A curved topology, as a characteristic of spherical nanoparticles, provides a template with fewer defects, and therefore higher surface anisotropy compared with the flat topology of cubes.[77] As surface anisotropy predominates over core anisotropy in determining H_c, spherical nanoparticles feature a higher H_c than cubic particles of similar volume. Other magnetic properties such as T_B, saturation magnetization (M_s) and remanent magnetization (M_r) are independent of the shape but dependent on size, with M_s and M_r, for instance, increasing with an increase in size.[77]

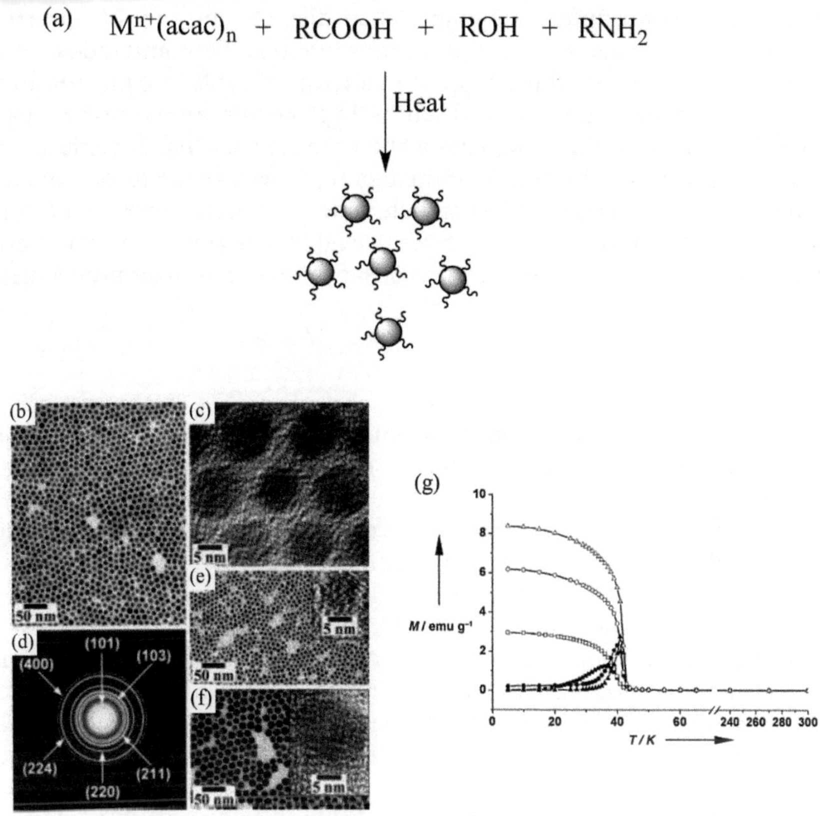

Figure 2.13 (a) General schematic for the synthesis of magnetic nanoparticles *via* thermal decomposition of an organometallic precursor. (b) TEM images of 10 nm Mn_3O_4 nanoparticles, (c) high resolution TEM images of 10 nm Mn_3O_4 nanoparticles, (d) SAED pattern of 10 nm Mn_3O_4 nanoparticles, (e) TEM images of 6 nm Mn_3O_4 nanoparticles, (f) TEM images of 15 nm Mn_3O_4 nanoparticles, and (g) size-dependent zero-field-cooling (filled symbols) and field-cooling (open symbols) magnetization of Mn_3O_4 nanoparticles (squares: 6 nm, circles: 10 nm, triangles: 15 nm). Reproduced from W. S. Seo, H. H. Jo, K. Lee *et al.*, Side-chain ferrocene-containing (meth)acrylate polymers: Synthesis and properties, *Journal of Polymer Science Part A: Polymer Chemistry*,[73] John Wiley and Sons. © 2011 Wiley Periodicals, Inc.

2.3 Conclusion

The foregoing discussion shows proof of the redox activity, photoactivity, and magnetic property of transition metals, which continuously lead materials scientists to explore them as structural and functional building blocks to construct macromolecules. For instance, the coordination chemistry enables the design of heterometallic systems *via* the selective affinity of different metals for different ligands. Their discrete and variable electronic configurations impact magnetic properties, photoactivity, and redox activity on these complexes and these properties are transferable into macromolecular systems through well-designed synthetic protocols. An interesting aspect of the chemistry of these complexes is their tunability, which is achievable by intelligent control of the nature of the ligand, as well as the oxidation states of the transition metals, and is transferable to macromolecules designed from these complexes. The subsequent chapters highlight how materials scientists employ these complexes to design functional macromolecules to meet specific applications.

References

1. F. A. Cotton, G. Wilkinson, C. A. Murillo and M. Bochmann, *Advanced Inorganic Chemistry*, John Wiley & Sons Inc., New York, United States, 1999.
2. C. Housecroft and A. G. Sharpe, *Inorganic Chemistry*, Pearson Education Limited, Harlow, United Kingdom, 2012.
3. G. L. Miessler, P. J. Fischer and D. A. Tarr, *Inorganic Chemistry*, Pearson Education Limited, Harlow, United Kingdom, 2013.
4. D. Shriver, M. Weller, T. Overton, F. Armstrong and J. Rourke, *Inorganic Chemistry*, W. H. Freeman & Company, New York, United States, 2014.
5. J. E. Huheey, E. A. Keiter and R. L. Keiter, *Inorganic Chemistry: Principle of Structure and Reactivity*, Prentice Hall, Upper Saddle River, United States, 1997.
6. A. S. Abd-El-Aziz, C. Agatemor and N. Etkin, *Biomaterials*, 2017, **118**, 27.
7. W.-Y. Wong, *Coord. Chem. Rev.*, 2005, **249**, 971.
8. International Union of Pure and Applied Chemistry, *Compendium of Chemical Terminology, the Gold Book*, http://goldbook.iupac.org/pdf/goldbook.pdf.
9. J. A. Lemire, J. J. Harrison and R. J. Turner, *Nat. Rev. Microbiol.*, 2013, **11**, 371.
10. K. L. Haas and K. J. Franz, *Chem. Rev.*, 2009, **109**, 4921.
11. R. G. Pearson, *J. Chem. Educ.*, 1968, **45**, 581.
12. R. G. Pearson, *J. Chem. Educ.*, 1968, **45**, 643.
13. T. Luo, L. Hsu, C. Su, C. Ueng, T. Tsai and K. Lu, *Inorg. Chem.*, 2007, **46**, 1532.
14. M. L. Workentine, J. J. Harrison, P. U. Stenroos, H. Ceri and R. J. Turner, *Environ. Microbiol.*, 2008, **10**, 238.
15. M. Elias, A. Wellner, K. Goldin-Azulay, E. Chabriere, J. A. Vorholt, T. J. Erb and D. S. Tawfik, *Nature*, 2012, **491**, 134.
16. V. Amendola, M. Boiocchi, V. Brega, L. Fabbrizzi and L. Mosca, *Inorg. Chem.*, 2010, **49**, 997.
17. V. Balzani, A. Juris, M. Venturi, S. Campagna and S. Serroni, *Chem. Rev.*, 1996, **96**, 759.
18. D. Astruc, *Acc. Chem. Res.*, 2000, **33**, 287.
19. D. R. Tyler, *Acc. Chem. Res.*, 1991, **24**, 325.

20. M. Nippe, R. S. Khnayzer, J. A. Panetier, D. Z. Zee, B. S. Olaiya, M. Head-Gordon, C. J. Chang, F. N. Castellano and J. R. Long, *Chem. Sci.*, 2013, **4**, 3934.
21. D. Astruc, J. R. Hamon, E. Roman and P. Michaud, *J. Am. Chem. Soc.*, 1981, **103**, 7502.
22. D. Astruc, J. R. Hamon, G. Althoff, E. Roman, P. Batail, P. Michaud, J. P. Mariot, F. Varret and D. Cozak, *J. Am. Chem. Soc.*, 1979, **101**, 5445.
23. J. R. Hamon, D. Astruc and P. Michaud, *J. Am. Chem. Soc.*, 1981, **103**, 758.
24. J. Ruiz, F. Ogliaro, J. Saillard, J. Halet, F. Varret and D. Astruc, *J. Am. Chem. Soc.*, 1998, **120**, 11693.
25. D. Astruc, E. Roman E, J. R. Hamon and P. Batail, *J. Am. Chem. Soc.*, 1979, **101**, 2240.
26. K. Moudgil, M. A. Mann, C. Risko, L. A. Bottomley, S. R. Marder and S. Barlow, *Organometallics*, 2015, **34**, 3706.
27. S. Trupia, T. E. Bitterwolf and W. E. Geiger, *Organometallics*, 2012, **31**, 6063.
28. S. K. Mohapatra, A. Fonari, C. Risko, K. Yesudas, K. Moudgil, J. H. Delcamp, T. V. Timofeeva, J. Brédas, S. R. Marder and S. Barlow, *Chem.–Eur. J.*, 2014, **20**, 15385.
29. S. Trupia, A. Nafady and W. E. Geiger, *Inorg. Chem.*, 2003, **42**, 5480.
30. J. C. Swarts, A. Nafady, J. H. Roudebush, S. Trupia and W. E. Geiger, *Inorg. Chem.*, 2009, **48**, 2156.
31. M. G. Hill, W. M. Lamanna and K. R. Mann, *Inorg. Chem.*, 1991, **30**, 4687.
32. A. Nafady, P. J. Costa, M. J. Calhorda and W. E. Geiger, *J. Am. Chem. Soc.*, 2006, **128**, 16587.
33. M. Schulze, A. Steffen and F. Würthner, *Angew. Chem., Int. Ed.*, 2015, **54**, 1570.
34. A. S. Abd-El-Aziz, C. Agatemor, N. Etkin, D. P. Overy and R. G. Kerr, *RSC Adv.*, 2015, **5**, 86421.
35. D. V. Yandulov and R. R. Schrock, *Science*, 2003, **301**, 76.
36. A. S. Abd-El-Aziz, E. K. Todd, R. M. Okasha, P. O. Shipman and T. E. Wood, *Macromolecules*, 2005, **38**, 9411.
37. G. Crosby, K. Hipps and W. Elfring Jr, *J. Am. Chem. Soc.*, 1974, **96**, 629.
38. G. A. Crosby, *Acc. Chem. Res.*, 1975, **8**, 231.
39. Y. Y. Chia and M. G. Tay, *Dalton Trans.*, 2014, **43**, 13159.
40. Z. C. Su, C. C. Zheng, G. Cheng, C.-M. Che and S. J. Xu, *J. Mater. Chem. C*, 2017, **5**, 4488.
41. M. Z. Shafikov, D. N. Kozhevnikov, M. Bodensteiner, F. Brandl and R. Czerwieniec, *Inorg. Chem.*, 2016, **55**, 7457.
42. D. Ma, H. He, K. Leung, D. S. Chan and C. Leung, *Angew. Chem., Int. Ed.*, 2013, **52**, 7666.
43. J. Heine and K. Müller-Buschbaum, *Chem. Soc. Rev.*, 2013, **42**, 9232.
44. U. Schatzschneider, *Eur. J. Inorg. Chem.*, 2010, **2010**, 1451.
45. C. Daniel, *Coord. Chem. Rev.*, 2015, **282**, 19.
46. H. Weissman, E. Shirman, T. Ben-Moshe, R. Cohen, G. Leitus, L. J. Shimon and B. Rybtchinski, *Inorg. Chem.*, 2007, **46**, 4790.

47. A. A. Rachford, S. Goeb and F. N. Castellano, *J. Am. Chem. Soc.*, 2008, **130**, 2766.
48. E. O. Danilov, A. A. Rachford, S. Goeb and F. N. Castellano, *J. Phys. Chem. A*, 2009, **113**, 5763.
49. W. Zhang, F. Zhang, Y. Wang, B. Song, R. Zhang and J. Yuan, *Inorg. Chem.*, 2017, **56**, 1309.
50. A. Bierhaus, T. Fleming, S. Stoyanov, A. Leffler, A. Babes, C. Neacsu, S. K. Sauer, M. Eberhardt, M. Schnölzer, F. Lasitschka, W. L. Neuhuber, T. I. Kichko, I. Konrade, R. Elvert, W. Mier, V. Pirags, I. K. Lukic, M. Morcos, T. Dehmer, N. Rabbani, P. J. Thornalley, D. Edelstein, C. Nau, J. Forbes, P. M. Humpert, M. Schwaninger, D. Ziegler, D. M. Stern, M. E. Cooper, U. Haberkorn, M. Brownlee, P. W. Reeh and P. P. Nawroth, *Nat. Med.*, 2012, **18**, 926.
51. L. Zhang, S. Yin, M. Pan, W. Liao, J. Zhang, H. Wang and C. Su, *J. Mater. Chem. A*, 2017, **5**, 9807.
52. J. Liao, Y. Chi, C. Yeh, H. Kao, C. Chang, M. A. Fox, P. J. Low and G. Lee, *J. Mater. Chem. C*, 2015, **3**, 4910.
53. M. Göppert-Mayer, *Ann. Phys.*, 1931, **401**, 273.
54. H. M. Kim and B. R. Cho, *Chem. Rev.*, 2015, **115**, 5014.
55. J. Jing, J. Chen, Y. Hai, J. Zhan, P. Xu and J. Zhang, *Chem. Sci.*, 2012, **3**, 3315.
56. D. Xie, J. Jing, Y. Cai, J. Tang, J. Chen and J. Zhang, *Chem. Sci.*, 2014, **5**, 2318.
57. Y. Hai, J. Chen, P. Zhao, H. Lv, Y. Yu, P. Xu and J. Zhang, *Chem. Commun.*, 2011, **47**, 2435.
58. F. Helmchen and W. Denk, *Nat. Methods*, 2005, **2**, 932.
59. W. Denk, J. H. Strickler and W. W. Webb, *Science*, 1990, **248**, 73.
60. W. Xu, J. Zuo, L. Wang, L. Ji and H. Chao, *Chem. Commun.*, 2014, **50**, 2123.
61. C. Xu and W. W. Webb, *J. Opt. Soc. Am. B*, 1996, **13**, 481.
62. A. R. West, *Basic Solid State Chemistry*, John Wiley & Sons Ltd, West Sussex, United Kingdom, 1999.
63. R. Valenzuela, *Magnetic Ceramics*, Cambridge University Press, Cambridge, United Kingdom, 1994.
64. X. Wang, C. Avendaño and K. R. Dunbar, *Chem. Soc. Rev.*, 2011, **40**, 3213.
65. N. A. Frey, S. Peng, K. Cheng and S. Sun, *Chem. Soc. Rev.*, 2009, **38**, 2532.
66. R. Boča, J. Miklovič and J. Titiš, *Inorg. Chem.*, 2014, **53**, 2367.
67. J. Titiš, J. Miklovič and R. Boča, *Inorg. Chem. Commun.*, 2013, **35**, 72.
68. D. Schweinfurth, J. Krzystek, M. Atanasov, J. Klein, S. Hohloch, J. Telser, S. Demeshko, F. Meyer, F. Neese and B. Sarkar, *Inorg. Chem.*, 2017, **56**, 5253.
69. J. Krzystek, D. C. Swenson, S. Zvyagin, D. Smirnov, A. Ozarowski and J. Telser, *J. Am. Chem. Soc.*, 2010, **132**, 5241.
70. W. Zhang, R. Ishikawa, B. Breedlove and M. Yamashita, *RSC Adv.*, 2013, **3**, 3772.
71. Y. Zhang, S. Gómez-Coca, A. J. Brown, M. R. Saber, X. Zhang and K. R. Dunbar, *Chem. Sci.*, 2016, **7**, 6519.

72. M. S. Fataftah, S. C. Coste, B. Vlaisavljevich, J. M. Zadrozny and D. E. Freedman, *Chem. Sci.*, 2016, **7**, 6160.
73. W. S. Seo, H. H. Jo, K. Lee, B. Kim, S. J. Oh and J. T. Park, *Angew. Chem., Int. Ed.*, 2004, **43**, 1115.
74. K. Woo, J. Hong, S. Choi, H. Lee, J. Ahn, C. S. Kim and S. W. Lee, *Chem. Mater.*, 2004, **16**, 2814.
75. S. Sun, H. Zeng, D. B. Robinson, S. Raoux, P. M. Rice, S. X. Wang and G. Li, *J. Am. Chem. Soc.*, 2004, **126**, 273.
76. S. Sun and H. Zeng, *J. Am. Chem. Soc.*, 2002, **124**, 8204.
77. Q. Song and Z. J. Zhang, *J. Am. Chem. Soc.*, 2004, **126**, 6164.
78. M. Unni, A. M. Uhl, S. Savliwala, B. H. Savitzky, R. Dhavalikar, N. Garraud, D. P. Arnold, L. F. Kourkoutis, J. S. Andrew and C. Rinaldi, *ACS Nano*, 2017, **11**, 2284.

CHAPTER 3

Synthesis of Transition-metal-containing Macromolecules

3.1 Introduction

Transition-metal-containing macromolecules are attractive materials because they integrate the desirable mechanical properties and processibility of organic polymers with the unique optoelectronic, catalytic and magnetic properties of transition metals.[1–6] Early attempts to synthesize well-defined high molecular weight macromolecules were unsuccessful due to ineffective synthetic methods and limited characterization techniques. Remarkable breakthroughs began in the 1990s,[4–6] which led to reports of high-molecular-weight, well-characterized transition-metal-containing polymers, thanks to the serendipity as well as the ingenuity of scientists that fostered innovative polymerization and characterization techniques. The ability to rationally metalate an organic polymer to realize a target function is now possible, though gaps exist that create space for more innovations. Strategies now abound to precisely install transition metals onto macromolecular frameworks *via* static covalent or dynamic coordinate bonds. Polymerization of transition-metal-containing monomers or post-synthesis metalation of macromolecules are state-of-the-art synthetic routes to access these macromolecules. This chapter focuses on advances in the design of metalated macromolecules within the context of selected examples.

Polymer Chemistry Series No. 27
Macromolecules Incorporating Transition Metals: Tackling Global Challenges
By Alaa S. Abd-El-Aziz, Christian Agatemor and Wai-Yeung Wong

Published by the Royal Society of Chemistry, www.rsc.org

3.2 Synthesis of Transition-metal-containing Macromolecules

3.2.1 Chain Polymerization of Transition-metal-containing Monomers

The first generation of transition-metal-containing macromolecules includes iron-containing polymers developed from ferrocene or ferrocene-containing vinyl, acrylate, or methacrylate monomers.[4] Pioneers in the field speculated on the direct coupling of the cyclopentadienyl (Cp) ligands in ferrocene to afford polymers *via* a free-radical chain polymerization mechanism.[7–10] A typical example is a peroxyl radical-initiated chain polymerization of ferrocene at 200 °C to yield a polymer with short-chain poly(ferrocenylene) segments formed *via* 1,2- or 1,3-homoannular and 1,1′-heteroannular coupling at the Cp ligand (Figure 3.1).[11–13] These early attempts yielded poorly defined, low-molecular-weight polymers with inferior hydrolytic and thermal stability but inspired the development of new monomers to access these polymers, which were targeted as functional materials.

The ability to facilely synthesize ferrocene derivatives, for instance, *via* reactions at the Cp ring (Scheme 3.1) enables the design of several ferrocene-containing monomers,[4] with some amenable to chain polymerization. Pittman *et al.* synthesized poly(ferrocenylethylene), poly(ferrocenylmethyl acrylate), and poly(ferrocenylmethyl methacrylate) *via* AIBN-catalyzed free-radical polymerization of vinylferrocene, ferrocenylmethyl acrylate, and ferrocenylmethyl methacrylate, respectively (Scheme 3.2).[14] They found a first-order reaction mechanism, which is typical of free-radical polymerization of methacrylate and methyl methacrylate monomers,[14] indicating limited interference from the ferrocenyl iron center on the polymerization mechanism. Gel permeation chromatography (GPC) revealed that the benzene-soluble fraction consists of mostly uncross-linked, non-uniform homopolymers. Vapor pressure osmometry data suggested that the polymers have moderate molecular weights with an absolute number average molecular weight (M_n) that varies from 6955–20 130 Da and 8650–35 500 Da for poly(ferrocenylmethyl acrylate) and poly(ferrocenylmethyl methacrylate), respectively.[14]

Polymerization of monomers with carbon–carbon double bonds remains the most explored of the chain polymerizable unsaturated monomers. Several well-established chemical reactions enable the grafting of organometallic

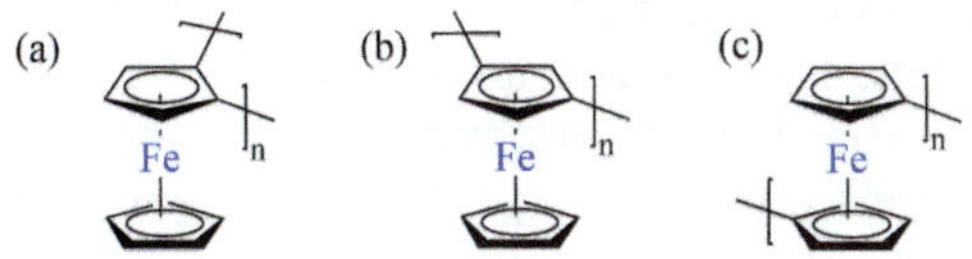

Figure 3.1 Poly(ferrocenylene) formed by (a) 1,2-homoannular (b) 1,3-homoannular and (c) 1,1′-heteroannular coupling at the cyclopentadienyl ring.

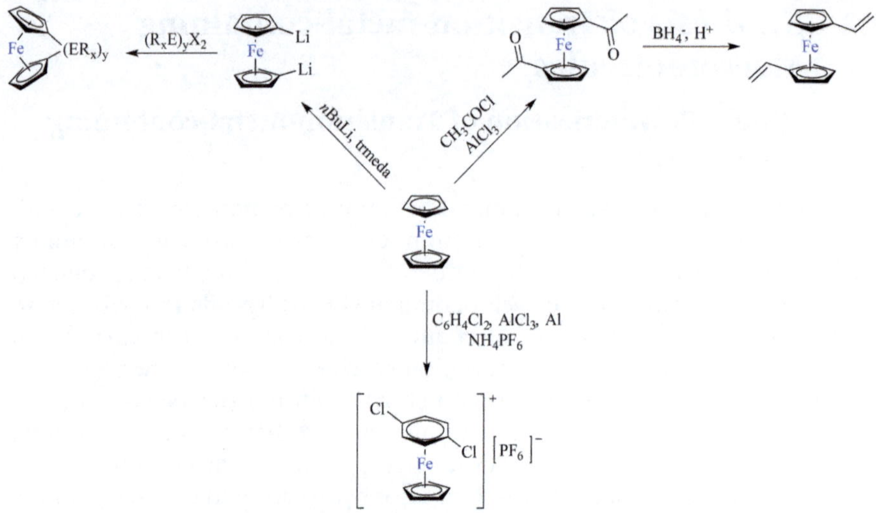

Scheme 3.1 Reactions of ferrocene to give polymerizable monomers.

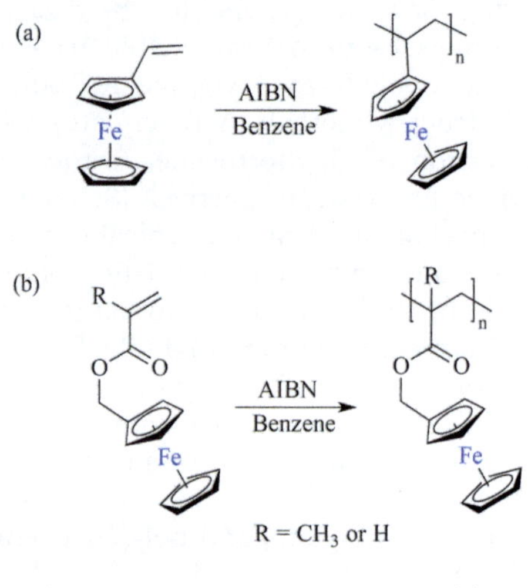

Scheme 3.2 AIBN-catalyzed radical polymerization of (a) vinylferrocene, and (b) ferrocenylmethyl acrylate, or ferrocenylmethyl methacrylate.

moieties or transition metal coordination complexes onto these carbon–carbon double-bond-containing monomers. To illustrate this, Chernyy *et al.* successfully grafted ferrocene onto a methyl methacrylate monomer *via* a series of reactions that includes Vilsmeier–Haack formylation, reduction, and acylation reactions (Scheme 3.3).[15] The ferrocenemethyl methacrylate monomer undergoes anionic chain polymerization, producing block copolymers of poly(ferrocenylmethyl methacrylate)-*block*-poly(methyl methacrylate) or poly(ferrocenylmethyl methacrylate)-*block*-poly(nonafluorohexyl

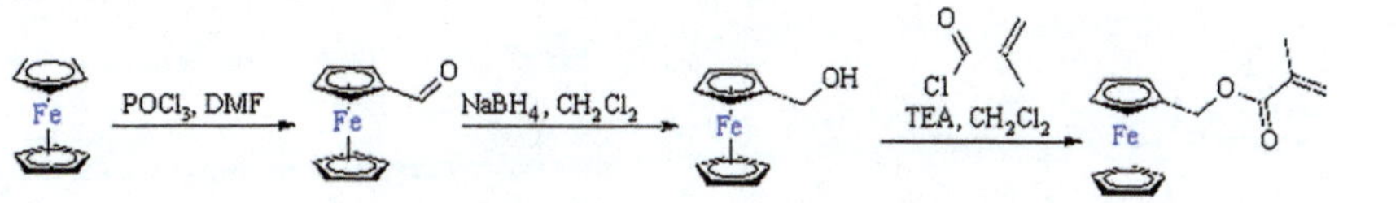

Scheme 3.3 Synthesis of ferrocenylmethyl methacrylate monomer.

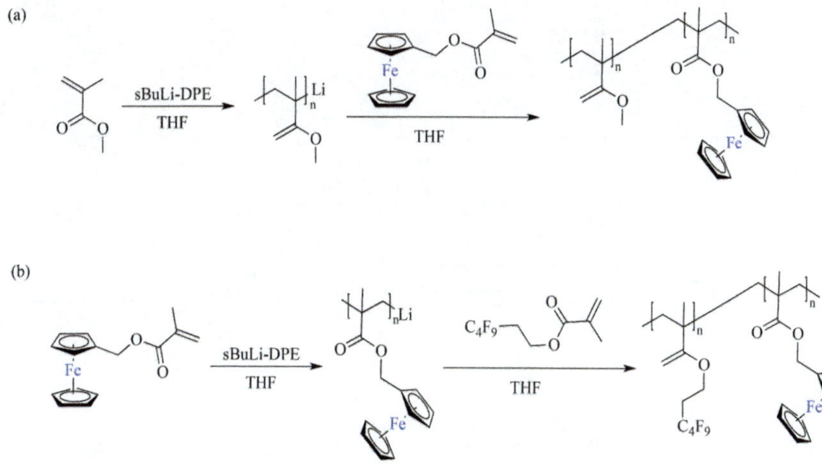

Scheme 3.4 Sequential anionic polymerization to produce (a) poly(ferrocenylmethyl methacrylate)-*block*-poly(methyl methacrylate) and (b) poly(ferrocenylmethyl methacrylate)-*block*-poly(nonafluorohexyl methacrylate).

methacrylate) (Scheme 3.4).[15] Polymerizability of the carbon–carbon double bond is contingent on the inductive and resonance effects of the substituent(s) around the bond, in this case, ferrocene. The substituent(s) modulate the electron-cloud density on the double bond, stabilizing or destabilizing the resulting radical, carbocation, or carbanion.[16] In these anionic chain polymerizations, a logical reaction sequence that involves the erstwhile polymerization of the methyl methacrylate block before the ferrocenylmethyl block was employed to obtain the poly(ferrocenylmethyl methacrylate)-*block*-poly(methyl methacrylate). This sequence is necessary since the ferrocenylmethyl substituent stabilizes the carbanion relative to a methyl methacrylate without a ferrocene. The overall effect is that the lithiated poly(ferrocenylmethyl methacrylate) is less nucleophilic than the lithiated poly(methyl methacrylate), and therefore, less reactive towards anionic polymerization. Conversely, to obtain poly(ferrocenylmethyl methacrylate)-*block*-poly(nonafluorohexyl methacrylate), Chernyy and coworkers polymerized the ferrocenylmethyl methacrylate in the first step and the nonafluorohexyl methacrylate in the final step because of the strong electrophilicity of a lithiated poly(nonafluorohexyl methacrylate).[15] This strategy affords ferrocene-containing copolymers with molecular weights in the range of 15 500–35 200 Da and dispersity (*Đ*) of 1.02–1.04. The anionic

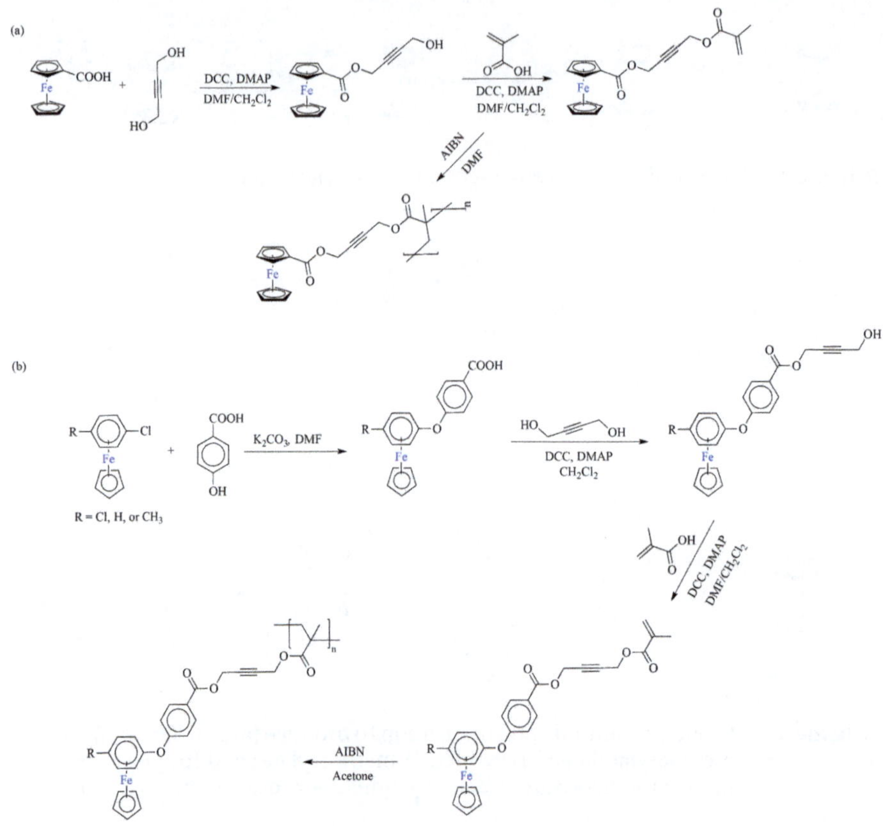

Scheme 3.5 Steglich esterification reaction affords polymerizable (a) ferrocene- and (b) η^6-dichlorobenzene-η^5-cyclopentadienyliron(II)-containing methacrylate monomers.

polymerization also enables the synthesis of a poly(ferrocenylmethyl methacrylate) homopolymer with a molecular weight of 45 800 Da and dispersity ($Đ$) of 1.15.

Nucleophilic acyl substitution reactions such as esterification and amidation are employed to graft organometallic moieties to (meth)acrylate monomers. Abd-El-Aziz *et al.* demonstrated this by designing ferrocene- and cationic η^6-dichlorobenzene-η^5-cyclopentadienyliron(II)-containing methacrylate monomers *via* the Steglich esterification reaction (Scheme 3.5).[17] The AIBN-catalyzed radical chain polymerization of these monomers requires dry DMF as the most suitable solvent and a monomer to initiator ratio of 7 : 1 as the optimal stoichiometry. The polymerization affords polymers with moderate molecular weights, with the M_n of the THF-soluble fractions of the demetalated cationic polymers being in the range of 17 800–23 200 Da. Also, a Steglich esterification reaction between a styrene moiety and a tertiary phosphine bearing a ferrocene and a ruthenocene moiety gives a polymerizable heterobimetallic monomer (Scheme 3.6).[18] AIBN-catalyzed radical chain

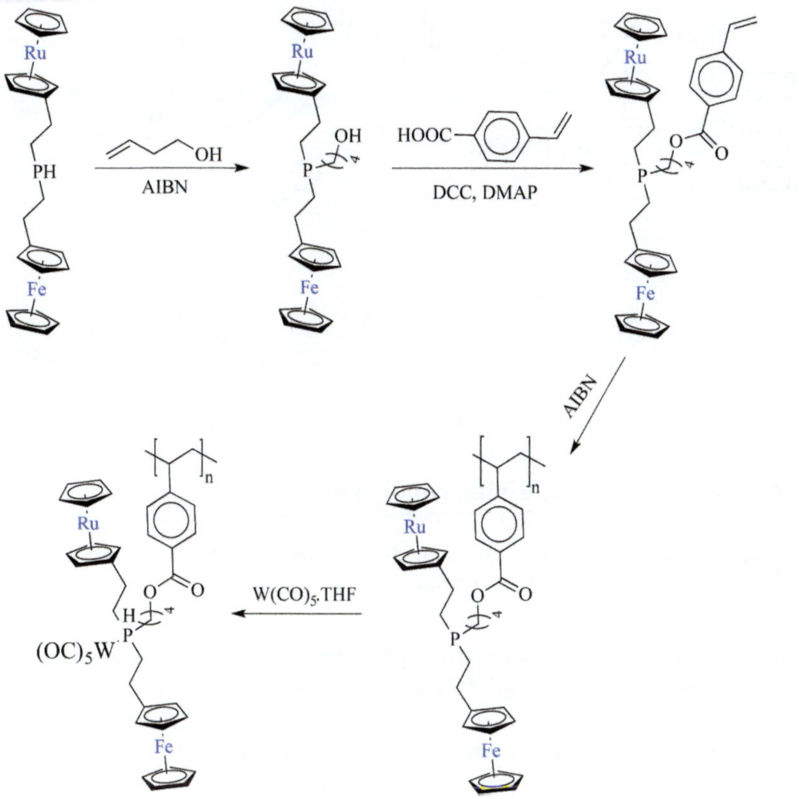

Scheme 3.6 Steglich esterification reaction affords a polymerizable Fe/Ru heterobimetallic styrene monomer.

polymerization of this monomer yielded a heterobimetallic polymer, which coordinates to $W(CO)_5$ to form a heterotrimetallic polymer (Scheme 3.6). ^1H NMR confirms successful polymerization of the monomer whereas ^{31}P NMR proves the coordination of $W(CO)_5$ to the phosphine ligand. The resulting polymers exhibit poor solubility profiles in common GPC solvents including H_2O and DMF, a challenge that limits GPC analysis to THF-soluble fractions.

Also, acylation is used to graft cobaltocenium cations onto methacrylate monomers to construct cobalt-containing polymers. As an example, Tang *et al.* designed a Co-containing monomer, 2-cobaltoceniumamidoethyl methacrylate hexafluorophosphate, by amidation reaction of cobaltocenium carboxylic acid hexafluorophosphate with 2-aminoethyl methacrylate hydrochloride (Scheme 3.7).[19,20] They copolymerized the monomer with a crosslinker, poly(ethylene glycol) dimethacrylate, in an AIBN-catalyzed radical chain polymerization to form a polymeric gel, whose rheological properties depend on the molar ratio of the crosslinker.[20]

Chain polymerization of monomeric transition metal coordination complexes provides an accessible route to metalated polymers. Lü and coworkers, for example, designed vinyl-containing coordination complexes in a

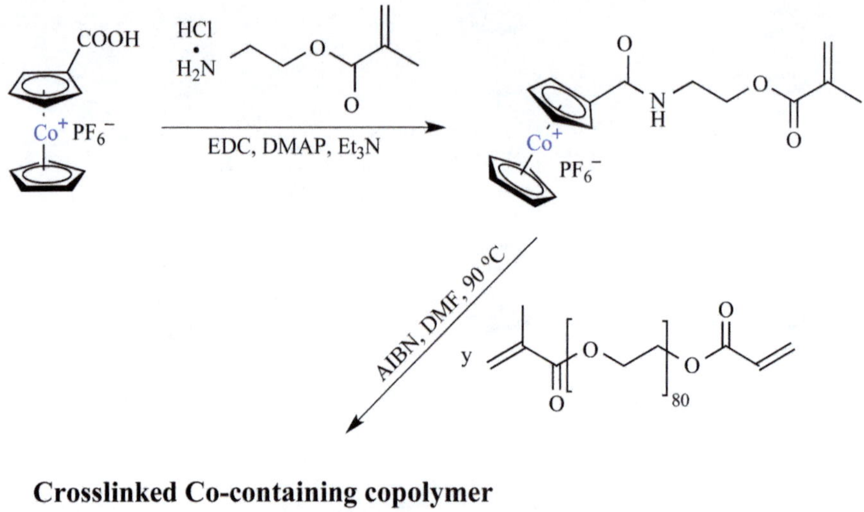

Scheme 3.7 Amidation reaction provides a polymerizable Co-containing methyl methacrylate monomer.

multistep reaction and successfully polymerized the complexes *via* AIBN-initiated chain polymerization.[21,22] The synthesis of a monomeric coordination complex entails a nucleophilic substitution reaction between a tripodal tetradentate ligand, tris(2-benzimidazolylmethyl)amine, and 1-(chloromethyl)-4-vinylbenzene, and the subsequent coordination of La^{3+}, Eu^{3+}, Tb^{3+}, or Gd^{3+} to the product (Scheme 3.8).[21] The vinyl group of these complexes is susceptible to AIBN-initiated chain polymerization, allowing the copolymerization of these monomeric complexes with methyl methacrylate monomer (Scheme 3.8). FT-IR, ^{1}H NMR and GPC data confirmed the success of the copolymerization reaction. GPC, for instance, revealed the formation of copolymers with close to uniform dispersity ($Đ = 1.03–1.07$) and molecular weights in the range of 54 391–78 149 Da.[21] Lü *et al.* also designed a vinyl-containing salen-type metal complex that undergoes AIBN-initiated copolymerization with methyl methacrylate monomer to produce triblock heterotrimetallic copolymers.[22]

As another strategy, the Tang group explored living radical polymerization techniques to polymerize transition-metal-containing methacrylate monomers.[23–29] In contrast with traditional radical and ionic chain polymerization mechanisms, living radical polymerization proceeds without chain-breaking reactions, and so could enable the design of block copolymers *via* sequential addition of different monomers.[16] Indeed, living radical polymerization extends the radical polymerization landscape to the design of high-molecular-weight, well-defined materials that feature narrow molecular weight distribution and are useful in microelectronics, nanotechnology, biomedical applications, and other emerging fields. Successful living radical polymerization requires a judicious choice of the appropriate monomer, initiator, activator catalyst, and deactivator as well as a fine-tuning

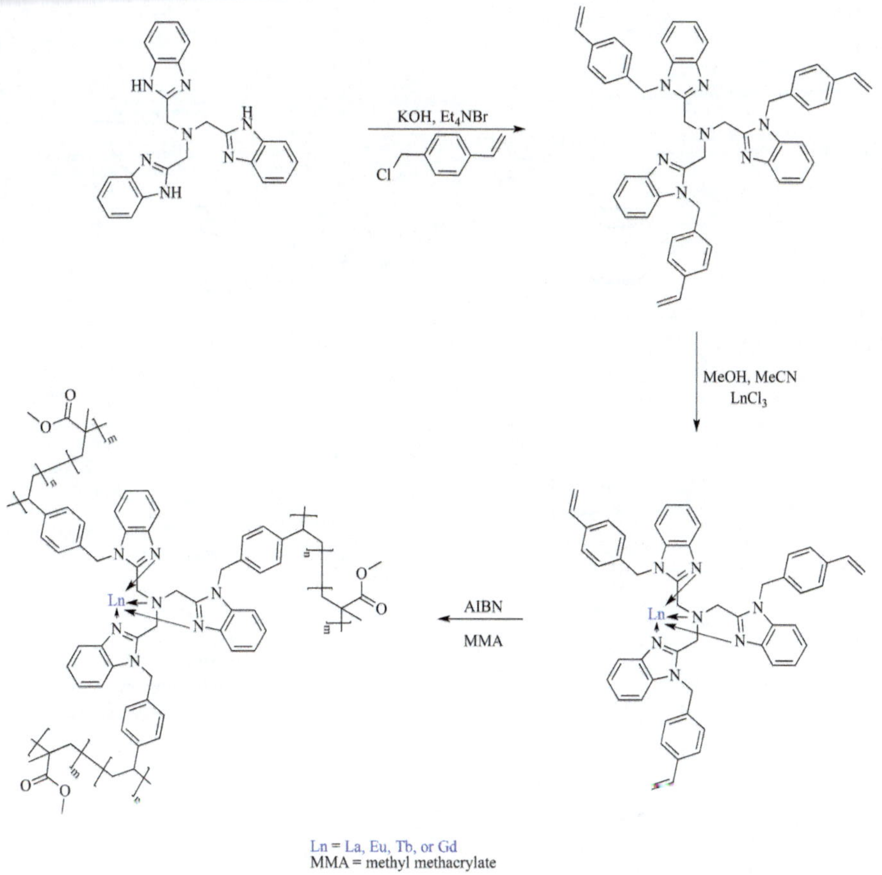

Scheme 3.8 Polymerization of monomeric coordination complexes.

of the components' concentration and reaction conditions to suppress chain-breaking reactions. Tang and coworkers successfully polymerized ferrocene-containing methacrylate monomers *via* atom transfer radical polymerization (ATRP), a living radical polymerization technique (Figure 3.2).[25] They found that in the series of acrylate monomers, as the length of the spacer between the ferrocene moiety and the double bond increases, the polymerization rate decreases (Figure 3.2).[25] This finding suggests that the steric effect of the bulky ferrocene moiety exerts no retarding effect on the polymerization rate since the monomers with longer alkyl spacers exhibit the slowest polymerization rates. Nonetheless, polymerizations of monomers with longer spacers are better controlled than those with shorter spacers as evidenced by their lower *Đ*, presumably because of slower termination rates. Grafting the ferrocene moiety to the acrylate monomer with a triazole linker obtained *via* a copper(I)-catalyzed azide–alkyne cycloaddition reaction (Scheme 3.2) resulted in a monomer that failed to undergo ATRP.[25] The

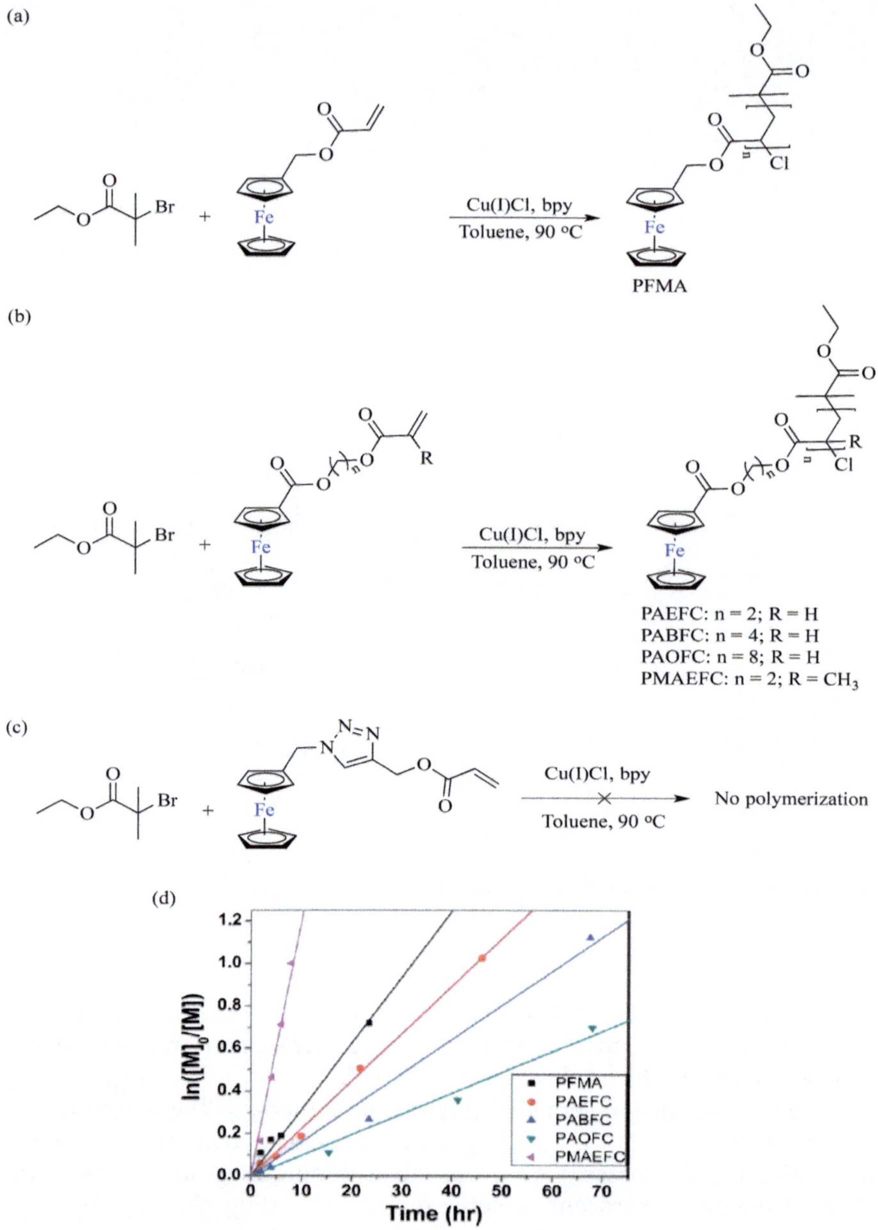

Figure 3.2 (a)–(c) Atom transfer radical polymerization of ferrocene-containing methacrylate monomers. (d) Effect of spacer length and structure on the rate of polymerization. Reproduced from C. G. Hardy, L. Ren, T. C. Tamboue and C. Tang, Side-chain ferrocene-containing methacrylate polymers: Synthesis and properties, *Journal of Polymer Science Part A: Polymer Chemistry*,[25] John Wiley and Sons, © 2011 Wiley Periodicals, Inc.

triazole ligand most likely either chelates the ATRP copper catalyst or exerts a deleterious steric effect on the monomer polymerizability.

Tang and coworkers observed that ATRP of ferrocene-containing methacrylate monomers proceeds faster than their acrylate analogs, achieving 63% conversion in eight hours compared with 64% in 46 hours realized with the acrylate.[25] In contrast, Ragogna and coworkers found that ATRP or the traditional AIBN-initiated radical polymerization of methacrylate-functionalized η^4-cyclobutadiene-η^5-cyclopentadienylcobalt(I) monomer fails but its acrylate analog successfully undergoes polymerization, although under stringent conditions (Scheme 3.9).[30] The nature of the substituent on the cyclobutadiene ring of the η^4-cyclobutadiene-η^5-cyclopentadienylcobalt(I) moiety or the linker connecting the moiety to the acrylate group impacts on the degree of polymerizability. Higher conversions are attainable when the cyclobutadiene ring has phenyl substituents and an ester linker compared with a system with methyl substituents and a ketone linker (Scheme 3.9).[30] Under similar conditions, ATRP of the monomer with phenyl substituents and an ester linker proceeds to 70% completion, compared with the 10% achieved with the monomer with methyl substituents and a ketone linker (Scheme 3.9).[30]

The AIBN-initiated radical polymerization of the acrylate-functionalized η^4-cyclobutadiene-η^5-cyclopentadienylcobalt(I) monomers requires stringent conditions, including prolonged polymerization times and high temperatures, to achieve good conversions.[30] Under these circumstances, the obtained polymers feature a low molecular weight (10 000–30 000 Da) and a non-uniform dispersity ($Đ = 1.31–2.09$).[30] Attempts to access high-molecular-weight polymers using another living polymerization technique—reversible addition-fragmentation transfer (RAFT) polymerization—failed, and a rigorous investigation of the system establishes that the steric effect of η^4-cyclobutadiene-η^5-cyclopentadienylcobalt(I) hinders chain propagation, but promotes chain termination and transfer reactions.[31] By copolymerizing the bulky acrylate-containing η^4-cyclobutadiene-η^5-cyclopentadienylcobalt(I) monomer with a less bulky methylacrylate monomer, Ragogna and coworkers mitigated the deleterious effect of the steric demand and realized a high-molecular-weight random copolymer with excellent control over molecular weight.[31] The living nature of the polymerization enables the design of triblock copolymers as demonstrated by the sequential addition of styrene to the diblock copolymer that functions as a macro RAFT agent to form the triblock copolymer.[31] The construction of a heterobimetallic block copolymer is possible by the sequential addition and copolymerization of a phosphonium-salt-functionalized styrene monomer that coordinates to a gold anion.[32]

Another illustrative exploration of living radical polymerization for the design of metal-containing macromolecules is exemplified by the recent work of Tang *et al.*, who successfully assembled Fe and Co homopolymers *via* photoinduced RAFT polymerization of Fe or Co-containing monomers, respectively (Figure 3.3).[23] RAFT polymerization of the Co-containing monomer, 2-cobaltocenium amidoethyl methacrylate hexafluorophosphate, proceeds slowly at room temperature, affording <20% conversion after 24 hours,

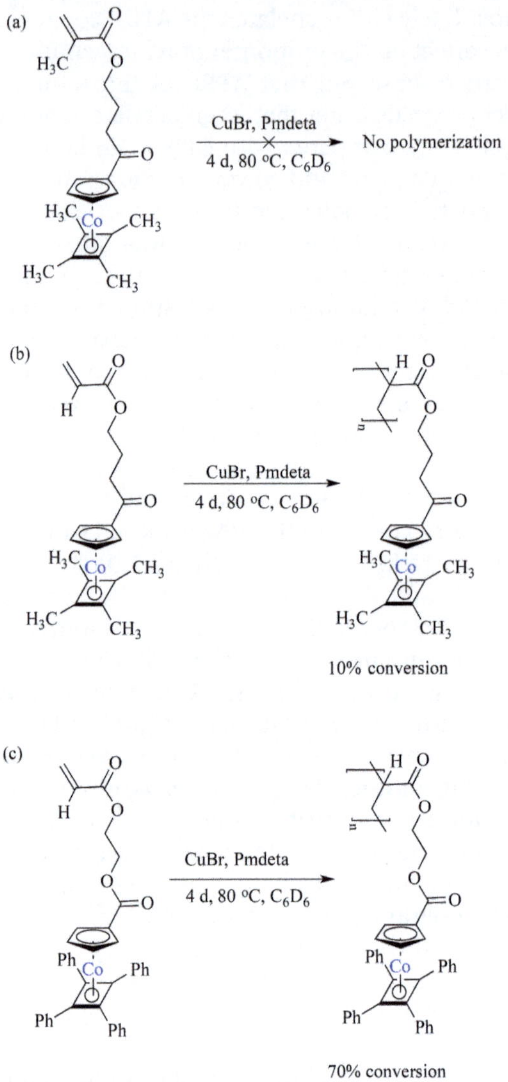

Scheme 3.9 Polymerizability of η[4]-cyclobutadiene-η[5]-cyclopentadienylcobalt(I)-functionalized methacrylate monomer.

presumably due to the low temperature, which ineffectively promotes polymerization.[23] Higher conversions are attainable at higher temperatures, with 60 °C and 90 °C yielding 43% and 64% conversions, respectively.[23] The nature of the RAFT agent impacts the degree of conversion of 2-cobaltocenium amidoethyl methacrylate hexafluorophosphate monomer as evidenced by the better conversions, for example 64% at 90 °C obtained with (dodecylthiocarbonothioylthio)-2-methylpropionic acid as the RAFT agent, compared with the 51% at 90 °C achieved with cumyl dithiobenzoate.[23] Kinetic studies proved that the polymerization is "living", as indicated by the linearity of the

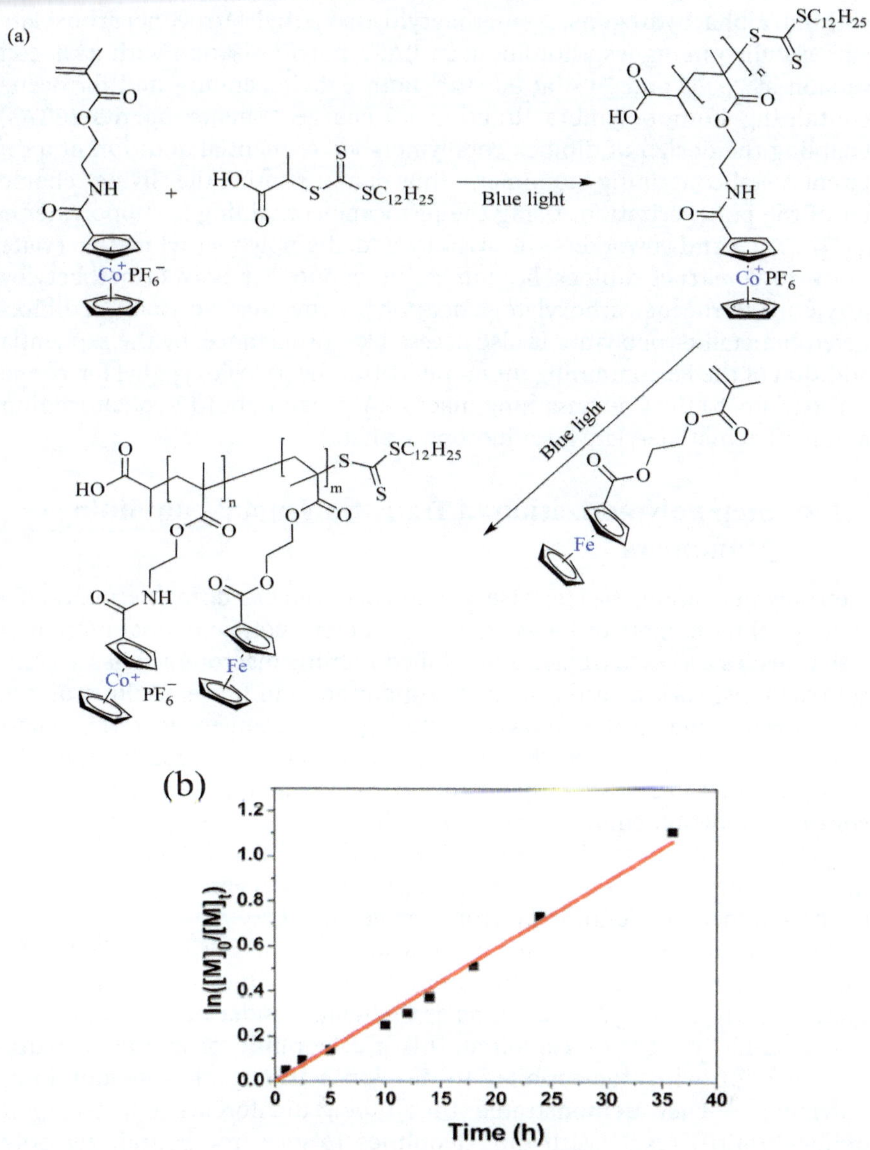

Figure 3.3 Photoinduced reversible addition–fragmentation chain transfer polymerization for the synthesis of Co-containing homopolymer and Co/Fe-containing heterobimetallic copolymer (a), and kinetics of polymerization (b). Reprinted from ref. 23. Copyright (2016) American Chemical Society.

plot of In([M]$_0$/[M]$_t$) *versus* irradiation time (Figure 3.3).[23] Photoinduced RAFT polymerization of ferrocenylmethyl methacrylate monomer is unsuccessful even at high temperature (90 °C) because of chain transfer side reactions that are driven by radical-initiated hydrogen abstraction from the methylene group in a position alpha to the ferrocene moiety.[23] A ferrocene monomer without alpha hydrogens, 2-(methacryloyloxy)ethyl ferrocenecarboxylate, successfully undergoes photoinduced RAFT polymerization with 28% conversion at 25 °C and 74% at 90 °C.[23] Both cobaltocenium- and ferrocene-containing homopolymers function as charge transfer agents (CTAs), enabling the design of diblock copolymers *via* sequential addition of a different metal-containing monomer,[23] thus demonstrating the "living" character of the polymerization. Using the ferrocene-containing homopolymer as a CTA, Tang and coworkers successfully added a poly(methyl methacrylate) block to construct diblock homometallic copolymer poly(2-(methacryloyloxy)ethyl ferrocene-carboxylate)-*block*-poly(methyl methacrylate). A diblock heterobimetallic copolymer is also accessible, for instance, by the sequential addition of the Fe-containing monomer, 2-(methacryloyloxy)ethyl ferrocenecarboxylate, to the Co-containing macro RAFT agent, poly(2-cobaltocenium amidoethyl methacrylate hexafluorophosphate).[23]

3.2.2 Step Polymerization of Transition-metal-containing Monomers

Step polymerization, the stepwise reaction between two different homo-polyfunctional monomers or between a single hetero-polyfunctional monomer, also provides access to transition-metal-containing macromolecules. Chemical reactions, such as esterification, amidation, and nucleophilic aromatic substitution, enable the stepwise coupling of monomers to form macromolecules. In step polymerization, high-molecular-weight macromolecules are only achievable at very high conversions, a characteristic that demands control of reaction equilibrium to form the product, as well as elimination of cyclization and other deleterious side reactions.[16] Regardless of these conditions, chemists have successfully explored this technique to construct transition-metal-containing macromolecules.

The coordination of metal complexes such as $Cr(CO)_3$, $Fe^+(C_5H_5)$, $Ru^+(C_5H_5)$, or $Mn^+(CO)_3$ to chloroarene ligands enables nucleophilic aromatic substitution (S_NAr) reactions at the ligand under mild conditions.[33] The Abd-El-Aziz group exploited this nucleophilic aromatic substitution reaction with nucleophiles to develop a library of iron-containing polymers.[2–4] They demonstrated that the η^6-dichloroarene-η^5-cyclopentadienyliron(II) reacts with dinucleophiles to give iron-containing poly(aromatic ethers), poly(thioethers) or poly(amines) under mild conditions (Schemes 3.10–3.12).[34–40] Structural diversity in available dinucleophiles is an asset they exploited to control the structure of the polymer, allowing fine-tuning of key fundamental properties such as solubility and thermal stability. For instance, a diphenolic acid, such as 4,4-bis(4-hydroxyphenyl) valeric acid, affords an iron-containing polymer that is soluble in a basic medium.[36] Also, alkylated nucleophiles yield polymers that are much more soluble than unalkylated ones.[36] These iron-containing polymers are more soluble in common laboratory solvents than their demetalated organic analogs, indicating that the iron centers enhance solubility.[36]

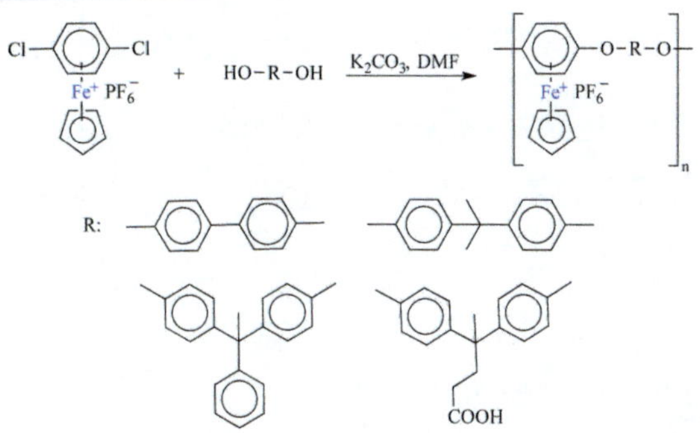

Scheme 3.10 Step polymerization *via* nucleophilic aromatic substitution reaction between η^6-dichloroarene-η^5-cyclopentadienyliron(II) and diphenolic nucleophiles.

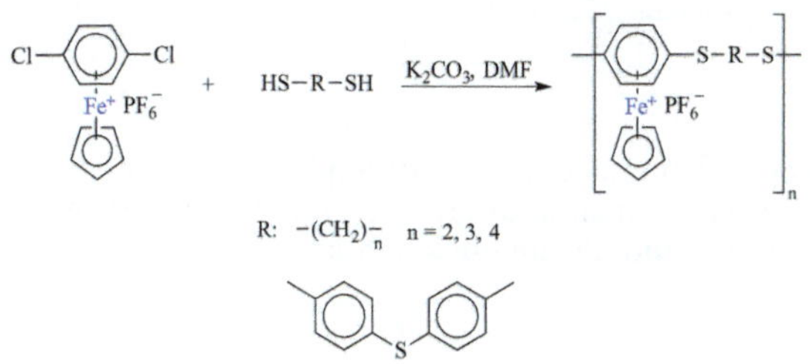

Scheme 3.11 Step polymerization *via* nucleophilic aromatic substitution reaction between η^6-dichloroarene-η^5-cyclopentadienyliron(II) and dithiol nucleophiles.

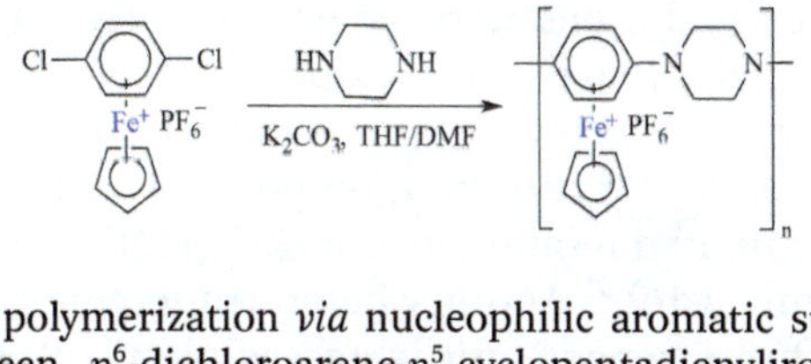

Scheme 3.12 Step polymerization *via* nucleophilic aromatic substitution reaction between η^6-dichloroarene-η^5-cyclopentadienyliron(II) and diamine nucleophiles.

Furthermore, the flexibility of the dinucleophile controls the solubility and monomer conversion as exemplified by the more soluble, more thermally stable, and higher-molecular-weight poly(thioether)s realized by changing from a 1,4-butanedithiol nucleophile to a more flexible 1,6-hexanedithiol or 1,8-octanedithiol.[36] With a judicious selection of dinucleophiles, Abd-El-Aziz and coworkers demonstrated the feasibility to synthetically access, under mild conditions, soluble iron-containing poly(aromatic sulfide) in excellent yield (92–96%) and polymers with alternating thioether, ether or amine backbones (Scheme 3.13).[36] As the cationic iron centers

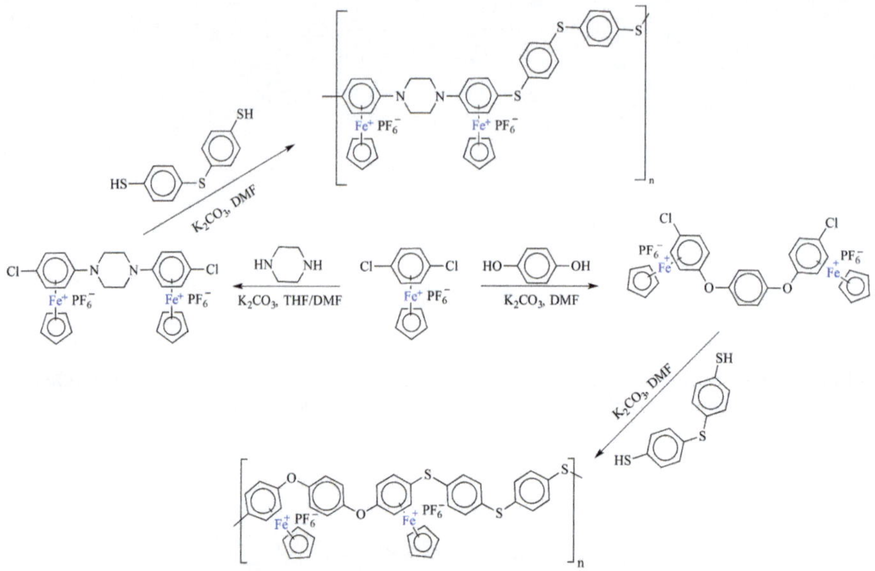

Scheme 3.13 Step polymerization *via* nucleophilic aromatic substitution reaction between η^6-dichloroarene-η^5-cyclopentadienyliron(II) and diphenolic, diamine, and dithiol nucleophiles.

interact with most GPC columns, it is a challenge to characterize the molecular weights of the iron-containing polymer using GPC. Nonetheless, photolytic demetalation affords organic polymers, which are compatible with most GPC columns, allowing estimation of the molecular weights.

Also, the η^6-dichloroarene-η^5-cyclopentadienyliron(II) mediates the facile synthesis of bifunctional monomers. As an example, a multistep synthesis that includes a η^6-dichloroarene-η^5-cyclopentadienyliron(II)-mediated S_NAr reaction provided access to terpyridine-containing organoiron monomers (Schemes 3.14 and 3.15).[41] A coordination polymerization reaction between the terpyridine-terminated monomers and Fe^{2+} or Ni^{2+} gives heterotrimetallic polymers (Scheme 3.16).[41] Another heterotrimetallic polymer with three metal centers within a repeat unit is accessible *via* S_NAr reaction of the η^6-dichloroarene-η^5-cyclopentadienyliron(II)-terminated monomer with diphenolic nucleophiles (Scheme 3.17).[41]

Polymetallaynes are an attractive class of conjugated polymers that incorporate a transition metal along the conjugated system and are rigorously investigated as semiconductors in optoelectronic devices. A widely-explored subclass, polyplatinynes, which incorporate platinum as the metal center, are synthetically accessible *via* Sonogashira-type cross-coupling step polymerization reactions between alkyne-terminated monomers and platinum chlorides. A structurally diverse library of polyplatinynes with distinctive optoelectronic properties has been developed *via* this cross-coupling polymerization by structurally modifying the aromatic spacer in the

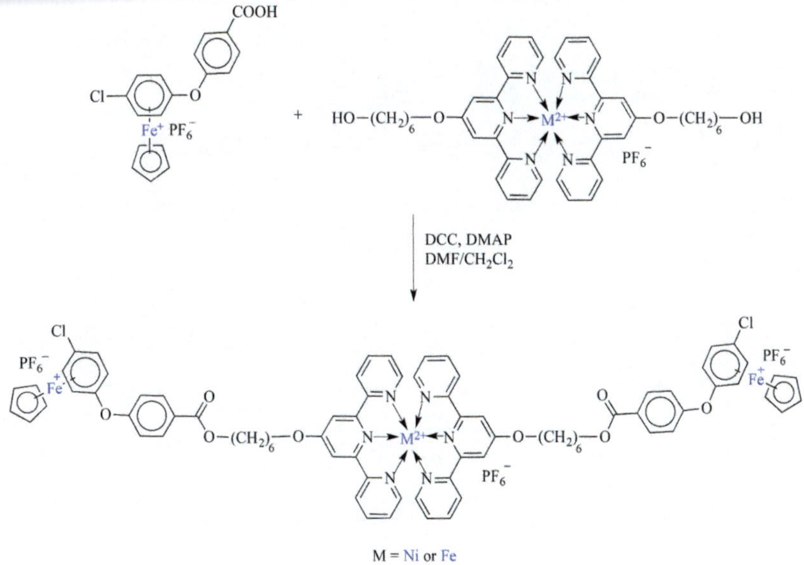

Scheme 3.14 Design of η^6-dichloroarene-η^5-cyclopentadienyliron(II)-terminated terpyridine-containing monomer.

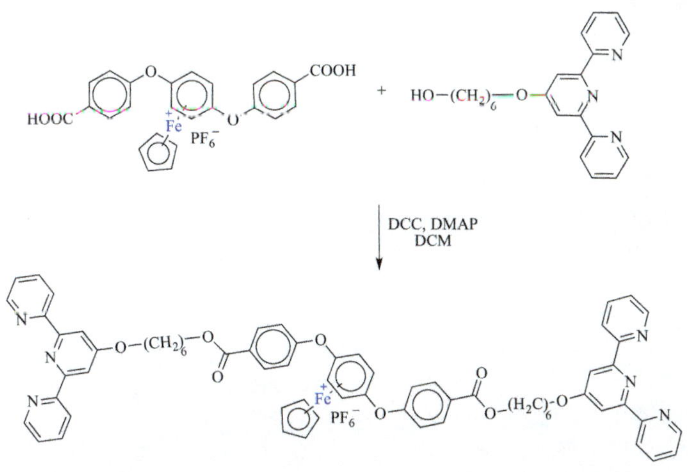

Scheme 3.15 Design of terpyridine-terminated η^6-dichloroarene-η^5-cyclopentadienyliron(II)-containing monomer.

alkyne-terminated monomers.[42–48] A clear demonstration of the possibility of using this polycondensation route to tune optoelectronic properties is evidenced by the decrease in the band gap (E_{gap}) of a polyplatinyne from 2.90 eV to 2.35 eV by changing from a benzene spacer to a more electron-rich anthracene spacer (Scheme 3.18).[45] Also, Wong *et al.* explored this polycondensation route to construct polyplatinynes of diverse colors and E_{gap},

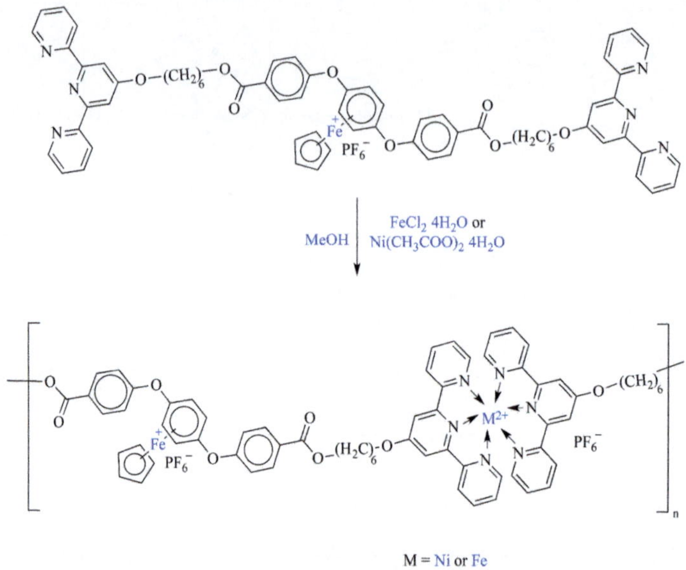

Scheme 3.16 Step polymerization *via* coordination of metals to terpyridine-terminated monomer.

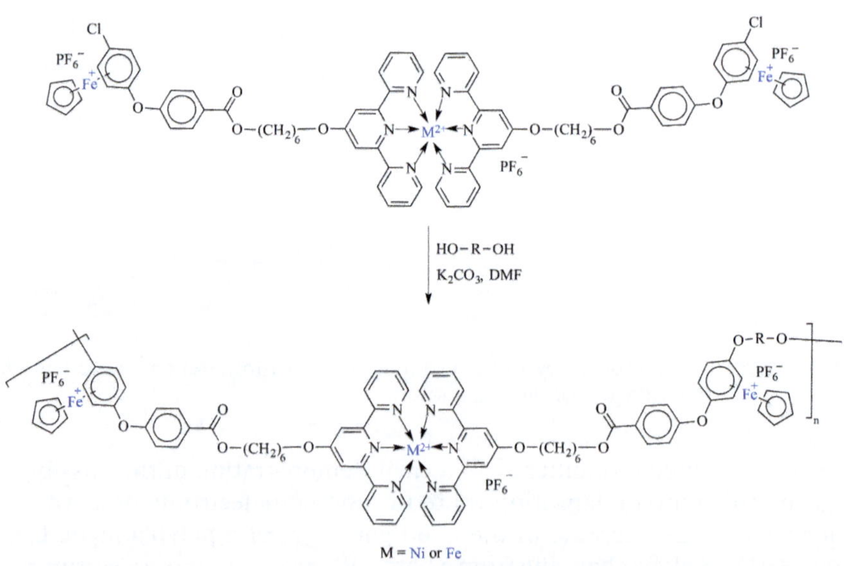

Scheme 3.17 Step polymerization *via* nucleophilic aromatic substitution reaction of η^6-dichloroarene-η^5-cyclopentadienyliron(II)-containing monomer.

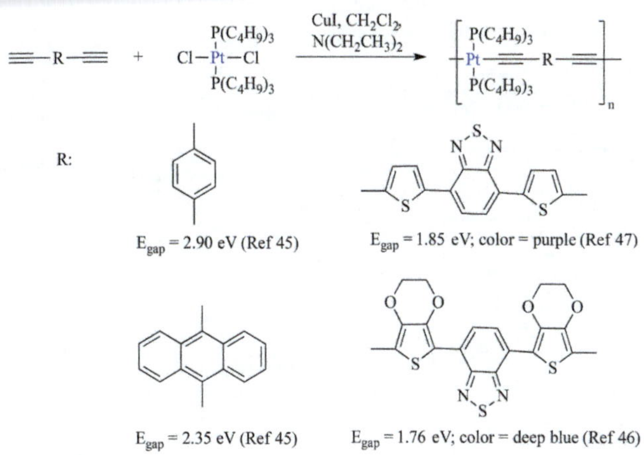

Scheme 3.18 Synthesis of polyplatinynes *via* Sonogashira-type step polymerization.

tuning the color and band gap of these polymers from deep blue to purple and 1.76 eV to 1.85 eV, by changing the aromatic spacer from 3,4-ethylenedioxythiophene-benzothiadiazole[46] to 4,7-di-2′-thienyl-2,1,3-benzothiadiazole,[47] respectively (Scheme 3.18). Copolymerization to achieve a heterometallic polymer is feasible through this polycondensation route. A graphical example is the stepwise cross-coupling reaction of three monomers, [$PtCl_2(P(C_4H_9)_3)_2$], alkyne-terminated iridium complex and alkyne-terminated bicarbazole, to obtain a yellow-emitting Pt/Ir heterobimetallic copolymer (Scheme 3.19).[44]

A Sonogashira-type cross-coupling reaction of substituted diiodobenzene with η^4-1,3-diethynylcyclobutadiene-η^5-cyclopentadienylcobalt(I) in the presence of an amine results in a Co-containing poly(*p*-phenylene-ethynylene) (Scheme 3.20).[49] The amine is a crucial component of the reaction since it controls the molecular weight and dispersity of the polymer. In the design of the Co-containing poly(*p*-phenylene-ethynylene), piperidine is the ideal amine, affording polymers with satisfactory molecular weights and dispersity, though the molecular weights are low and dispersity is high.[49] High-molecular-weight Co-containing poly(*p*-phenylene-ethynylene)s are synthetically obtainable *via* acyclic diyne metathesis (ADIMET) reactions (Scheme 3.20).[49] In an archetypal ADIMET polycondensation, a catalyst formed *in situ* metathesized dipropynylated benzene to give high-molecular-weight polymers, with the chain propagation being driven by the elimination of 2-butyne.[49] Efficient ADIMET reactions run for a long time (~16 hours) and at high temperatures (~150 °C), necessary conditions for forming the catalyst from a combined [$Mo(CO)_6$] and 4-chlorophenol or 4-trifluoromethylphenol mixture.[49] ADIMET polycondensation produces higher-molecular-weight polymers in good yields and with better purity compared with those obtained from Sonogashira-type polycondensation.[49]

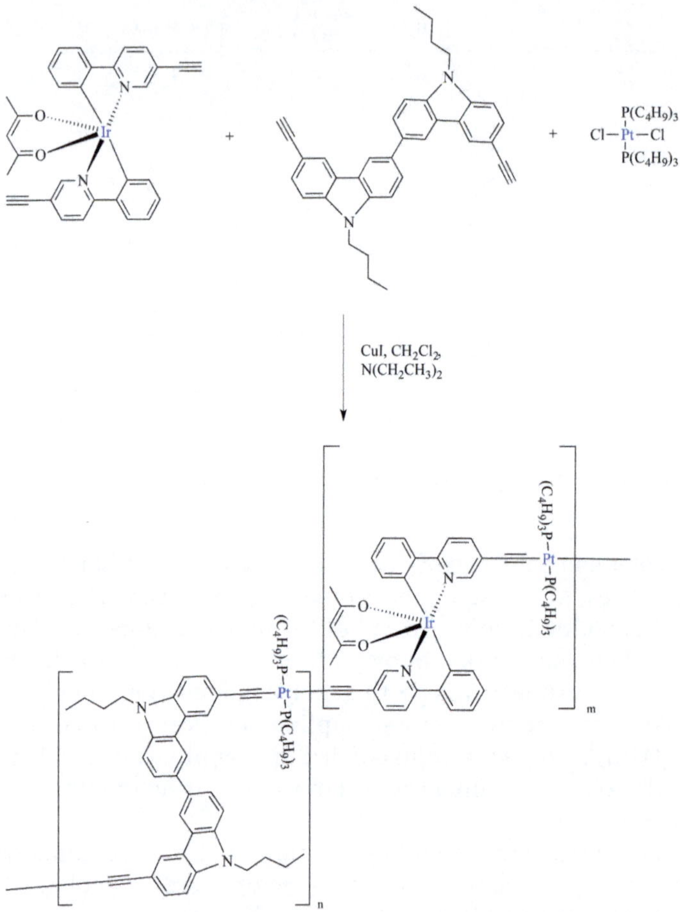

Scheme 3.19 Copolymerization using a Sonogashira-type coupling reaction.

Sonogashira-type coupling reactions remain a useful synthetic protocol to design organometallic monomers for polycondensation reactions. A Sonogashira-type coupling of *cis*-[Ru(dppe)$_2$Cl$_2$] and 1,4-dibutoxy-2-ethynyl-5-iodobenzene yields a Ru-containing conjugated monomer, which undergoes Sonogashira-type polycondensation with an alkyne-terminated BODIPY derivative (Scheme 3.21).[50] Also, a brominated thiophene organocobalt molecule, functioning as the core for a star polymer, is synthetically accessible *via* a sequential two-step reaction that includes Sonogashira coupling.[51] A subsequent Suzuki coupling reaction between the brominated organocobalt complex and 2-(pinacol ester boronic acid)-5,2′-bithiophene successfully yields a four-arm star organocobalt oligothiophene.[51]

The Suzuki coupling reaction is an established polycondensation reaction that also enables the design of metalated conjugated polymers. As an example, an iridium-containing poly(9,9-dioctylfluorene-*co*-tris(2-phenylpyridine)) is

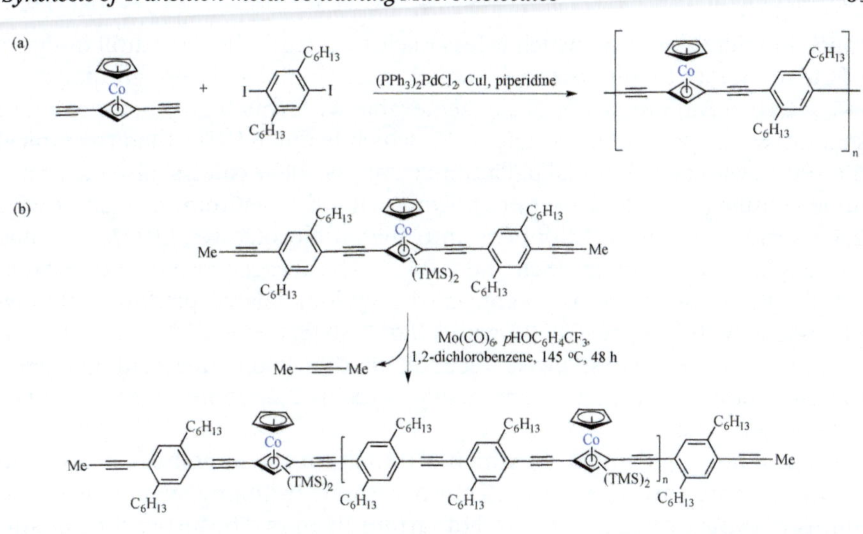

Scheme 3.20 Synthesis of Co-containing poly(*p*-phenylene-ethynylene) *via* (a) Sonogashira-type polycondensation and (b) ADIMET polycondensation.

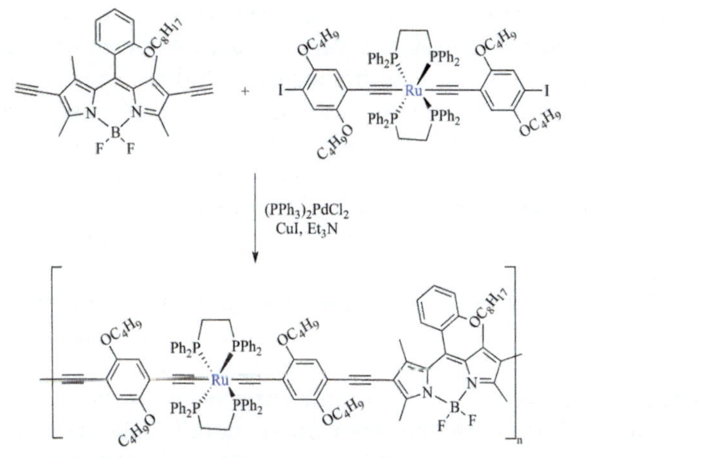

Scheme 3.21 Synthesis of a Ru-containing conjugated polymer *via* Sonogashira-type polycondensation.

constructed *via* Suzuki polycondensation reaction between 9,9-dioctyl-2,7-dibromofluorene, (5-bromo-2-(4-bromophenyl)pyridine))Ir(III)bis(2-phenylpyridine), and 9,9-dioctylfluorene-2,7-bis(trimethyleneborate).[52] Also, Fréchet *et al.* employed the Suzuki coupling reaction to polymerize platinum(II) (5-bromo-2-(5-bromothiophen-2-yl)-1,3-thiazolato-*N*,$C^{2'}$) ((2*Z*)-3-hydroxy-1,3-bis[(4-octyloxy)phenyl]prop-2-en-1-onato-*O*,*O*) with 9,9-dioctylfluorene-2,7-bis(trimethylborate) to obtain a platinum-containing conjugated polymer (Scheme 3.22).[53] As the platinum-containing monomer degrades under the polycondensation conditions, it is essential to replace the conventional $Pd(PPh_3)_4$ catalyst with a much bulkier $Pd(P^tBu_3)_2$, which is less likely to perturb the structural integrity of the monomer to initiate the degradation. High-molecular-weight polymers (M_n = 24 000 Da) are synthetically achievable with this highly active catalyst that drives the polymerization at 40 °C, which is much lower than the typical 80–100 °C with conventional palladium catalysts.[53] The catalyst also facilitates Stille coupling polycondensation, polymerizing the platinum monomer with 2,5-bis(trimethylstannyl)thiophene into high-molecular-weight (M_n = 45 000 Da) conjugated polymers (Scheme 3.22).[53] The Swager group used Suzuki coupling polycondensation to construct a cyclometalated platinum(II) polyphenylene with a M_n of 12 000 Da and $Đ$ of 2.50 (Scheme 3.23).[54] The polymerization involves the stepwise reaction of dibromophenylpyridine-ligated pyridine monomer with a commercially available 2,7-diboronic ester of fluorene (Scheme 3.23).

Coordination polymers, in which a transition metal coordinates to the polymer framework, are synthetically accessible by taking advantage of the ability of transition metals to bind to various ligands. The terpyridine ligand is reputed for its ability to coordinate to different metal ions, a characteristic that enables the construction of coordination polymers of distinctive architecture and with different metal centers. Indeed, the assorted coordination geometries of transition metals coupled with the ability to synthetically access terpyridine ligands of diverse structures permit the design of various terpyridine-based macromolecules including linear metallopolymers, and coordination nanosheets and nanowires. This was exemplified in the works of Abd-El-Aziz *et al.*, who constructed linear terpyridine-based metallopolymers (Schemes 3.16 and 3.17)[41] and the collaborative work of the Wong and Nishihara groups, who developed a library of terpyridine-based coordination nanosheets and nanowires.[55–65] By changing the terpyridine ligand from 1,3,5-tris[4-(4′-2,2′:6′,2″-terpyridyl)phenyl]benzene to tris[4-(4′-2,2′:6′,2″-terpyridyl)phenyl]amine or the metal salt from $Zn(BF_4)_2$ to $ZnSO_4$, Wong and coworkers produced three unique multilayer nanosheets (Scheme 3.24) *via* a liquid/liquid interfacial bottom-up synthetic protocol.[56] They also explored the liquid/liquid interfacial synthetic protocol and the dynamic and spontaneous coordinate bonding between a three-armed dipyrrin ligand in CH_2Cl_2 and zinc(II) acetate in H_2O to construct a multilayer bis(dipyrrinato)zinc(II) coordination nanosheet.[57] Controlling the concentration of the dipyrrin ligand in CH_2Cl_2 tunes the thickness of the nanosheet from 6 to 800 nm and the number of layers from 5 to 670 layers.[57] An air/liquid interfacial synthetic protocol that utilizes a very small amount of a diluted CH_2Cl_2 solution of the dipyrrin ligand gives a single or a few layers of bis(dipyrrinato)zinc(II) nanosheets.[57] The bottom-up liquid/liquid interfacial protocol is adaptable to diverse ligands, allowing the functionality of the nanosheet to be rationally tailored to an application by selecting an appropriate functional ligand. As an example, the photoelectric property of the bis(dipyrrinato)zinc(II) nanosheet is improved by replacing the three-armed dipyrrin ligand with a four-armed zinc(II) porphyrin-cored dipyrrin ligand.[55] A concentrated solution of the four-armed ligand in CH_2Cl_2/C_5H_5N

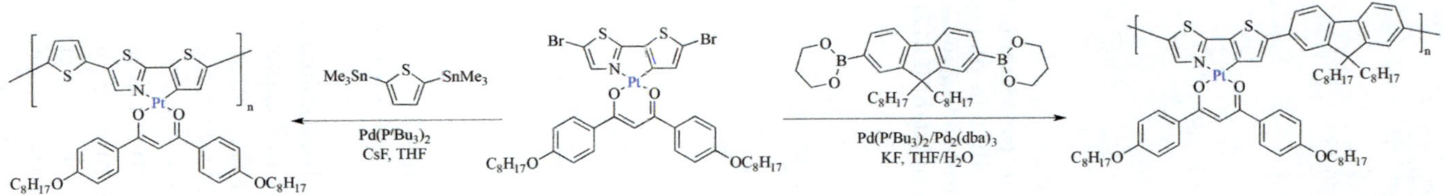

Scheme 3.22 Cyclometalated platinum polymer *via* Suzuki (right) and Stille (left) coupling polycondensation.

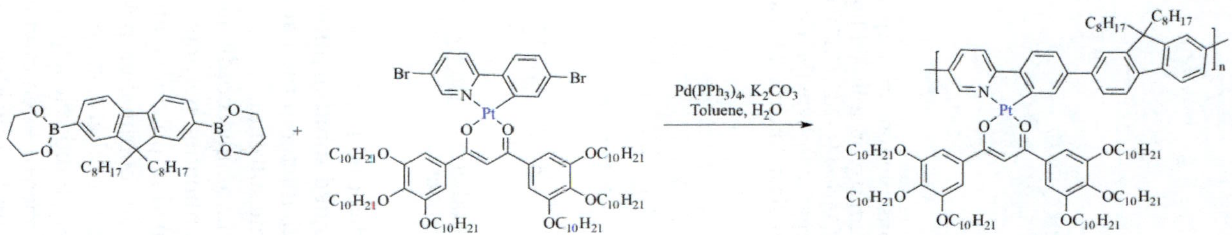

Scheme 3.23 Cyclometalated platinum polymer *via* Suzuki coupling polycondensation.

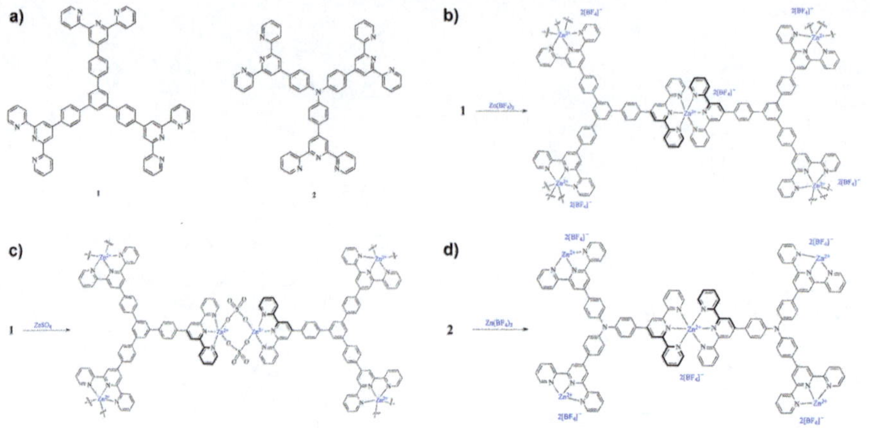

Scheme 3.24 Construction of terpyridine–zinc nanosheets *via* coordination chemistry in a liquid/liquid interfacial bottom-up method (a–d). (**1**) 1,3,5-tris[4-(4′-2,2′:6′,2″-terpyridyl)phenyl]benzene, (**2**) tris[4-(4′-2,2′:6′,2′-terpyridyl)phenyl]amine. Reprinted from ref. 56. Copyright (2017) American Chemical Society.

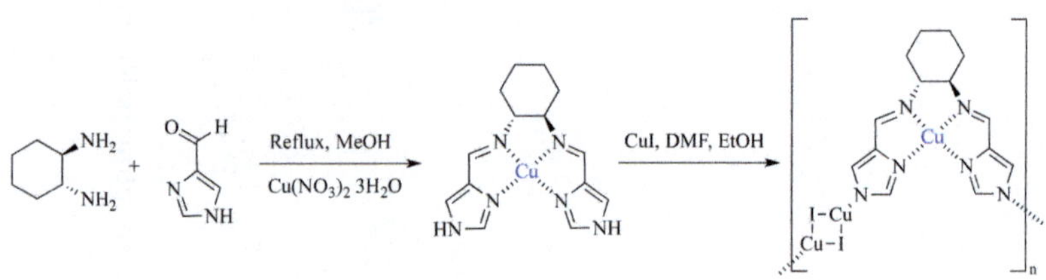

Scheme 3.25 Solvothermal synthesis of a Cu-salen polymer *via* polycondensation reaction.

forms a microsheet of 1100 ± 47 nm with an aqueous solution of zinc(II) acetate but a diluted solution affords three-layer or two-layer nanosheets.[55] The polymeric character of the nanosheet is evidenced by its insolubility in water and organic solvents.

Salen ligands favorably coordinate to metals, a property that is exploited in the design of coordination polymers. Solvothermal synthesis is a well-explored route to coordination polymers, and is well-demonstrated in the construction of salen-based coordination polymers including a bimetallic salen-based coordination polymer that contains a copper(II) salen molecular framework linked by copper(I) iodide clusters (Scheme 3.25). A one-pot reaction afforded the air-stable copper(II) salen monomer, which reacts with two molar equivalents of CuI salt under solvothermal conditions for 24 hours to give red crystals of the coordination polymer (Scheme 3.25). Also, a solvothermal reaction of *N*,*N*′-bis(4-carboxysalicylidene)ethylenediamine with $MnCl_2{\cdot}(H_2O)_4$ and $Ln(NO_3)_3{\cdot}n(H_2O)$ (Ln = Nd, Eu, Gd, Dy, or Tb) produced bimetallic salen-based

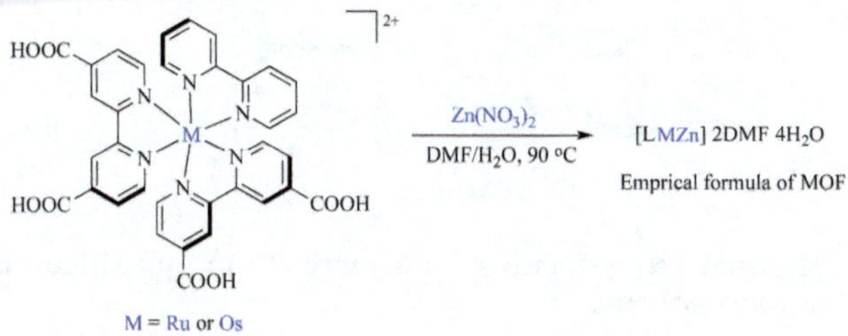

Scheme 3.26 Solvothermal synthesis of a metal–organic framework (MOF) *via* polycondensation reaction.

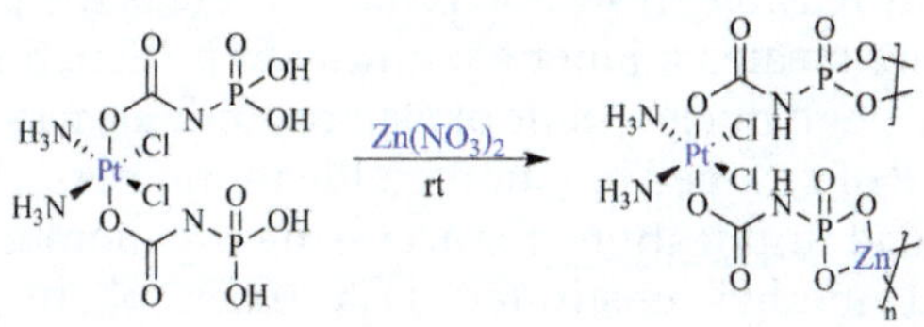

Scheme 3.27 Room temperature synthesis of a nanoscale coordination polymer *via* polycondensation reaction.

1D coordination polymers.[66] The lanthanide ions coordinate to the carboxylate group of the salen ligand, linking the Mn-salen framework into a 1D chain structure. Solvothermal reactions are also relevant in the synthesis of an attractive class of coordination polymers, metal–organic frameworks (MOFs), as demonstrated by Lin *et al.* in their design of Ru and Os-derived MOFs.[67,68] They obtained a Ru/Os-derived MOF by reacting {M[4,4′-bpy-$(COOH)_2]_2$bpy}$(PF_6)_2$ (M = Ru or Os; bpy = 2,2′-bipyridine) with $Zn(NO_3)_2$ in DMF and H_2O at 90 °C for three days (Scheme 3.26).[67] The Ru-MOF can be doped by adding a varying amount of {Os[4,4′-bpy-$(COOH)_2]_2$bpy}$(PF_6)_2$ to the {Ru[4,4′-bpy-$(COOH)_2]_2$bpy}$(PF_6)_2$ and $Zn(NO_3)_2$ reaction mixture. Low temperature synthesis of coordination polymers is feasible, as demonstrated by Lin and coworkers, who obtained nanoscale coordination polymers at room temperature.[69–72] For instance, a vigorously agitated microemulsion of a cisplatin or oxaliplatin prodrug with $Zn(NO_3)_2$ with the Na salt of 1,2-dioleoyl-*sn*-glycero-3-phosphoric acid in a Triton X-100/1-hexanol/cyclohexane/water mixture at room temperature for 30 minutes gives the nanoscale coordination polymer nanoparticles (Scheme 3.27).[70–72]

3.2.3 Ring-opening Polymerization of Transition-metal-containing Monomers

Ring-opening polymerization (ROP) is another mode of polymerization investigated by academic researchers, which is of interest in industry. Efficient ROP depends on both thermodynamic and kinetic factors; the thermodynamic element relies on the stability of the cyclic monomer and the

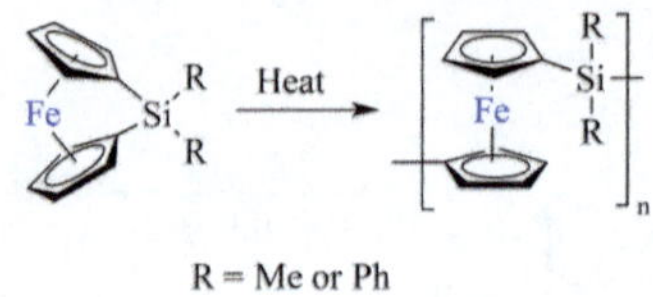

Scheme 3.28 Thermal ring-opening polymerization of silicon-bridged [1] ferrocenophane.

resulting polymer while the kinetic element depends on the presence of a pathway for ring opening and subsequent polymerization. Generally, ROP is thermodynamically feasible in strained rings, whereas the presence of a heteroatom in the ring creates a kinetic pathway to initiate and propagate the polymerization through nucleophilic or electrophilic attacks.[16] Manners *et al.* pioneered the ROP of strained organometallic monomers, developing a synthetic route of broad implications for accessing organometallic polymers.[73] A silicon-bridged methyl-substituted [1]ferrocenophane organometallic monomer, accessible *via* reaction of dilithioferrocene·TMEDA (TMEDA = tetramethylethylenediamine) with Me_2SiCl_2 at −78 °C, undergoes ROP when heated at 130 °C to produce a high-molecular-weight polymer (M_n = 340 000 Da) (Scheme 3.28).[74] The phenylated monomer, obtained by replacing Me_2SiCl_2 with Ph_2SiCl_2, also undergoes ROP at 200–235 °C to afford an insoluble organometallic polymer.[74]

The synthetic method to these strained organometallic monomers is flexible, and therefore, allows the substitution of the Si bridge with other bridges including Sn, Ge, P, Ni, Pd, or Pt bridges.[73,75–77] Replacing R_2SiCl_2 with R_2GeCl_2 (R = methyl, ethyl or phenyl), for instance, affords Ge-bridged [1]ferrocenophanes[77] just as $^tBu_2SnCl_2$ and Mes_2SnCl_2 (Mes = 2,4,6-trimethylphenyl) produced the Sn-bridged analogs.[76] From a polymerization perspective, a rich library of synthetically accessible structurally diverse monomers provides several options to access polymers of diverse structures and properties. Similar to the Si-bridged [1]ferrocenophane, the Ge-bridged and Sn-bridged [1]ferrocenophanes undergo thermally induced ROP to produce high-molecular-weight polymers.[76,77] The Ge-bridged methyl-substituted [1]ferrocenophane readily polymerizes in the melt at 90 °C to produce high-molecular-weight poly(ferrocenylgermane) (M_n up to 850 000 Da).[77] Replacing the methyl substituent on Ge with ethyl or phenyl groups increases the onset temperature for ROP from 90 °C to 120 °C or 230 °C, respectively.[77] The ROP of the Sn-bridged methyl-substituted [1]ferrocenophane occurs in the melt at 150 °C and in toluene solution at room temperature.[76] The solution ROP affords a polymer with a higher molecular weight (M_n = 560 000 Da) compared with the melt ROP (M_n = 83 000 Da). The choice of solvent has an impact on the degree of polymerization as evidenced by solution polymerization of the Sn-bridged methyl-substituted [1]ferrocenophane in $CHCl_3$, which produced lower-molecular-weight polymers (M_n = 4700 Da) compared with polymerization in toluene. The nature of the substituent on the bridging atom affects the onset temperature of melt ROP. For instance, with the Mes-substituted [1]ferrocenophane, ROP commences at a higher temperature and takes a longer time compared with the ^{t}butyl-substituted [1]ferrocenophane.

While the release of ring strain that accompanied the thermal ROP of [1]ferrocenophanes is thermodynamically favored, a kinetic pathway, usually created by a labile site, is indispensable to the feasibility of the polymerization. Indeed, the nature of the bridging atom contributes to the polymerizability of the strained ring, with thermal ROP realized with Si-, Ge-, Sn-, P-, Se-, and S-bridged [1]ferrocenophanes[76–82] but is yet to be achieved with Ni-, Pd-, and Pt-bridged [1]ferrocenophanes[75] despite the tilt angles between the cyclopentadienyl ligands of the non-polymerizable [1]ferrocenophanes being intermediate between the polymerizable [1]ferrocenophanes.

ROP of the bridged [1]ferrocenophanes to high-molecular-weight polymers is also promoted using transition-metal-based catalysts, γ-irradiation, and a cationic initiator as well as *via* a living anionic ROP route.[73,83–85] Late transition metal complexes, such as those of Pt, Pd, and Rh, are typical catalytic systems that initiate the ROP of Si-bridged [1]ferrocenophane.[79,86–90] With the Pt complex, $Pt(1,5\text{-cod})_2$ (cod = cyclooctadiene), mechanistic evidence proved that colloidal Pt(0) is the primary catalyst, and that oxidative-addition and reductive-elimination or σ-bond metathesis reactions at the surface of Pt(0) drives the ROP under a heterogeneous condition.[89] The nature of the bridging atom influences the activity of the catalyst in the transition-metal-catalyzed polymerization, with Karstedt's catalyst inhibiting the ROP of alkyne-substituted Si-bridged[91] or ^{t}Bu-substituted Sn-bridged [1]ferrocenophane[76] but promoting that of Me-substituted Si-bridged [1]ferrocenophane at room temperature.[90] In contrast with thermal ROP of unsymmetrical Si-bridged [1]ferrocenophane, where the Si–Cp bond is randomly cleaved to produce a regioirregular polymer, transition-metal-catalyzed ROP ensures control over the microstructure to give a regioregular polymer.[90] Also, it is possible to synthetically access architecture other than linear polymers, as demonstrated with the design of a star-shaped polymer when the Pt-catalyzed ROP is carried out in the presence of cyclotetrasiloxane.[90] Cyclic polymers and oligomers can be generated from Si-bridged [1]ferrocenophanes on irradiation in the presence of a labile Lewis acid, such as 4,4′-dimethyl-2,2′-bipyridine, to cleave the Fe–Cp bond.[92] Control over the ratio of cyclic polymers to cyclic oligomers is achievable by varying the temperature and concentration, which ultimately allows for the synthesis of high-molecular-weight cyclic polymers (M_n = 48 800 Da).[92]

ROP of other strained metallocenophanes other than ferrocenophanes is achievable and provides a route to metalate polymers with different transition metals. Anionic ROP of the 19-electron dicarba[2]cobaltocenophane with $Li[^tBuC_5H_4]$ as an initiator produced oligocobaltocenium chloride with nine repeat units (Scheme 3.29).[93] Thermal ROP is also feasible with this monomer, which, when heated at 140 °C for 1 hour, produced polycobaltocenium nitrate (Scheme 3.29).[93] Also, the thermally-induced

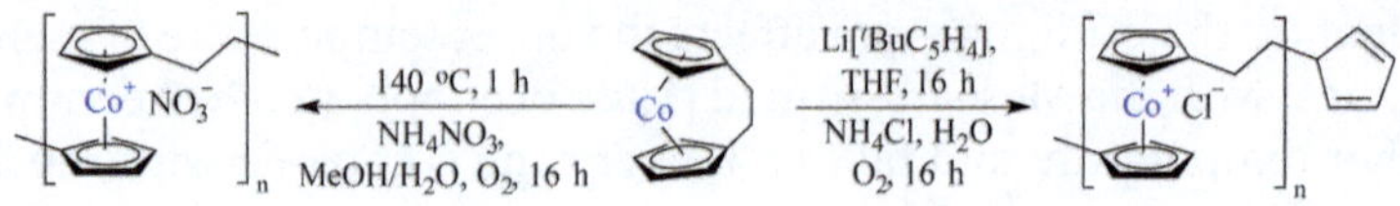

Scheme 3.29 Ring-opening polymerizations of [2]cobaltocenophane.

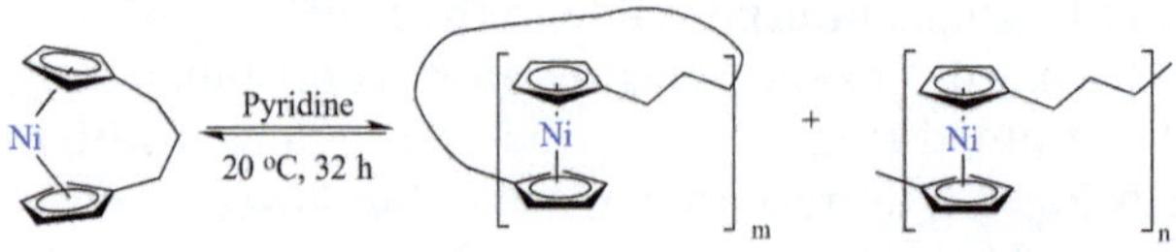

Scheme 3.30 Ring-opening polymerizations of [3]nickelocenophane.

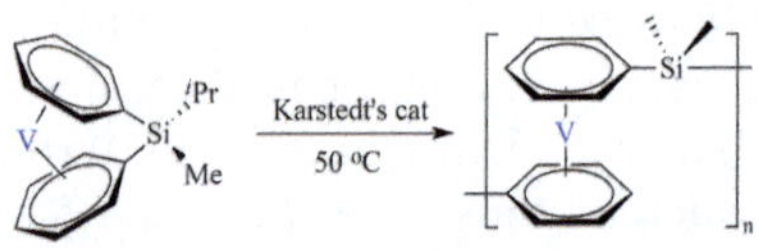

Scheme 3.31 Ring-opening polymerization of [3]vanadocenophane.

ROP of tricarba[3]nickelocenophane at 80 °C for 20 hours affords an insoluble material that swells in organic solvents, suggesting a cross-linked or high-molecular-weight polymer.[94] Stirring tricarba[3]nickelocenophane in pyridine at 20 °C for 32 hours releases the ring strain, driving polymerization into cyclic poly(nickelocenylpropylene) and linear polynickelocene *via* cleavage of the Ni–Cp bond (Scheme 3.30).[94,95] In a coordinating or polar solvent, the lability of the Ni–Cp bond in the polynickelocene triggers a dynamic behaviour that results in depolymerization at low concentration or elevated temperature.[95] This behavior contrasts with that of polymetallocenes derived from ferrocene or cobaltocene, where the strong metal–Cp bond is covalent and therefore static. Pt-catalyzed ROP of Si-bridged bis(benzene) vanadium (Scheme 3.31),[96] as well as a heteroleptic Si-[97] and tin-bridged [1]trovacenophane,[98] is possible.

Ring strain drives cycloalkenes, such as norbornenes, to undergo ring-opening metathesis polymerization (ROMP) in the presence of a transition-metal-based initiator,[16] producing polymers of commercial interest. Researchers graft transition-metal-containing moieties onto cycloalkenes using well-established chemistries, then polymerize these metalated monomers *via* ROMP to access functional materials with wide-ranging implications in diverse fields. This is exemplified in the work of Tew's group, who explored Mitsunobu coupling to graft ferrocene onto 10-oxa-4-azatricyclo[5.2.1.0^{2,6}]dec-8-ene-3,5-dione, a norbornene derivative, to produce an Fe-containing monomer (Scheme 3.32).[99] The Mitsunobu coupling also enables the functionalization of the norbornene derivative with a terminal alkyne, which reacts with $Co_2(CO)_8$ to afford a Co-containing monomer (Scheme 3.32).[99,100] These Co- and Fe-containing monomers, in the presence

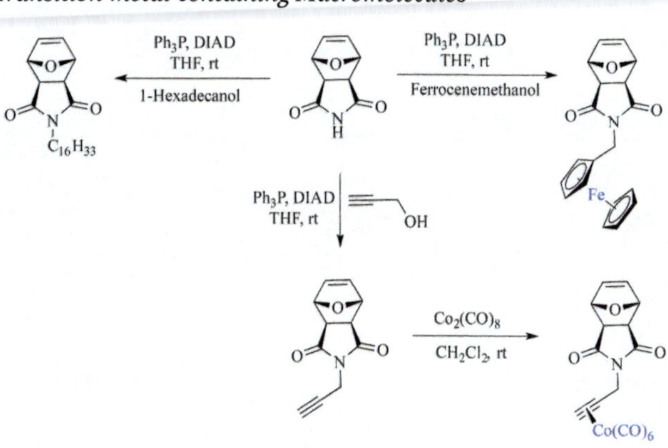

Scheme 3.32 Design of transition-metal-containing norbornenes.

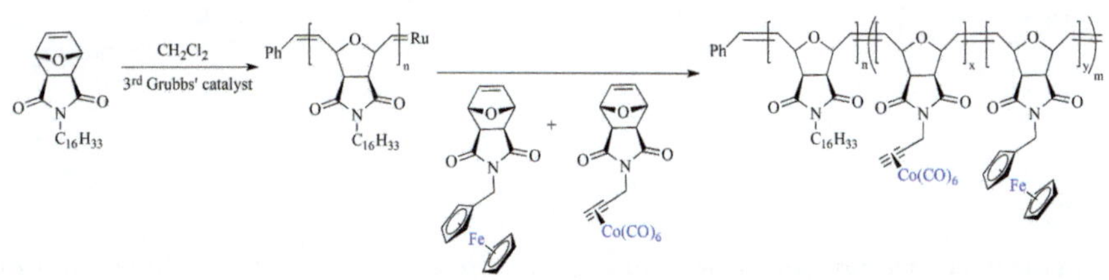

Scheme 3.33 Ring-opening metathesis polymerization of ferrocene- and cobalt-containing monomers.

of a third-generation Grubbs' catalyst, are copolymerized with another norbornene derivative, yielding a diblock homometallic[100] or triblock heterobimetallic copolymer[99] (Scheme 3.33), with useful self-assembling properties. The presence of a metal center in the monomers has no adverse effects on the polymerization efficiency, because the reactions are well-controlled, and affords high-molecular-weight (M_n = 100 000 Da) and nearly uniform dispersity ($Đ$ = 1.05) triblock copolymers.[99]

Instead of a heterometallic block copolymer with different metal centers in each repeat unit, it is feasible to construct a heterometallic polymer with different metal centers in the same repeat unit, a design strategy that enriches the polymer with metal centers. Abd-El-Aziz *et al.* successfully demonstrated this by using facile reactions that include Steglich esterification to graft a $Co_2(CO)_6$ moiety and ferrocene or η^6-arene-η^5-cyclopentadienyliron(II) onto a norbornene derivative (Scheme 3.34).[101,102] Usually, $Co_2(CO)_6$ moieties are incorporated into most polymers, post-polymerization, but in this scenario, as in other alkyne-functionalized norbornene derivatives, the Grubbs' catalyst reacts with the alkyne triple bond, an undesirable metathesis reaction that inspires the pre-polymerization grafting of $Co_2(CO)_6$ onto the monomers to protect the alkyne bonds. The monomers successfully undergo ROMP in

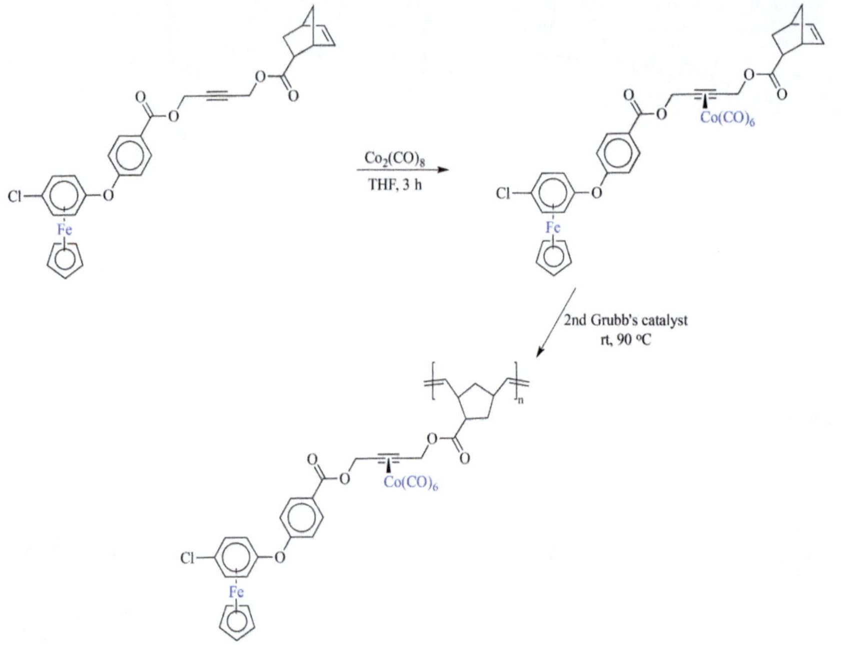

Scheme 3.34 Design *via* a ring-opening metathesis polymerization route to an iron and cobalt-containing heterobimetallic polymer.

the presence of Grubbs' catalyst to yield redox-active polymers with weight average molecular weights up to 69 000 Da.[101]

Evidently, the chemistry of the metalated norbornene monomer is critical to the efficiency of ROMP, with, for instance, different solvents enabling the ROMP depending on the polarity of the monomer. Astruc *et al.* found CH_2Cl_2 as the ideal solvent for the ROMP of neutral ferrocene- or pentamethylferrocene-containing norbornene monomers and DMF as the only solvent that works for the ROMP of cationic cobaltocenium-containing norbornene monomers.[103] The time required to achieve 100% monomer conversion also depends on the monomer, with the pentamethylferrocene-containing norbornene monomer reaching 100% conversion in 10 minutes, the ferrocene analog in 15 minutes and up to several days depending on the polymer size, and the cobaltocenium analog in 30 minutes.[103] The slow ROMP of the cobaltocenium monomer in DMF is due to the coordination of the solvent to the ruthenium center of the Grubbs' catalyst, a phenomenon that competes with the initiation step. The three monomers afford polymers that differ in their solubility profiles, a property that influences the efficiency of the ROMP process. It is pertinent to consider the properties of metal-containing norbornene monomers in developing copolymerization routes that aim to incorporate all monomers within the copolymer framework. Indeed, Astruc and coworkers judiciously selected the appropriate solvent to realize the copolymerization of the three monomers in a one-pot three-step sequential

ROMP.[103] For instance, they achieved the ROMP of a pentamethylferrocene-containing norbornene monomer in dry CH_2Cl_2 to 100% conversion in 10 minutes in the first step. Next, a DMF solution of the cobaltocenium-containing norbornene monomer was added and polymerized to 100% conversion in a DMF–CH_2Cl_2 mixture, followed by the addition and polymerization of the ferrocene-containing monomer in the DMF–CH_2Cl_2 mixture in the final step (Scheme 3.35).[103] Altering the sequence in which the monomers are added offers an avenue to control the macrostructure, the precise arrangement of the blocks in the copolymer, to access new materials. The ROMP of the pentamethylferrocene-containing norbornene monomer in dry CH_2Cl_2 in the first step, the ferrocene-containing monomer also in dry CH_2Cl_2 in the second step, and the cobaltocenium-containing norbornene monomer in the DMF–CH_2Cl_2 mixture in the final step yields a polymer with a distinctive architecture (Scheme 3.35).[103]

Depending on the catalyst, a controlled and living ROMP is possible, especially with a third-generation Grubbs' catalyst.[103–108] Also, the polymerization tolerates a broad range of monomers, including those incorporating transition metal coordination complexes. A norbornene monomer grafted onto a coordination complex of nickel(II) in Goedken's macrocycle undergoes ROMP in the presence of a third generation Grubbs' catalyst, producing a Ni-containing polymer with the efficiency and characteristics of the polymerization depending on the reaction solvent (Figure 3.4).[109] For instance, the ROMP fails in DMF, affording low-molecular-weight (M_n = 16 500 Da) non-uniform ($Đ$ = 2.69) polymer in THF and higher-molecular-weight (M_n = 24 100 Da) nearly uniform ($Đ$ = 1.12) polymer in CH_2Cl_2, which suggests better control in the latter solvent. The linearity of the plot of $\ln([M]_0/[M])$ *vs.* polymerization time ($[M]_0$ = initial monomer concentration; $[M]$ = monomer concentration at different times) as well as increase in M_n as a function of time confirms the controlled character of the ROMP in CH_2Cl_2 (Figure 3.4). Side reactions and/or quenching are probable at a high monomer–catalyst feed ratio because $Đ$ increases and the relationship between feed ratio and degree of polymerization deviates from linearity at a high ratio (Figure 3.4). The authors concluded from these findings that ROMPs of these transition-metal-containing monomers are controlled but may lack living character.

3.2.4 Post-synthesis Metalation of Macromolecules with Transition Metal Moieties

The installation of transition metal moieties on macromolecules is also achievable post-synthesis using well-established chemistries. It is pertinent that a functional handle exists in the macromolecule to graft the transition metal moieties. Also, care is needed to ensure the structural and functional integrity of the macromolecule by avoiding adverse reaction conditions such as extreme pH or redox conditions that can trigger depolymerization. Post-synthesis metalation is a perfect approach to install transition metal

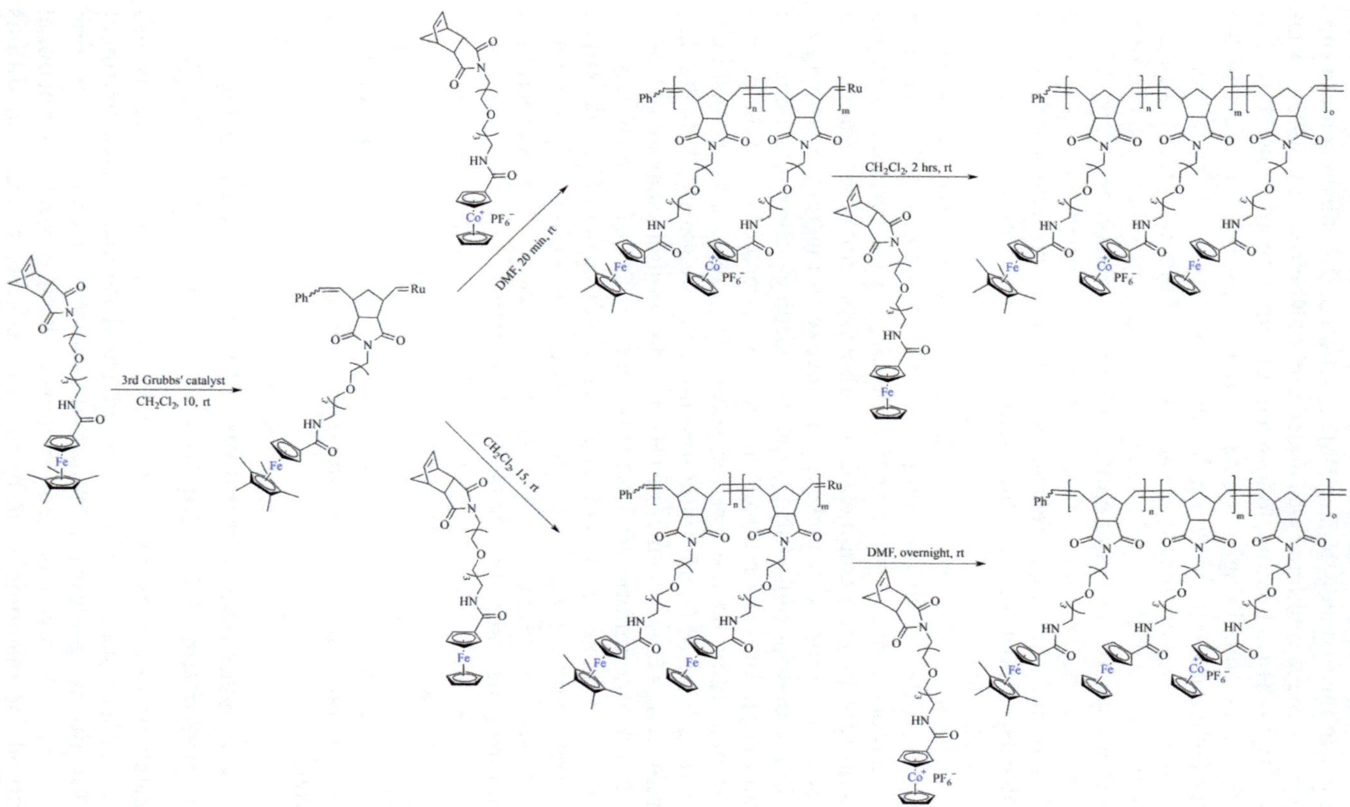

Scheme 3.35 Control of reaction conditions to attain successful ring-opening metathesis polymerization of metallocene-functionalized norbornenes.

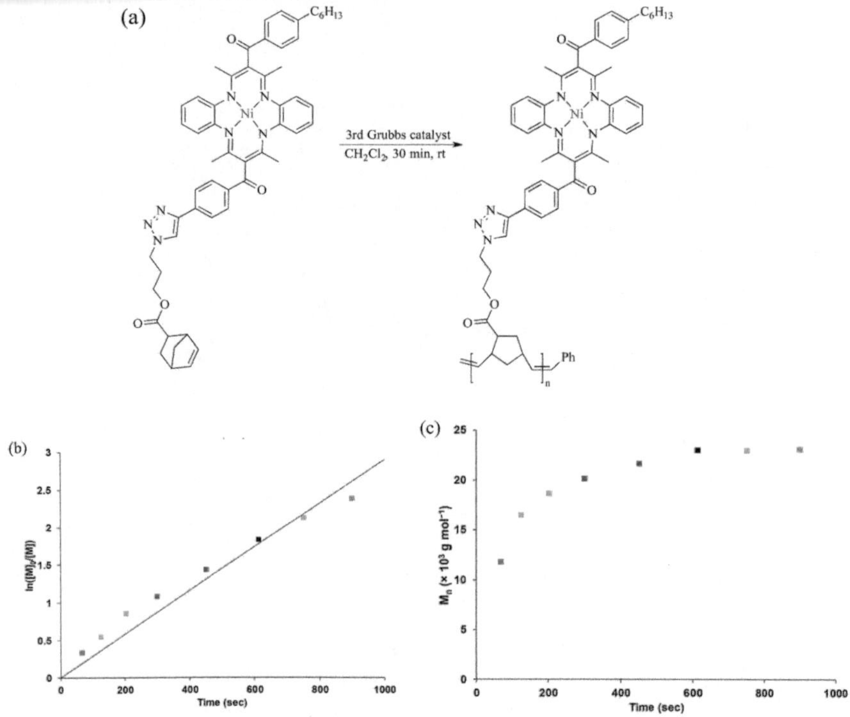

Figure 3.4 (a) Ring-opening metathesis polymerization of a nickel(II) complex-containing norbornene monomer. (b) Plots of $[M]_0/[M]$ *vs.* time; and (c) number average molecular weight (M_n) *vs.* time. Reproduced from ref. 109 with permission from The Royal Society of Chemistry.

moieties onto the periphery of dendrimers constructed *via* the divergent synthetic route. Starting from a multifunctional core, a dendrimer can be assembled outwards and finally decorated with transition metals at the outer termini or periphery. Abd-El-Aziz *et al.* explored this route to construct Fe/Co heterobimetallic dendrimers by building the dendrimers from a pentaerythritol core with organoiron branches and functionalizing the dendrimer with alkyne bonds at the outer termini to react with $Co_2(CO)_8$ moieties (Scheme 3.36).[110] The eventual reaction of the alkyne-functionalized organoiron dendrimers with $Co_2(CO)_8$ in a DMF–THF mixture occurs under benign conditions that maintain the structural and functional integrity of the dendrimers, though the resulting Fe/Co heterobimetallic dendrimers are less soluble than the starting organoiron dendrimer precursors. Post-synthesis installation of $Co_2(CO)_6$ into a macromolecule is also feasible in alkyne-functionalized linear polymers as demonstrated by Abd-El-Aziz's group with poly(methyl methacrylate)s (Scheme 3.37)[17] and Grubbs' group with poly(4-(phenylethynyl)styrene) block copolymers (Scheme 3.38).[111–114] Post-synthesis metalation with $Co_2(CO)_6$ is mostly carried out in dry THF or toluene, but in case of insolubility of the precursor

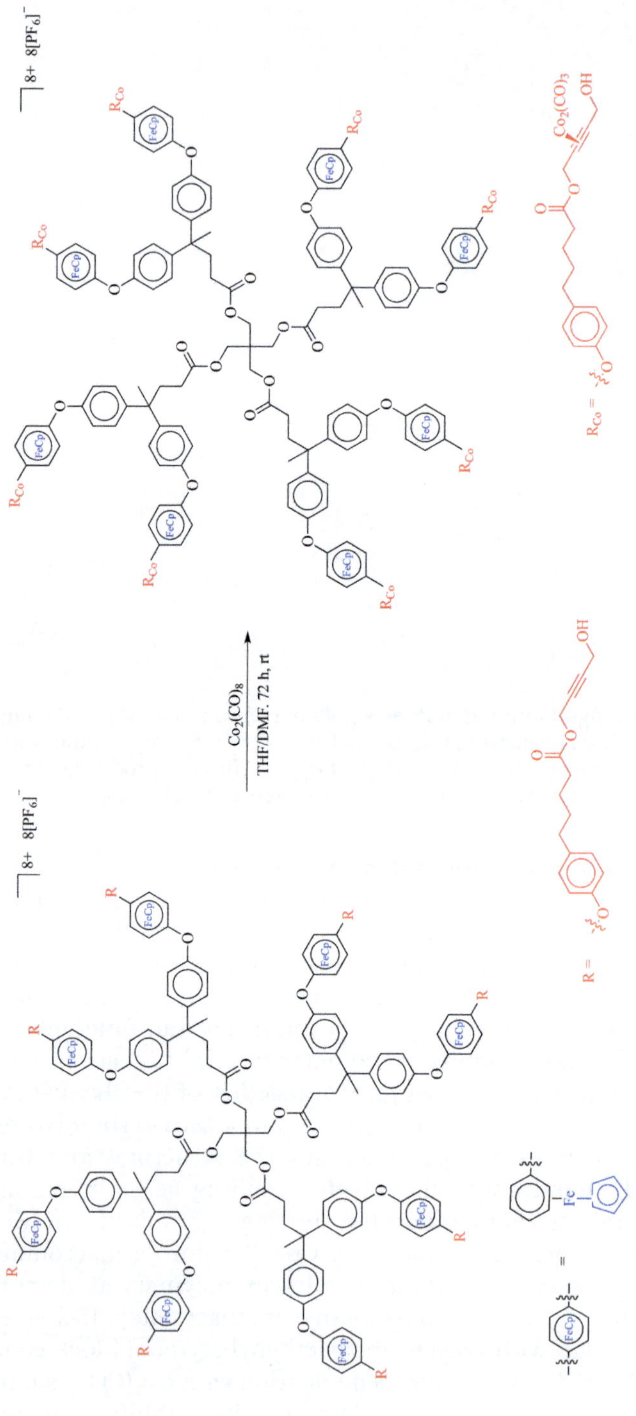

Scheme 3.36 Post-synthesis metalation of an iron-containing dendrimer.

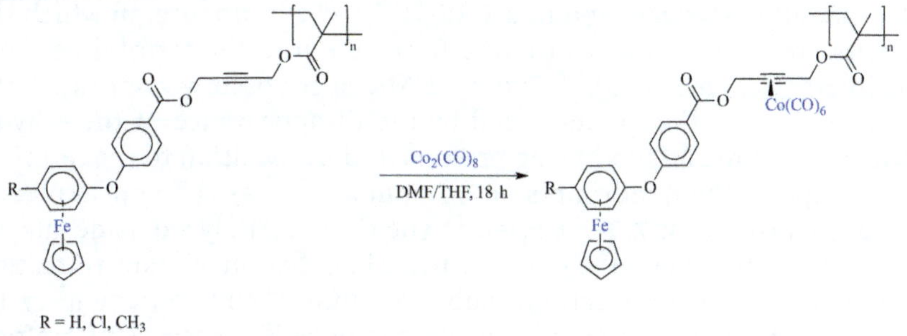

Scheme 3.37 Post-synthesis metalation of an iron-containing poly(methyl methacrylate).

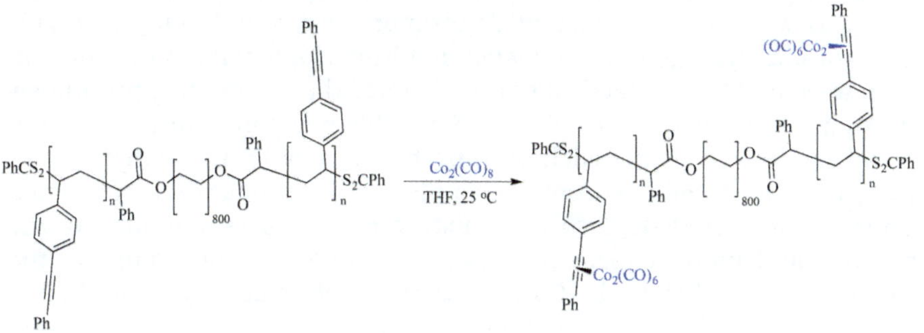

Scheme 3.38 Post-synthesis metalation of a polystyrene copolymer.

macromolecules in these solvents, metalation is still achievable in a solvent mixture that ensures solubility of precursors and products. Indeed, Abd-El-Aziz *et al.* metalated three generations of organoiron dendrimers as well as organoiron poly(methyl methacrylates) in a THF–DMF solvent mixture by adding a dry DMF solution of the macromolecules to a dry THF solution of $Co_2(CO)_8$ at room temperature.[17,110] While it may be difficult to realize 100% metalation of the alkyne bonds in a macromolecule, a very high degree of metalation, greater than 90%, is achievable by using a slight excess of molar equivalent of $Co_2(CO)_8$ to alkyne bond[111–114] or prolonging the reaction time.[17,110] Refluxing the reaction mixture also enables efficient metalation, though with a slight decomposition of the cobalt carbonyl moieties.[111–114]

The click reaction, especially copper-catalyzed azide–alkyne cycloaddition to 1,2,3-triazole, is a powerful, versatile and functional-group-tolerating chemistry that enables the facile metalation of macromolecules, post-synthesis. Astruc and coworkers explored the copper-catalyzed azide–alkyne cycloaddition to metalate several dendrimers and linear polymers post-synthesis.[115–118] As an example, Astruc and coworkers reacted azido-terminated dendrimers with ethynylbiferrocene in the presence of

$CuSO_4$–sodium ascorbate and in a THF–H_2O solvent mixture, in which the reactants are soluble, to incorporate ferrocene into the termini of these dendrimers (Scheme 3.39).[115] Proton NMR spectroscopy confirmed the success of the reaction as indicated by the disappearance of the ethynyl proton at 2.7 ppm and methylene protons in the α-position to the terminal azido groups in the dendrimers at 2.8 ppm as well as the appearance of the triazole protons at 7.0–7.1 ppm.[115] The copper-catalyzed azide–alkyne cycloaddition also allows the post-synthesis installation of many transition metal moieties, including cationic cobaltocenium[116] and η^6-arene-η^5-cyclopentadienyliron(II) complexes[118] on macromolecules. Successful grafting of these cationic organometallic moieties is contingent on reaction conditions, such as an appropriate solvent mixture, that ensure solubility of the reactants as well as the products. To illustrate this, the cationic ethynylcobaltocenium and η^6-ethynylmesitylene-η^5-cyclopentadienyliron(II) complexes, as well as the metalated dendrimers, are soluble only in highly polar solvents but the non-metalated dendrimer precursor is insoluble in these solvents,[116,118] a practicality that dictates the use of an appropriately adjusted solvent mixture to achieve a favorable reaction. Thus, Astruc and coworkers performed the click reaction between cationic η^6-ethynylmesitylene-η^5-cyclopentadienyliron(II) complex and a nonaazido-terminated dendrimer in a DMF–H_2O solvent mixture that ensures solubility of the reactants and product (Scheme 3.39).[118] Multiple triazole groups in the macromolecules chelate the Cu(I) catalyst, which is usually derived from $CuSO_4$ and ascorbic acid, a chemistry that decreases the reaction rates and product yields. Thus, a stoichiometric rather than a catalytic amount of Cu catalyst is required to ensure efficiency of the reaction, but the excess Cu(I) contaminates the resulting macromolecule.[116] Recently, a new Cu(I) catalytic system, [Cu(I)-(hexabenzyl)tren]Br, was designed and shown to ensure efficient click reactions in macromolecules.[117] Astruc and coworkers employed this catalyst to drive the click reaction between ethynylbiferrocene and azido-functionalized linear poly(methylstyrene) to produce poly(biferrocenylmethylstyrene) in 97% yield.

It is attractive to explore mild chemistries to metalate macromolecules post-synthesis as these approaches ensure the integrity of the final product. As an example, cobaltocenium- or rhodicenium-terminated dendrimers are accessible *via* a post-synthesis metathesis ion exchange reaction between the sodium salt of a polybenzoate-terminated dendrimer and cobaltocenium or rhodicenium chloride.[119] The same chemistry enables the design of a water-soluble rhodicenium polystyrene sulfonate with 330 ± 30 units of rhodicenium ions.[119] Also, a mild uncatalyzed hydroamination reaction allows the metalation of poly(amido amine) (PAMAM) dendrimers without by-products.[120] Indeed, a hydroamination reaction between an ethynylcobaltocenium cation and primary or secondary amino groups in a PAMAM dendrimer at 35 °C for two days produces *trans*-enamine-cobaltocenium-terminated dendrimers in 90% yield. A Fe/Co-containing heterometallic dendrimer is also accessible *via* a hydroamination reaction between a PAMAM dendrimer

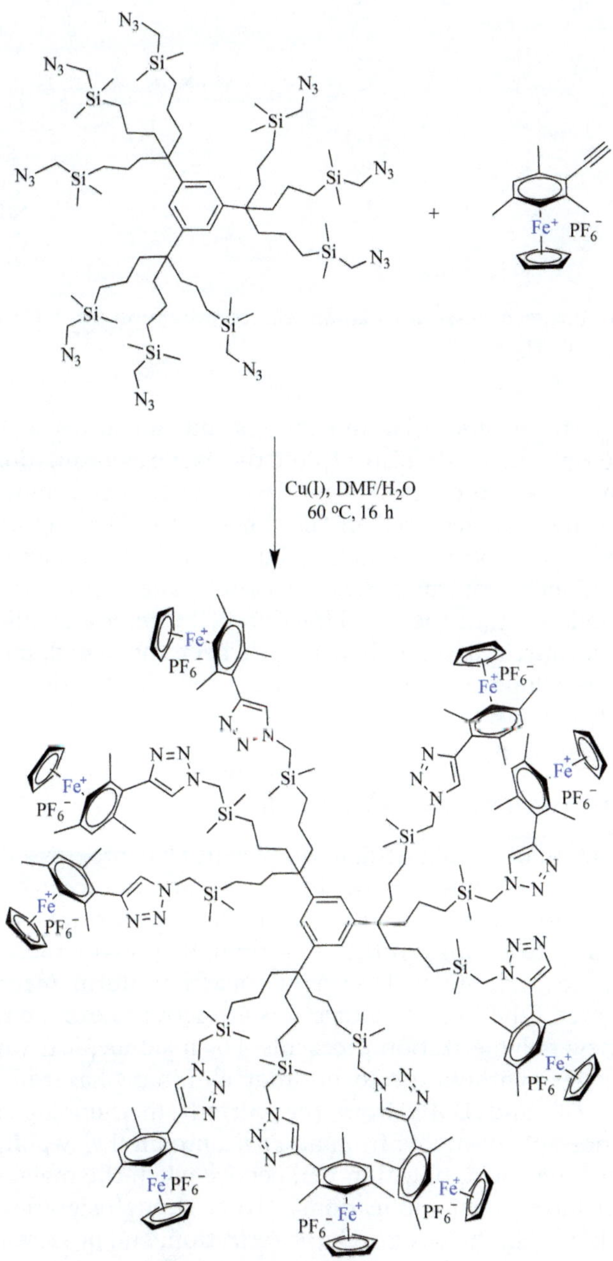

Scheme 3.39 Post-synthesis metalation of a dendrimer *via* a copper-catalyzed azide–alkyne cycloaddition reaction.

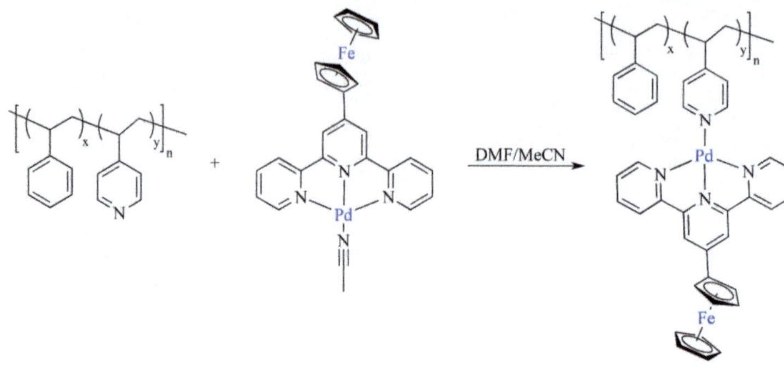

Scheme 3.40 Post-synthesis metalation *via* coordination of Pd to pyridine in polystyrene.

containing ferrocenylmethylamino groups and an ethynylcobaltocenium cation. Materials chemists also exploit the facile coordination chemistry of transition metals to organic ligands to metalate macromolecules, post-synthesis. Wong *et al.* accessed an Fe/Pd heterometallic diblock copolymer by coordinating a ferrocene-containing Pd-terpyridine complex to the pyridyl groups of a diblock copolymer (Scheme 3.40).[121] The coordination chemistry can be controlled to tune the metal fraction in the heterometallic copolymer. Also, some porphyrin-based polymers have been metalated, post-synthesis, to produce metalloporphyrin polymers *via* facile coordination of the metals to the pyrrole ligands.[122]

3.3 Conclusion and Outlook

The development of metalated macromolecules has progressed to an era of remarkable achievements. Today, chemistries abound to incorporate transition metals into macromolecules *via* chain, step or ring-opening polymerizations as well as post-synthesis. Although most of these traditional routes produce high-molecular-weight, nearly uniform macromolecules with the desired functionality intact, it is attractive to expand the frontiers to explore new polymerization protocols. For instance, it is worthwhile to explore greener synthetic protocols that eliminate hazardous reagents, such as CH_2Cl_2 and DMF, from the current methodologies. Such an advance is not only desirable from an environmental viewpoint, but most importantly from the application perspective when the macromolecule is designed for biomedical applications. Also, although significant progress has been achieved in the design, characterization, and processing of transition-metal macromolecules, solubility remains an issue in some cases, and it precludes detailed characterization and use of these functional materials. With persistent effort, most of these problems will be addressed, and transition-metal-containing macromolecules will emerge as the "holy grail" of functional materials.

References

1. C.-L. Ho, Z.-Q. Yu and W.-Y. Wong, *Chem. Soc. Rev.*, 2016, **45**, 5264.
2. A. S. Abd-El-Aziz and E. A. Strohm, *Polymer*, 2012, **53**, 4879.
3. A. S. Abd-El-Aziz, C. Agatemor and N. Etkin, in *Functional Materials Based on Metal-containing Polymers in Functional Metallosupramolecular Materials*, ed. J. G. Hardy and F. H. Schacher, Royal Society of Chemistry Smart Materials Series, 2015, pp. 87–119.
4. A. S. Abd-El-Aziz, C. Agatemor and N. Etkin, *Macromol. Rapid Commun.*, 2014, **35**, 513.
5. C. G. Hardy, J. Zhang, Y. Yan, L. Ren and C. Tang, *Prog. Polym. Sci.*, 2014, **39**, 1742.
6. G. R. Whittell, M. D. Hager, U. S. Schubert and I. Manners, *Nat. Mater.*, 2011, **10**, 176.
7. V. Weinmayr, *J. Am. Chem. Soc.*, 1955, **77**, 3009.
8. A. N. Nesmeyanov, V. A. Sazonova and V. N. Drozd, *Dokl. Akad. Nauk SSSR*, 1959, **126**, 1004.
9. S. I. Goldberg and D. W. Mayo, *Chem. Ind.*, 1959, 671.
10. O. A. Nesmeyanova and E. G. Perevalova, *Dokl. Akad. Nauk SSSR*, 1959, **126**, 1007.
11. V. V. Korshak, S. L. Sosin and V. P. Alekseeva, *Dokl. Akad. Nauk SSSR*, 1960, **132**, 360.
12. A. N. Nesmeyanov, V. V. Korshak, V. V. Voevodski, N. S. Kochetkova, S. L. Sosin, R. B. Materikova, T. N. Bolotnikova, V. M. Chibrikin and N. M. Bazhin, *Dokl. Akad. Nauk SSSR*, 1961, **137**, 1370.
13. H. Rosenberg and E. W. Neuse, *J. Organomet. Chem.*, 1966, **6**, 76.
14. C. U. Pittman Jr, J. Lai, D. Vanderpool, M. Good and R. Prado, *Macromolecules*, 1970, **3**, 746.
15. S. Chernyy, Z. Wang, J. J. K. Kirkensgaard, A. Bakke, K. Mortensen, S. Ndoni and K. Almdal, *J. Polym. Sci., Part A: Polym. Chem.*, 2017, **55**, 495.
16. G. Odian, *Principles of Polymerization*, John Wiley & Sons, 2004.
17. A. S. Abd-El-Aziz, D. J. Winram, P. O. Shipman, C. L. Rock, M. S. Vandel and B. O. Patrick, *Macromol. Chem. Phys.*, 2012, **213**, 2136.
18. A. R. Kenaree and J. B. Gilroy, *Organometallics*, 2017, **36**, 2483.
19. J. Zhang, Y. P. Chen, K. P. Miller, M. S. Ganewatta, M. Bam, Y. Yan, M. Nagarkatti, A. W. Decho and C. Tang, *J. Am. Chem. Soc.*, 2014, **136**, 4873.
20. J. Zhang, J. Yan, P. Pageni, Y. Yan, A. Wirth, Y. Chen, Y. Qiao, Q. Wang, A. W. Decho and C. Tang, *Sci. Rep.*, 2015, **5**, 11914.
21. W. Feng, S. Yin, M. Pan, H. Wang, Y. Fan, X. Lü and C. Su, *J. Mater. Chem. C*, 2017, **5**, 1742.
22. L. Liu, H. Li, P. Su, Z. Zhang, G. Fu, B. Li and X. Lü, *J. Mater. Chem. C*, 2017, **5**, 4780.
23. P. Yang, P. Pageni, M. P. Kabir, T. Zhu and C. Tang, *ACS Macro Lett.*, 2016, **5**, 1293.
24. J. Zhang, L. Ren, C. G. Hardy and C. Tang, *Macromolecules*, 2012, **45**, 6857.

25. C. G. Hardy, L. Ren, T. C. Tamboue and C. Tang, *J. Polym. Sci., Part A: Polym. Chem.*, 2011, **49**, 1409.
26. J. Zhang, Y. Yan, J. Chen, W. M. Chance, J. Hayat, Z. Gai and C. Tang, *Chem. Mater.*, 2014, **26**, 3185.
27. L. Ren, C. G. Hardy and C. Tang, *J. Am. Chem. Soc.*, 2010, **132**, 8874.
28. L. Ren, J. Zhang, C. G. Hardy, D. Doxie, B. Fleming and C. Tang, *Macromolecules*, 2012, **45**, 2267.
29. Y. Yan, J. Zhang, Y. Qiao and C. Tang, *Macromol. Rapid Commun.*, 2014, **35**, 254.
30. P. Chadha and P. J. Ragogna, *Chem. Commun.*, 2011, **47**, 5301.
31. M. Hadadpour, Y. Liu, P. Chadha and P. J. Ragogna, *Macromolecules*, 2014, **47**, 6207.
32. M. Hadadpour, J. Gwyther, I. Manners and P. J. Ragogna, *Chem. Mater.*, 2015, **27**, 3430.
33. A. S. Abd-El-Aziz, D. A. Armstrong, S. Bernardin and H. M. Hutton, *Can. J. Chem.*, 1996, **74**, 2073.
34. A. S. Abd-El-Aziz, K. M. Epp, C. R. de Denus and G. Fisher-Smith, *Organometallics*, 1994, **13**, 2299.
35. A. S. Abd-El-Aziz, A. A. Abdelghani, J. K. Pearson, M. K. Awad, D. P. Overy and R. G. Kerr, *Macromol. Chem. Phys.*, 2016, **217**, 987.
36. A. Abd-El-Aziz, E. Todd and G. Ma, *J. Polym. Sci., Part A: Polym. Chem.*, 2001, **39**, 1216.
37. A. S. Abd-El-Aziz, C. R. de Denus, M. J. Zaworotko and L. R. MacGillvray, *J. Chem. Soc., Dalton Trans.*, 1995, 3375.
38. A. S. Abd-El-Aziz, E. K. Todd, R. M. Okasha, P. O. Shipman and T. E. Wood, *Macromolecules*, 2005, **38**, 9411.
39. A. S. Abd-El-Aziz and C. P. de Denus, *J. Chem. Soc., Chem. Commun.*, 1994, 663.
40. A. S. Abd-El-Aziz and K. M. Epp, *Polyhedron*, 1995, **14**, 957.
41. A. S. Abd-El-Aziz, J. L. Pilfold, B. Z. Momeni, A. J. Proud and J. K. Pearson, *Polym. Chem.*, 2014, **5**, 3453.
42. M. Younus, A. Köhler, S. Cron, N. Chawdhury, M. R. Al-Mandhary, M. S. Khan, J. Lewis, N. J. Long, R. H. Friend and P. R. Raithby, *Angew. Chem., Int. Ed.*, 1998, **37**, 3036.
43. W. Wong, X. Wang, Z. He, K. Chan, A. B. Djurišić, K. Cheung, C. Yip, A. M. Ng, Y. Y. Xi, C. S. Mak and W. Chan, *J. Am. Chem. Soc.*, 2007, **129**, 14372.
44. Z. Huang, B. Liu, J. Zhao, Y. He, X. Yan, X. Xu, G. Zhou, X. Yang and Z. Wu, *RSC Adv.*, 2015, **5**, 36507.
45. M. S. Khan, M. R. Al-Mandhary, M. K. Al-Suti, F. R. Al-Battashi, S. Al-Saadi, B. Ahrens, J. K. Bjernemose, M. F. Mahon, P. R. Raithby, M. Younus, N. Chawdhury, A. Köhler, E. A. Marseglia, E. Tedesco, N. Feeder and S. J. Teat, *Dalton Trans.*, 2004, 2377.
46. W.-Y. Wong, X. Wang, H.-L. Zhang, K.-Y. Cheung, M.-K. Fung, A. B. Djurišić and W.-K. Chan, *J. Organomet. Chem.*, 2008, **693**, 3603.
47. W.-Y. Wong, X.-Z. Wang, Z. He, A. B. Djurišić, C.-T. Yip, K.-Y. Cheung, H. Wang, C. S. K. Mak and W.-K. Chan, *Nat. Mater.*, 2007, **6**, 521.

48. P.-Y. Ho, S.-C. Yiu, D. Y. Wu, C.-L. Ho and W.-Y. Wong, *J. Organomet. Chem.*, 2017, **849–850**, 4–9.
49. W. Steffen, B. Köhler, M. Altmann, U. Scherf, K. Stitzer, H. zur Loye and U. H. F. Bunz, *Chem.–Eur. J.*, 2001, **7**, 117.
50. S. Zhang, Y. Sheng, G. Wei, Y. Quan, Y. Cheng and C. Zhu, *J. Polym. Sci., Part A: Polym. Chem.*, 2014, **52**, 1686.
51. G. H. Bertrand, L. Tortech, V. Gandon, C. Aubert and D. Fichou, *Chem. Commun.*, 2014, **50**, 8663.
52. G. L. Schulz and S. Holdcroft, *Chem. Mater.*, 2008, **20**, 5351.
53. T. A. Clem, D. F. J. Kavulak, E. J. Westling and J. M. J. Freéchet, *Chem. Mater.*, 2009, **22**, 1977.
54. S. W. Thomas III, S. Yagi and T. M. Swager, *J. Mater. Chem.*, 2005, **15**, 2829.
55. R. Sakamoto, T. Yagi, K. Hoshiko, S. Kusaka, R. Matsuoka, H. Maeda, Z. Liu, Q. Liu, W.-Y. Wong and H. Nishihara, *Angew. Chem., Int. Ed.*, 2017, **56**, 3526.
56. T. Tsukamoto, K. Takada, R. Sakamoto, R. Matsuoka, R. Toyoda, H. Maeda, T. Yagi, M. Nishikawa, N. Shinjo, S. Amano, T. Iokawa, N. Ishibashi, T. Oi, K. Kanayama, R. Kinugawa, Y. Koda, T. Komura, S. Nakajima, R. Fukuyama, N. Fuse, M. Mizui, M. Miyasaki, Y. Yamashita, K. Yamada, W. Zhang, R. Han, W. Liu, T. Tsubomura and H. Nishihara, *J. Am. Chem. Soc.*, 2017, **139**, 5359.
57. R. Sakamoto, K. Hoshiko, Q. Liu, T. Yagi, T. Nagayama, S. Kusaka, M. Tsuchiya, Y. Kitagawa, W.-Y. Wong and H. Nishihara, *Nat. Commun.*, 2015, **6**, 6713.
58. R. Sakamoto, S. Katagiri, H. Maeda and H. Nishihara, *Coord. Chem. Rev.*, 2013, **257**, 1493.
59. H. Maeda, R. Sakamoto, Y. Nishimori, J. Sendo, F. Toshimitsu, Y. Yamanoi and H. Nishihara, *Chem. Commun.*, 2011, **47**, 8644.
60. H. Maeda, R. Sakamoto and H. Nishihara, *Coord. Chem. Rev.*, 2017, **346**, 139.
61. H. Nishihara, K. Kanaizuka, Y. Nishimori and Y. Yamanoi, *Coord. Chem. Rev.*, 2007, **251**, 2674.
62. S. Katagiri, R. Sakamoto, H. Maeda, Y. Nishimori, T. Kurita and H. Nishihara, *Chem.–Eur. J.*, 2013, **19**, 5088.
63. Y. Nishimori, K. Kanaizuka, M. Murata and H. Nishihara, *Chem.–Asian J.*, 2007, **2**, 367.
64. T. Kurita, Y. Nishimori, F. Toshimitsu, S. Muratsugu, S. Kume and H. Nishihara, *J. Am. Chem. Soc.*, 2010, **132**, 4524.
65. M. Miyachi, M. Ohta, M. Nakai, Y. Kubota, Y. Yamanoi, T. Yonezawa and H. Nishihara, *Chem. Lett.*, 2008, **37**, 404.
66. A. Bhunia, M. A. Gotthardt, M. Yadav, M. T. Gamer, A. Eichhöfer, W. Kleist and P. W. Roesky, *Chem.–Eur. J.*, 2013, **19**, 1986.
67. C. A. Kent, B. P. Mehl, L. Ma, J. M. Papanikolas, T. J. Meyer and W. Lin, *J. Am. Chem. Soc.*, 2010, **132**, 12767.
68. D. Liu, R. C. Huxford and W. Lin, *Angew. Chem., Int. Ed.*, 2011, **123**, 3780.

69. W. J. Rieter, K. M. Pott, K. M. L. Taylor and W. Lin, *J. Am. Chem. Soc.*, 2008, **130**, 11584.
70. C. He, D. Liu and W. Lin, *Biomaterials*, 2015, **36**, 124.
71. C. Poon, C. He, D. Liu, K. Lu and W. Lin, *J. Controlled Release*, 2015, **201**, 90.
72. D. Liu, C. Poon, K. Lu, C. He and W. Lin, *Nat. Commun.*, 2014, **5**, 4182.
73. R. L. Hailes, A. M. Oliver, J. Gwyther, G. R. Whittell and I. Manners, *Chem. Soc. Rev.*, 2016, **45**, 5358.
74. D. A. Foucher, B. Z. Tang and I. Manners, *J. Am. Chem. Soc.*, 1992, **114**, 6246.
75. I. Matas, G. R. Whittell, B. M. Partridge, J. P. Holland, M. F. Haddow, J. C. Green and I. Manners, *J. Am. Chem. Soc.*, 2010, **132**, 13279.
76. F. Jäkle, R. Rulkens, G. Zech, D. A. Foucher, A. J. Lough and I. Manners, *Chem.–Eur. J.*, 1998, **4**, 2117.
77. D. A. Foucher, M. Edwards, R. A. Burrow, A. J. Lough and I. Manners, *Organometallics*, 1994, **13**, 4959.
78. R. Rulkens, D. P. Gates, D. Balaishis, J. K. Pudelski, D. F. McIntosh, A. J. Lough and I. Manners, *J. Am. Chem. Soc.*, 1997, **119**, 10976.
79. D. L. Zechel, K. C. Hultzsch, R. Rulkens, D. Balaishis, Y. Ni, J. K. Pudelski, A. J. Lough, I. Manners and D. A. Foucher, *Organometallics*, 1996, **15**, 1972.
80. C. H. Honeyman, D. A. Foucher, F. Y. Dahmen, R. Rulkens, A. J. Lough and I. Manners, *Organometallics*, 1995, **14**, 5503.
81. E. K. Sarbisheh, J. E. Flores, B. J. Anderson, J. Zhu and J. Muller, *Organometallics*, 2017, **36**, 2182.
82. J. K. Pudelski, R. Rulkens, D. A. Foucher, A. J. Lough, P. M. MacDonald and I. Manners, *Macromolecules*, 1995, **28**, 7301.
83. H. Bhattacharjee and J. Müller, *Coord. Chem. Rev.*, 2016, **314**, 114.
84. E. K. Sarbisheh, H. Bhattacharjee, M. P. T. Cao, J. Zhu and J. Muller, *Organometallics*, 2017, **36**, 614.
85. V. Bellas and M. Rehahn, *Angew. Chem., Int. Ed.*, 2007, **46**, 5082.
86. Y. Ni, R. Rulkens, J. K. Pudelski and I. Manners, *Macromol. Rapid Commun.*, 1995, **16**, 637.
87. N. P. Reddy, H. Yamashita and M. Tanaka, *J. Chem. Soc., Chem. Commun.*, 1995, 2263.
88. I. Manners, *Chem. Commun.*, 1999, 857.
89. K. Temple, F. Jäkle, J. Sheridan and I. Manners, *J. Am. Chem. Soc.*, 2001, **123**, 1355.
90. P. Gómez-Elipe, R. Resendes, P. M. Macdonald and I. Manners, *J. Am. Chem. Soc.*, 1998, **120**, 8348.
91. W. Y. Chan, A. J. Lough and I. Manners, *Organometallics*, 2007, **26**, 1217.
92. D. E. Herbert, J. B. Gilroy, W. Y. Chan, L. Chabanne, A. Staubitz, A. J. Lough and I. Manners, *J. Am. Chem. Soc.*, 2009, **131**, 14958.
93. U. F. J. Mayer, J. B. Gilroy, D. O'Hare and I. Manners, *J. Am. Chem. Soc.*, 2009, **131**, 10382.

94. S. Baljak, A. D. Russell, S. C. Binding, M. F. Haddow, D. O'Hare and I. Manners, *J. Am. Chem. Soc.*, 2014, **136**, 5864.
95. R. A. Musgrave, A. D. Russell, D. W. Hayward, G. R. Whittell, P. G. Lawrence, P. J. Gates, J. C. Green and I. Manners, *Nat. Chem.*, 2017, **9**, 743.
96. H. Braunschweig, C. J. Adams, T. Kupfer, I. Manners, R. M. Richardson and G. R. Whittell, *Angew. Chem., Int. Ed.*, 2008, **47**, 3826.
97. C. J. Adams, H. Braunschweig, M. Fuß, K. Kraft, T. Kupfer, I. Manners, K. Radacki and G. R. Whittell, *Chem.–Eur. J.*, 2011, **17**, 10379.
98. H. Braunschweig, A. Damme, S. Demeshko, K. Dück, T. Kramer, I. Krummenacher, F. Meyer, K. Radacki, S. Stellwag-Konertz and G. R. Whittell, *J. Am. Chem. Soc.*, 2015, **137**, 1492.
99. Y. Zha, H. D. Thaker, R. R. Maddikeri, S. P. Gido, M. T. Tuominen and G. N. Tew, *J. Am. Chem. Soc.*, 2012, **134**, 14534.
100. Z. M. AL-Badri, R. R. Maddikeri, Y. Zha, H. D. Thaker, P. Dobriyal, R. Shunmugam, T. P. Russell and G. N. Tew, *Nat. Commun.*, 2011, **2**, 482.
101. A. S. Abd-El-Aziz, D. J. Winram, P. O. Shipman and L. Bichler, *Macromol. Rapid Commun.*, 2010, **31**, 1992.
102. A. S. Abd-El-Aziz, R. M. Okasha, L. J. May and J. Hurd, *J. Polym. Sci., Part A: Polym. Chem.*, 2006, **44**, 3053.
103. H. Gu, R. Ciganda, P. Castel, J. Ruiz and D. Astruc, *Macromolecules*, 2016, **49**, 4763.
104. T. Choi and R. H. Grubbs, *Angew. Chem., Int. Ed.*, 2003, **115**, 1785.
105. H. Gu, R. Ciganda, R. Hernandez, P. Castel, P. Zhao, J. Ruiz and D. Astruc, *Macromol. Rapid Commun.*, 2016, **37**, 630.
106. H. Gu, R. Ciganda, R. Hernandez, P. Castel, P. Zhao, J. Ruiz and D. Astruc, *Macromolecules*, 2015, **48**, 6071.
107. Y. Wang, A. Rapakousiou and D. Astruc, *Macromolecules*, 2014, **47**, 3767.
108. L. Ren, J. Zhang, C. G. Hardy, S. Ma and C. Tang, *Macromol. Rapid Commun.*, 2012, **33**, 510.
109. J. A. Paquette, A. R. Kenaree and J. B. Gilroy, *Polym. Chem.*, 2017, **8**, 2164.
110. A. S. Abd-El-Aziz, C. Agatemor, N. Etkin and R. Bissessur, *J. Mater. Chem. C*, 2017, **5**, 2268.
111. L. B. Sessions, L. A. Mîinea, K. D. Ericson, D. S. Glueck and R. B. Grubbs, *Macromolecules*, 2005, **38**, 2116.
112. B. Jiang, W. L. Hom, X. Chen, P. Yu, L. C. Pavelka, K. Kisslinger, J. B. Parise, S. R. Bhatia and R. B. Grubbs, *J. Am. Chem. Soc.*, 2016, **138**, 4616.
113. B. Jiang, D. Nykypanchuk, M. K. Endoh, X. Chen, B. Qian, K. Kisslinger, T. Koga, J. B. Parise and R. B. Grubbs, *Macromolecules*, 2016, **49**, 853.
114. L. A. Mîinea, L. B. Sessions, K. D. Ericson, D. S. Glueck and R. B. Grubbs, *Macromolecules*, 2004, **37**, 8967.
115. A. Rapakousiou, R. Djeda, M. Grillaud, N. Li, J. Ruiz and D. Astruc, *Organometallics*, 2014, **33**, 6953.
116. A. Rapakousiou, Y. Wang, C. Belin, N. Pinaud, J. Ruiz and D. Astruc, *Inorg. Chem.*, 2013, **52**, 6685.

117. A. Rapakousiou, C. Deraedt, J. Irigoyen, Y. Wang, N. Pinaud, L. Salmon, J. Ruiz, S. Moya and D. Astruc, *Inorg. Chem.*, 2015, **54**, 2284.
118. A. Rapakousiou, Y. Wang, R. Ciganda, J. Lasnier and D. Astruc, *Organometallics*, 2014, **33**, 3583.
119. A. Rapakousiou, L. Herrer Jimenez, Y. Wang, C. Mouche and D. Astruc, *Organometallics*, 2014, **33**, 1259.
120. Y. Wang, A. Rapakousiou, J. Ruiz and D. Astruc, *Chem.–Eur. J.*, 2014, **20**, 11176.
121. Z. Meng, G. Li, N. Zhu, C.-L. Ho, C.-W. Leung and W.-Y. Wong, *J. Organomet. Chem.*, 2017, **849–850**, 10–16.
122. Z. Dai, Q. Sun, X. Liu, C. Bian, Q. Wu, S. Pan, L. Wang, X. Meng, F. Deng and F. Xiao, *J. Catal.*, 2016, **338**, 202.

CHAPTER 4

Tackling the Global Energy Crisis with Transition-metal-containing Macromolecules

4.1 Introduction

Human survival and civilization considerably depend on sustainable, low-cost energy. The invention of the steam engine ushered in the industrial revolution with its radical transition from animal-powered machines to fossil-fuel-powered machines that have immensely increased industrial production and improved our quality of life. Human demand for energy is ever increasing, being driven by a growing population, rising economies, and improvements in the standard of living in many parts of the world.[1–3] The International Energy Agency projects a 48% increase in global energy consumption from 2012 to 2040.[1] Currently, fossil fuels make up 80% of the global energy consumption and will continue as the major players in the global energy landscape, with liquid fuels, natural gas, and coal projected to account for 78% of the global total energy consumption in 2040 (Figure 4.1).[1,3] Environmental concerns, national security issues, dwindling fossil fuel reserves, and socio-political instability in fossil-fuel-rich regions of the world challenge policymakers and researchers to seek competitive, sustainable, renewable, and clean alternative energy.[4–7]

Alternative energy sources that are environmentally benign and renewable are becoming the gold standard of future energy technologies. Indeed, the 2015 Paris climate accord, which deals with climate change, urges parties to chart and accelerate actions that reduce the emission of greenhouse gases such as carbon dioxide, suggesting energy-efficient technologies and

Polymer Chemistry Series No. 27
Macromolecules Incorporating Transition Metals: Tackling Global Challenges
By Alaa S. Abd-El-Aziz, Christian Agatemor and Wai-Yeung Wong

Published by the Royal Society of Chemistry, www.rsc.org

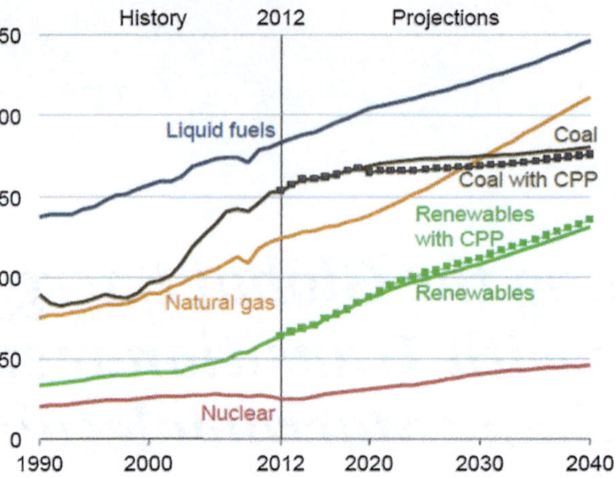

Figure 4.1 Contribution of energy sources to total world energy consumption with projection to 2040. Dotted line: projected effect of the US Clean Power Plan. Source: United States Energy Information Agency (US EIA) *International Energy Outlook 2016*, US EIA Washington DC USA (2016).[3]

renewable energy as options.[2,8,9] At present, global energy is harnessed from three fundamental reactions/processes.[10] The most common one harnesses the energy in chemical bonds to generate electric current. Typically, this involves chemical, electrochemical or photochemical reactions as exemplified by the chemical oxidation of hydrocarbons to produce heat or electrical energy. Another one is the nuclear fission of heavy nuclei, which releases an enormous amount of energy from the nucleus. This low-carbon technology contributed 2.3 trillion kWh to global electricity generation in 2012 and is projected to supply 4.5 trillion kWh of electricity in 2040.[1] The third process exploits energy from mechanical/geothermal processes, such as the wind, falling water or geothermal energy, and is increasingly employed to supply power to a significant proportion of the world population. They are the largest contributors to the rapid increase in renewable energy generation.[1]

This chapter focuses on efforts by the materials chemistry community to efficiently harness the energy release in the breaking and making of chemical bonds by enhancing energy or electron transfer processes. Indeed, the efficiency of these processes is better and more safely controlled using state-of-the-art chemistry toolkits than the energy from nuclear reactions and mechanical/geothermal processes. Currently, chemical oxidation of reduced substances, such as hydrocarbons and biomass, remains a more efficient option in harnessing the energy of chemical bonds than electrochemically- and light-driven processes. Nonetheless, chemical oxidation of reduced substances is unsustainable and notorious for its high carbon footprint, as demonstrated by the lifecycle emission of 490 g CO_2 eq per kWh from natural-gas-powered energy technologies *versus* 41 g CO_2 eq per kWh from rooftop solar photovoltaic devices.[11] The challenge, therefore, is to achieve sustainability, mitigate the carbon footprint of the chemical oxidation reactions, or boost the efficiency of the electrochemically- and light-driven energy technologies. Towards this end, enzymes or microorganisms are recruited to catalyze the oxidation process as demonstrated in biofuel cells, where transition-metal-containing macromolecules mediate the electron transfer process. Also, transition-metal-containing macromolecules are investigated as solar energy antennas and photocatalysts to enhance the efficiency of electrochemically- and light-driven energy technologies. Furthermore, hydrogen is proposed as an alternative to fossil fuels because its oxidation product, water, is environmentally benign; however, production and storage of hydrogen remain a problem.[10,12] Transition-metal-containing macromolecules are emerging platforms for the efficient generation and storage of hydrogen.[13–17] Another strategy employed to tackle the energy crisis focuses on energy-efficient products, electron reservoirs, and redox mediators. In this aspect, transition-metal-containing macromolecules are proven effective precursors for the design of energy-efficient products, as well as molecular batteries and wires. This chapter discusses the applications of transition-metal-containing macromolecules as solar energy antennas, catalysts, redox mediators in biofuels, platforms for hydrogen generation and storage, and precursors for the design of energy-saving products, molecular batteries, and wires.

4.2 Biofuel Cells: Transition-metal-containing Macromolecules as Redox Mediators

Fuel cells are emerging technologies that convert fuel or chemical energy to electrical energy mainly *via* electrochemical oxidation of carbon monoxide or hydrogen.[5,10,18,19] Notwithstanding their impressive performance, fuel cells are far from being a commercial success given their high cost. At the heart of fuel cells is the electrocatalyst system that is comprised of the anode, where electro-oxidation of fuels such as hydrogen gas or glucose takes place, and the cathode, where electro-reduction of oxygen takes place (Figure 4.2). Currently, the electrocatalyst system poses a technical and economic bottleneck to the performance and cost-competitiveness of this promising technology. The first issue is the high cost of the electrocatalyst system, which is based on platinum group catalysts; and the second is the low efficiency of oxygen reduction at the cathode.[18] Furthermore, platinum, the archetypal electrocatalyst for oxygen reduction, is limited in its scope of practical applications because it functions optimally only at pH 0 and 50 °C.[20] The need for a secure supply of hydrogen gas, which is currently plagued with production and storage problems,[10] makes the sustainability of fuel cells questionable. Biofuel cells utilize biological substances, such as glucose, to generate electrical energy in a process catalyzed by enzymes or microorganisms. The renewability and biodegradability of these biocatalysts and biofuels make the technology clean and sustainable. Nonetheless, the energy efficiency of

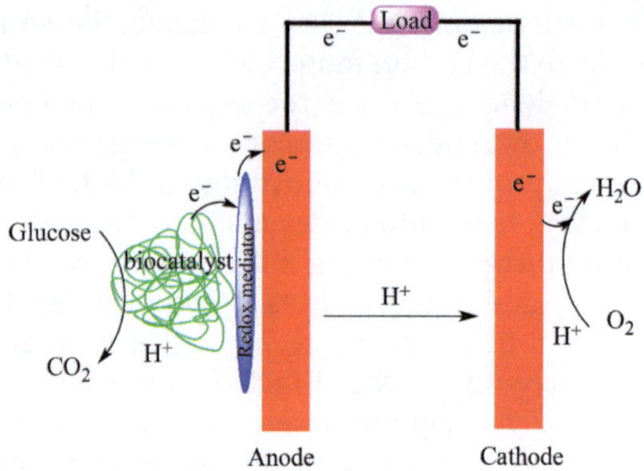

Figure 4.2 Biofuel cell components.

biofuel cells is low, due to poor electron transfer between the biocatalysts and the electrodes.[19]

To develop more efficient biofuel cells, researchers recruit redox-active transition metal complexes as redox mediators to shuttle electrons between the biocatalyst and the electrode. Calvo *et al.*, for instance, explored an osmium (Os) complex, [Os(bpy)$_2$Cl(pyCOH)]Cl (where bpy is bipyridine, and py is pyridine), to mediate electron transfer between the electrodes and the enzyme in a biofuel cell (Figure 4.3).[21] The cell consists of gold electrodes, laccase enzyme to catalyze the four-electron reduction of oxygen to water, acetate buffer, and a $NaClO_4$ supporting electrolyte. In the absence of the Os complex in the biofuel cell, no redox wave was found, indicating inefficient electron transfer between the electrode and the enzyme. The addition of the Os complex to the argon-purged biofuel cell results in a reversible Os^{3+}/Os^{2+} redox wave. Also, when added to an oxygen saturated biofuel cell, a catalytic wave results due to the reduction of Os^{3+} generated from the laccase reduction of oxygen (Figure 4.3).[21]

It is critical to immobilize all components of the catalytic chain on the electrode to improve the efficiency of biofuel cells. Also, practical applications, especially *in vivo* applications, require biofuel cells to function at physiological conditions. Heller *et al.* described biofuel cells that consist of enzyme-coated carbon fiber electrodes and operate on glucose at pH 7.2[20,22–26] and 37 °C.[22,25,26] Under these physiological conditions, the cells continuously shuttle electrons between the enzyme and the electrode, generating 0.9 J of electrical energy in a week.[22] The performance is due to the ability of a redox-active osmium-containing polymer (Figure 4.4) to wire the redox centers of the enzyme to the electrodes, enhancing the efficiency of electron transport between the enzyme and the electrodes.[22] In the biofuel cell, the anodic electrocatalytic system consists of a film made of 60% redox-active polymer, 5% poly(ethylene glycol) diglycidyl ether cross-linker and 35% glucose oxidase enzyme that catalyzes the oxidation of glucose to gluconolactone. The cathodic system consists of 48.5%

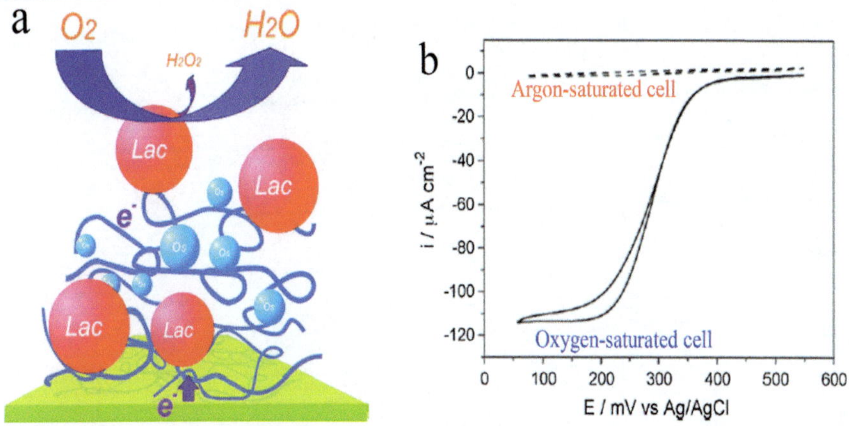

Figure 4.3 (a) Illustration of an osmium-complex-mediated electron transfer in a biofuel cell. (b) Cyclic voltammograms showing the catalytic current from Os^{3+} reduction in argon-saturated (dashed line) or oxygen-saturated media. (a) Reprinted with permission from ref. 27. Copyright (2010) American Chemical Society. (b) Reprinted from *Electrochimica Acta,* **54**, R. Szamocki, V. Flexer, L. Levin *et al.*, Oxygen cathode based on a layer-by-layer self-assembled laccase and osmium redox mediator, 1970–1977.[21] Copyright (2009), with permission from Elsevier.

polyacrylamide-*co*-poly(*N*-vinylimidazole) coordinated to redox-active $[Os(4,4'\text{-dichloro-}2,2'\text{-bipyridine})_2Cl]^{+/2+}$, 6.9% of the cross-linker, and 44.6% bilirubin oxidase that catalyzes the reduction of oxygen to water. Electrostatic interactions between the polyanionic enzymes and the polycationic redox-active polymers prevent phase separation of the catalytic chain components. The polymer-designed strategy includes a long flexible spacer that ensures fast electron transfer. Also, the use of a *N,N'*-dialkyl-2,2'-bis-imidazole ligand in place of a 2,2'-bipyridine ligand shifts the redox potential from +0.36 V to −0.19 V *vs* Ag/AgCl, allowing the operation of the anodic voltage at −0.1 V, which is closer to the −0.36 V redox potential of FAD/FADH cofactor in glucose oxidase. With this biofuel cell, the current densities for glucose oxidation and oxygen reduction show plateaus at 10 and 9.5 $\mu A\ mm^{-2}$, respectively. Also, the cell continuously operates for a week under physiological conditions at 0.52 V, generating 4.3 $\mu W\ mm^{-2}$ power density and losing only ~6% of power per day (Figure 4.4).[22]

The nature of the biocatalyst dictates the products of the electro-reduction of oxygen. For instance, using the bilirubin oxidase enzyme, which is wired to a carbon electrode by a copolymer of polyacrylamide-*co*-poly(*N*-vinylimidazole) coordinated to a redox-active osmium complex, $[Os\text{-}(4,4'\text{-dichloro-}2,2'\text{-bipyridine})_2Cl]^{+/2+}$, Heller *et al.* found that water is the exclusive product of the four-electron enzymatic reduction of oxygen in the biofuel cell.[25] However, Calvo *et al.* proved the contrary in a biofuel cell consisting of laccase enzyme wired to gold electrodes by a polymer coordinated to a redox-active osmium bipyridine complex.[27] They found a trace amount of

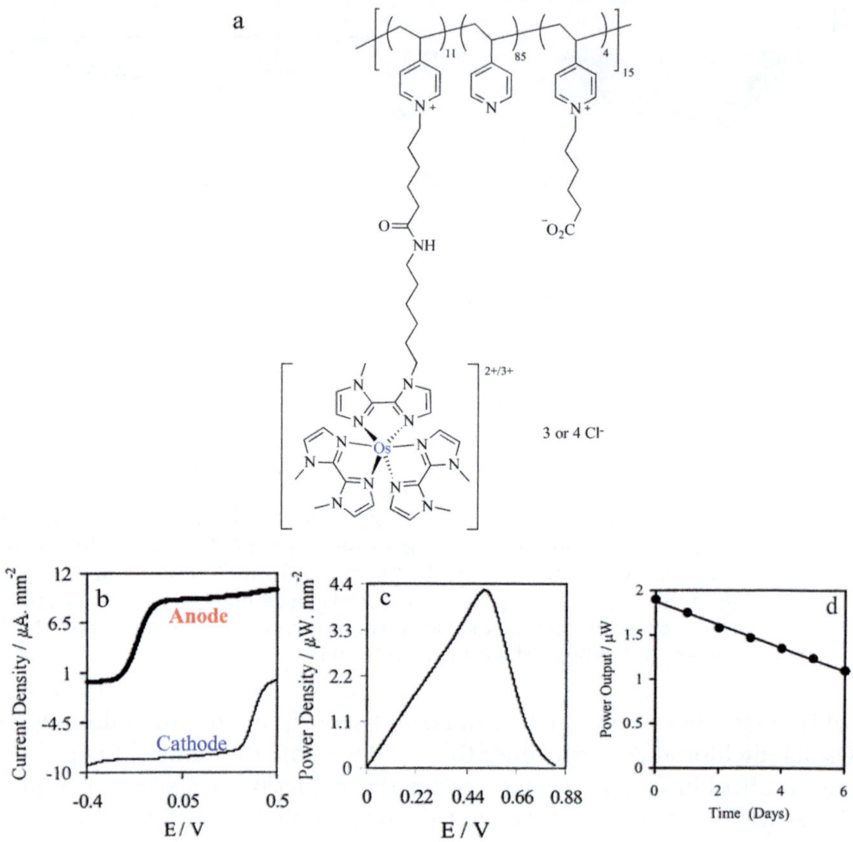

Figure 4.4 (a) Schematic of a redox-active osmium-containing polymer used as a mediator in a biofuel cell. (b) Current density at the anode (thick line) and cathode (thin line). (c) Power density of the biofuel cell. (d) Change in power density of the cell over a week. Reprinted with permission from ref. 22. Copyright (2002) American Chemical Society.

hydrogen peroxide along with the major product, water, during the enzymatic reduction of oxygen.[27] Calvo and coworkers linked the hydrogen peroxide to the enzymatic activity, providing convincing evidence to show that the electrodes and osmium mediator play no role in generating the peroxide, which decreases the catalytic reduction current.[27] Also, Heller and coworkers observed that biofuel cells incorporating an osmium-based mediator in the catalytic chain more efficiently reduce oxygen to water than cells without a mediator (Figure 4.5).[25] Evidently, a cell with a mediator-wired catalytic chain reaches oxygen-diffusion-defined collection efficiency at an overpotential of −0.3 V at pH 7.2 phosphate buffered saline in contrast with a platinum catalyzed cell that reaches efficiency at an overpotential of −0.5 V in 0.5 M H_2SO_4 and 0.65 V at pH 7.2 (Figure 4.5).[25]

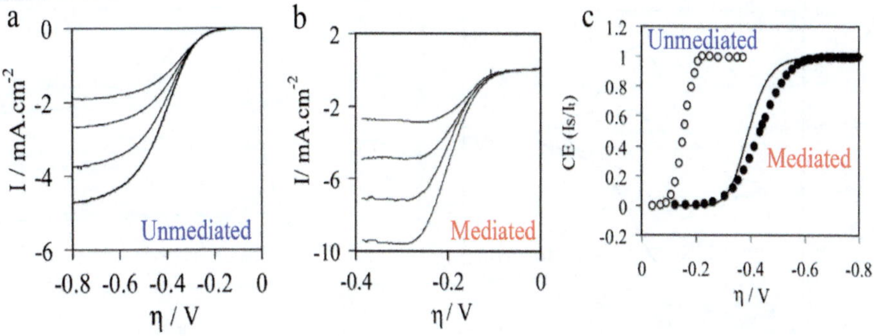

Figure 4.5 Effect of a redox-active polymer mediator on the performance of biofuel cell. (a) and (b) Effect of overpotential on current density for rotating electrode at different speeds (from top to bottom: 500, 1000, 2000, and 4000 rpm). (c) Effect of overpotential on collection efficiency. Reprinted with permission from ref. 25. Copyright (2003) American Chemical Society.

The ratio of the mediator to enzyme defines the current limit associated with each step in the four-step electron transfer process during reduction of oxygen in the biofuel cell. Increasing the ratio of mediator to enzyme increases the current limit flowing from the electrode to (step I) and through (step II) the mediator but decreases the limit flowing from the enzyme to (step IV) the oxygen molecule.[25] Also, concurrently increasing the specific activity of the enzyme and increasing the percentage of the redox mediator increases the current limit in steps I, II, and IV.[25] In contrast to platinum-catalyzed fuel cells that operate optimally in halide-free media, biofuel cells that use transition-metal-containing polymer mediators in their catalytic chain perform extremely well in the presence of a chloride ion.[25] The mobility of the redox-active centers in the mediator tunes the electron transfer between the enzyme and the electrode. As an example, incorporating a flexible 13-atom spacer between the polymer backbone and the redox-active osmium complex enhances electron transfer, compared with the use of mediators without spacers.[20,26] Switching from a mediator without a spacer to one with a spacer increases the apparent electron diffusion coefficient from $(3.4 \pm 0.4) \times 10^{-9}$ cm^2 s^{-1} to $(5.8 \pm 0.5) \times 10^{-6}$ cm^2 s^{-1}.[26] Cyclic voltammetry and polarization experiments performed with a mediator-modified anode in a biofuel cell reveal the impact of a spacer on the efficiency of electron transfer and current density (Figure 4.6).[26] Evidently, incorporating a spacer decreases peak separation in the CV wave and increases the current density of the biofuel cell.

The length of the spacer is an important parameter to control the efficiency of biofuel cells. Indeed, current density increases from 222 ± 19 to 980 ± 51 μA cm^{-2} when the length of the spacer in the mediator is decreased from six carbon atoms to three in a ferrocene-containing polymer mediator (Figure 4.7).[28] Other redox-active transition metals are capable of mediating

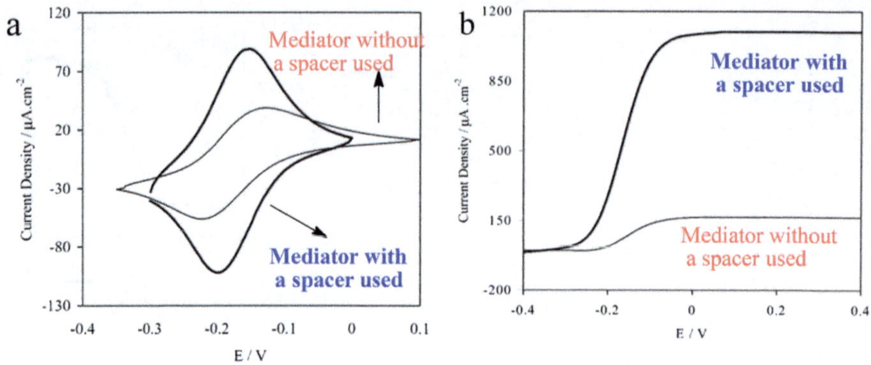

Figure 4.6 (a) Cyclic voltammograms, and (b) current density of electrode wired to an enzyme with a mediator containing a 13-carbon atom spacer (bold line) and without a spacer (fine line). Reprinted with permission from ref. 26. Copyright (2003) American Chemical Society.

electron transfer in biofuel cells, as exemplified by the redox-active iron center of ferrocene. When covalently bonded to a polymer backbone, ferrocene is an effective mediator in the catalytic chain of a biofuel cell.[28–31] For instance, ferrocenylalkyl-containing linear poly(ethylenimine) (Figure 4.7) wires glucose oxidase to a gold cathode *via* a layer-by-layer assembling technique, enabling electron transfer between enzymes and electrodes.[28] Using glucose as the fuel and platinum as the cathode, Glatzhofer and coworkers generated a power density of 86 ± 3 μW cm^{-2} with the ferrocene mediator/enzyme-modified anode at pH 7.[28] They increased the power density to 149 ± 7 μW cm^{-2} by lowering the pH to 5 (Figure 4.7).[28] The layer-by-layer assembling technique better controls cell design parameters, such as electrode thickness, enzyme loading, mediator loading, and the nature of the surface charge to tune the efficiency of transition-metal-mediated biocatalysis in biofuel cells.[21,27,28] Glatzhofer and coworkers, for instance, tuned current density by modifying the electrode with different numbers of enzyme/mediator layers.[28]

4.3 Harnessing Solar Energy with Transition-metal-containing Macromolecules

4.3.1 Photovoltaic Cells

Sunlight is the most abundant source of energy on earth and a promising solution to our energy crisis. At the zenith in a cloudless sky, the sun radiates an estimated 46 700 quads of energy per year in the continental United States alone.[10] This amount exceeds the 97.4 quads of energy consumed in the United States in 2016,[32] and the 815 quads projected to be consumed globally in 2040.[3,33] Unfortunately, the intensity of sunlight

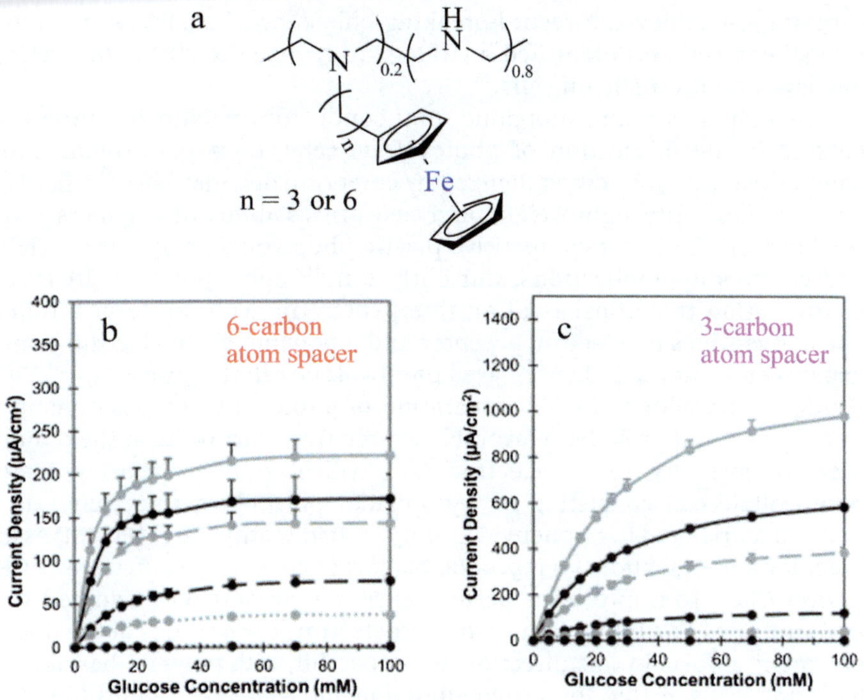

Figure 4.7 (a) Schematic of a ferrocene-containing polymer mediator with varying length of spacer between the ferrocene and polymer backbone. (b) and (c) Dependence of current density on the length of the spacer. (b): 6-Carbon atom spacer; (c): 3-carbon atom spacer, and on the number of enzyme/mediator layers on the electrode, *viz.*, current density increases with the number of layers: 1, 2, 4, 8, 12, 16. Reprinted with permission from ref. 28. Copyright (2016) American Chemical Society.

considerably varies with the time and latitude, and worse still, converting sunlight to electrical energy is inefficient and costly.[10] To exploit the energy of the sun, researchers and policymakers are pursuing strategies that will enhance the efficiency of photovoltaic cells and make these cells cost-competitive. For instance, the United States Department of Energy's SunShot Initiative 2020 aims at a competitive solar industry by 2020, targeting an unsubsidized Levelized Cost of Energy (LCOE) of 0.06 USD per kWh by 2020 for utility-scale photovoltaic cells.[34] Impressively, the collective effort of stakeholders is yielding fruit in the United States, as exemplified by the reduction in the LCOE from 0.27 USD per kWh in 2010 to 0.07 USD per kWh in 2016.[34] Remarkably, an unsubsidized LCOE is offered as low as 0.03 USD per kWh and 0.05 USD per kWh in Dubai and Saudi Arabia, respectively, where sunlight is intense and abundant.[35] The commendable push towards competitiveness is complemented by scientists' efforts to boost the efficiency of photovoltaic cells. Recently, researchers at Kaneka

Corporation achieved a record-breaking efficiency of 26.3% with a 180.4 cm^2 silicon cell and identified a strategy to realize the 29.1% theoretical efficiency limit for silicon cells.[36]

Although silicon and inorganic semi-conductors remain frontline precursors in the fabrication of photovoltaic cells, their performance and competitiveness are now challenged by emerging polymer-based cells. The low cost, flexibility, light weight, and easy processibility of polymers allow the fabrication of cost-competitive plastic photovoltaic cells for specialty and commodity applications. Currently, a bulk heterojunction (BHJ) cell configuration that consists of an interpenetrating composite of a fullerene derivative as an electron acceptor and a polymer as an electron donor ensures efficiency in polymer-based photovoltaic cells (Figure 4.8).[37,38] This configuration allows efficient separation of photogenerated excitons into free charges and provides a favorable charge transport route at the donor–acceptor interface to the electrodes. A typical state-of-the-art polymer photovoltaic cell consisting of regioregular poly(3-hexylthiophene) as a donor material and [6,6]-phenyl-C_{61}-butyric acid methyl ester (PCBM) as an acceptor material affords a reproducible power conversion efficiency (PCE) of over 6%.[37] To improve the performance, researchers have adopted several strategies, including tuning the optoelectronic properties of the donor polymer.[38] Efforts in this direction are impactful, with polymer-based cells featuring PCEs of 10% for a single-junction and ~12% for a multi-junction configuration being reported.[39,40]

An understanding of how a BHJ cell functions is helpful in developing a strategy to improve its performance. In brief, a BHJ cell operates in four steps: absorption of light and generation of an exciton, diffusion of the exciton to a donor–acceptor interface, photoinduced separation of the exciton into free charges at the donor–acceptor interface, and transport of charge to the collecting electrodes.[37,38,40–43] Any of these steps can be influenced to tune the efficiency of the cell, but exciton diffusion length and charge separation affect PCE more than the others.[37] A strategy is to increase the exciton diffusion length, and facilitate charge separation to reduce charge recombination. As a triplet excited charge transfer state with a long lifetime increases exciton diffusion length, and eventually promotes charge separation,[40–43] it is logical to presume that the formation of a triplet excited state would enhance the cell efficiency. It is established that heavy atoms, such as transition metals, allow singlet–triplet intersystem crossing *via* spin–orbit coupling to enable the formation of the triplet excited charge transfer state.[37,38,40–43] Thus, the integration of transition metal complexes within BHJ cells should favor the formation of a triplet excited state. Figure 4.8 illustrates the mechanism of photoinduced electron transfer between a triplet excited donor and an acceptor to generate a triplet exciton that favorably undergoes charge separation as the recombination process is spin-forbidden and the long lifetimes of triplet excited states ensure that the exciton reaches the donor–acceptor interface.[42] Indeed, recent results validate the enhanced performance of polymer BHJ solar cells when a

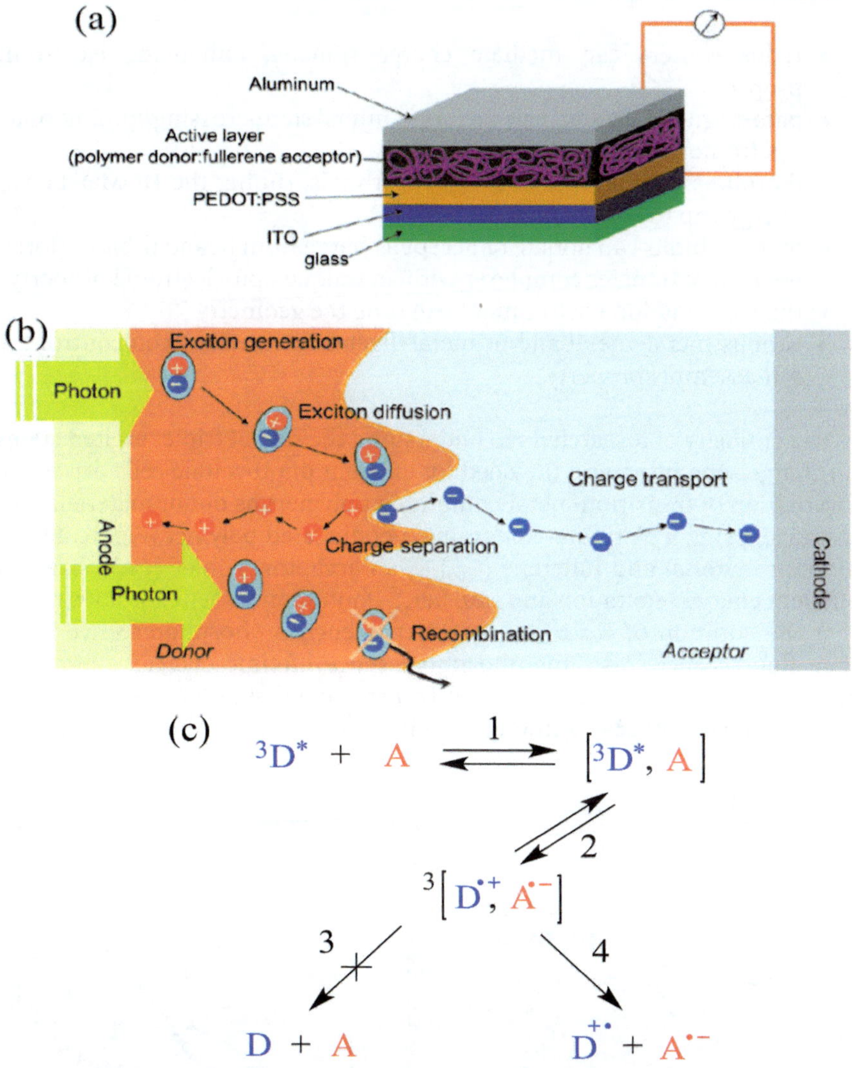

Figure 4.8 (a) Standard configuration of a bulk heterojunction (BHJ) photovoltaic cell; (b) simplified operation steps in a BHJ cell; (c) dynamics of charge separation and transfer in a triplet excited state donor (1: diffusion, 2: electron transfer to ion radical; 3: spin-forbidden charge recombination; 4: favored cage escape of ion radical escape). (a) and (b) Reprinted with permission from ref. 38. Copyright (2010) American Chemical Society. (c) Adapted from ref. 42 with permission from The Royal Society of Chemistry.

transition-metal-containing polymer is used as an electron donor.[44] It is also important to note that the presence of transition metals may endow the polymer with certain attributes including the following:[38]

- redox centers can mediate charge transfer, enhancing electronic property
- paramagnetic centers can facilitate intersystem crossing, tuning optoelectronic property
- d-orbitals can interact with ligand orbitals, tuning the HOMO–LUMO energy gap to control charge transport
- metal orbitals can donate or accept a charge from ligand orbitals, forming charge transfer complexes with an unique optoelectronic property
- the coordination environment can tune the geometry
- various metal–metal and/or metal–ligand interactions can control the self-assembly property.

The curiosity of researchers to understand the role of triplet excited states on charge separation and the need for efficient photovoltaic cells drives the exploration of transition-metal-containing polymers as donor materials. For instance, using a platinum-containing π-conjugated polymer (Figure 4.9) as a donor material and fullerene (C_{60}) as an acceptor, Köhler *et al.* observed efficient charge separation and transfer.[45] Photoluminescence studies reveal that the addition of C_{60} to the polymer quenches phosphorescence more than fluorescence. Also, the phosphorescence lifetime decreases after the addition of C_{60}, indicating that the triplet excited state is involved in the charge transfer process. Although the binding energy of the triplet exciton

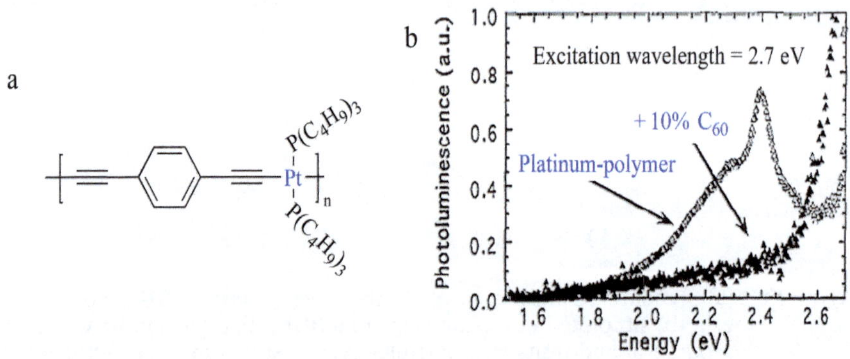

Figure 4.9 (a) Schematic of a platinum-containing polymer, donor material for the fabrication of a photovoltaic cell. (b) Quenching of the phosphorescence of a polymer film (open triangles) and polymer + 10% C_{60} blend film (solid triangles). Reprinted from *Synthetic Metals*, 77, A. Köhler, H. F. Wittmann, R. H. Friend *et al.*, Enhanced photocurrent response in photocells made with platinum-poly-yne/C_{60} blends by photoinduced electron transfer, 147–150.[45] Copyright (1996), with permission from Elsevier.

is high (~0.8 eV), the long lifetime (2.1 μs) associated with the triplet excited state ensures favorable charge transfer to the C_{60} acceptor.[45] This results in a photovoltaic cell that generates current at a photocurrent quantum yield of 1–2%.[45]

Schanze *et al.* also demonstrated the beneficial role of transition metals in photovoltaic cells.[40–43] Using a platinum-containing π-conjugated polymer (Figure 4.10), Schanze and coworkers showed that platinum-induced formation of the triplet excited state facilitates photoinduced charge transfer to enhance the efficiency of photovoltaic cells.[40–43] They observed the quenching of phosphorescence from the polymer, and unequivocally proved the existence of the triplet excited state of the polymer, the absence of an energy transfer quenching pathway, and the presence of electron transfer to a fullerene acceptor.[42] The photovoltaic cell fabricated with this platinum-containing polymer as a donor material affords an impressive incident photon-to-current efficiency of 0.16–0.27%, depending on the thickness of the photoactive layer (Figure 4.10).[42]

Platinum-containing π-conjugated alkynyl polymers are well-investigated in the pursuit for efficient polymer-based BHJ cells. The platinum atom, *via* the heavy atom effect, promotes singlet–triplet state intersystem crossing to form a triplet excited state of the donor.[37,38,46] Also, the polymer structure allows sufficient overlap of the d-orbitals of the platinum atoms with the p-orbitals of the alkynyl carbon atoms to strengthen delocalization of π-electrons and promote intrachain charge transport.[37] Also, the bulky PBu_3 coordinated to the platinum atom in this acetylide polymer impedes aggregation and enhances solubility,[37] a desirable property from the processing perspective.

Other transition metals exhibit the heavy-metal-induced formation of a triplet exciton. For instance, an iridium pyridine complex coordinated to a poly(fluorene-*co*-phenylpyridine) framework (Figure 4.11) generates a triplet

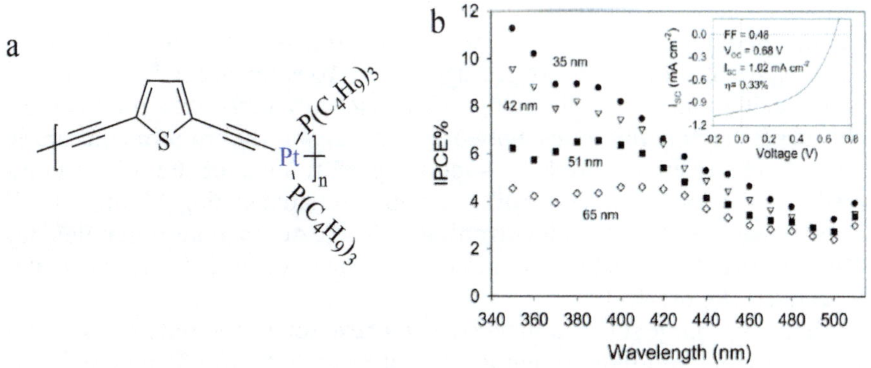

Figure 4.10 (a) Schematic of a platinum-containing polythiophene. (b) Incident photon-to-current efficiency *vs* wavelength as a function of the thickness of the photoactive layer. Reproduced from ref. 42 with permission from The Royal Society of Chemistry.

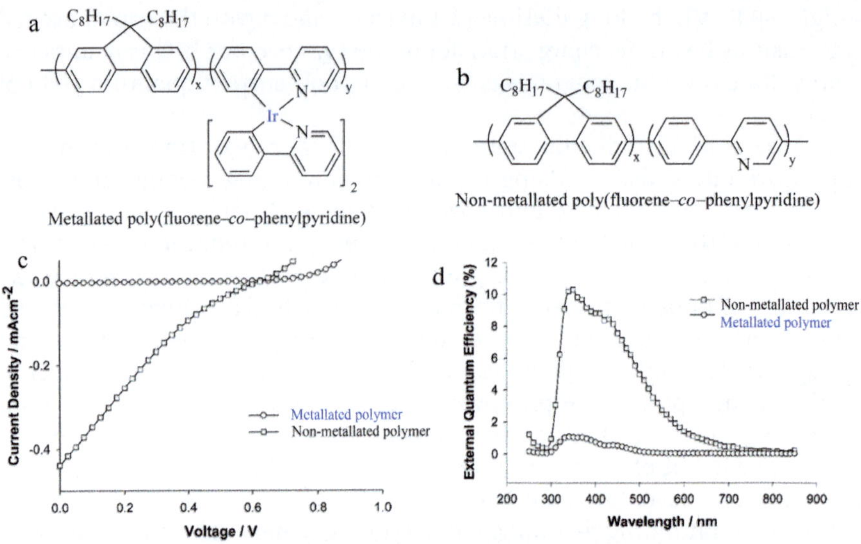

Figure 4.11 Schematic of (a) an iridium-containing (metalated) polymer and (b) non-metalated polymer; (c) current density and (d) quantum efficiency of the metalated and non-metalated polymers. Reprinted with permission from ref. 44. Copyright (2008) American Chemical Society.

exciton to promote charge separation and transfer, eventually increasing external quantum efficiency from 1.1 to 10.3%.[44] Photoluminescence studies with the polymer films reveal that energy transfer from the polymer backbone to the iridium complex quenches the fluorescence of the poly(fluorene-*co*-phenylpyridine) backbone at 422 nm and introduces phosphorescence into the metalated polymer (Figure 4.11). Also, blending the iridium-containing polymer with an acceptor, [6,6]-phenyl C_{61} butyric acid methyl ester, to form a film quenches the phosphorescence (Figure 4.11). The authors concluded that the phosphorescence quenches by electron transfer from the triplet excited state of the polymer to the acceptor.[44] To validate the benefit of the triplet excited state on enhancing efficiency, they fabricated photovoltaic cells using the poly(fluorene-*co*-phenylpyridine) and the iridium-containing analog.[44] The results proved the superior performance of the device fabricated with the iridium polymer over its non-metalated analog (Figure 4.11).[44] For instance, the short circuit current, PCE, and external quantum efficiency (EQE) increase from 0.01 to 0.44 mA cm^{-2}, 0.002 to 0.07%, and 1.1 to 10.3%, respectively (Figure 4.11).[44]

Enhancing charge separation with a triplet exciton generated from a triplet excited polymer donor offers a modest improvement to the efficiency of polymer photovoltaic cells. However, this efficiency is inferior to that obtained with state-of-the-art devices fabricated from organic polymer semi-conductors such as poly(3-hexylthiophene). Theoretical calculations predict that an optimized energy level and electron transfer process improve efficiency to

~8% and that a lower band gap polymer is needed to exceed ~6% efficiency.[47] Indeed, the theoretical limit of an optimized polymer/fullerene photovoltaic cell is predicted to be close to 11%; however, state-of-the-art transition-metal-containing polymer donors/fullerene cells afford much lower efficiencies.[47] Wong *et al.* explored a strategy to optimize the electron transfer process while lowering the band gap of the polymer donor to enhance the efficiency of polymer-based cells.[47] The strategy is inspired by the n-i-p-i superlattice quantum well structure of some inorganic semi-conductors and involves the construction of a π-conjugated polymer with alternating electron-donating thiophene rings and electron-accepting benzothiadiazole segments separated by neutral platinum and alkynyl functional units (Figure 4.12).[47] The extended delocalization of π-electrons throughout the polymer chain lowers the optical band gap of the polymer to 1.85 eV compared with that of electron-deficient benzothiadiazole (2.20 eV) and electron-rich bithienyl (2.55 eV).[47] Also, the small Stokes shift, short emission lifetime, absence of vibronic structure, and weak temperature dependence of the luminescence imply reduced singlet–triplet intersystem crossing.[47] The results corroborate the findings that high band gap polymers are phosphorescent, emitting from the triplet state, whereas low band gap polymers are fluorescent, emitting from the singlet state.[48,49] Nonetheless, transient absorption spectroscopy gives evidence of the generation of the triplet state with a lifetime in the microsecond time range, but the triplet yield is reduced due to quenching of the singlet state by ground-state donor–acceptor charge transfer interactions.[47] In this scenario, the triplet excited state does not contribute to charge separation, rather the strong donor–acceptor charge transfer excited state

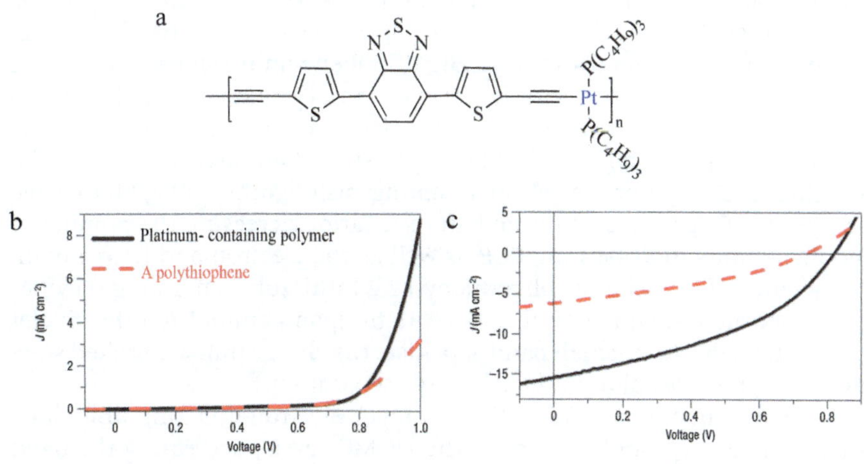

Figure 4.12 (a) Schematic of a platinum-containing π-conjugated polymer with over 4.1% power conversion efficiency. Current density of platinum-containing polymer and poly(3-hexylthiophene) in (b) the dark and (c) simulated AM1.5 solar irradiation. Reprinted by permission from Macmillan Publishers Ltd: *Nature Materials* (ref. 47), copyright (2007).

promotes the photoinduced charge separation. The platinum-containing polymer shows improved absorption of solar radiation, and in some cases, absorbing in the entire visible region and even above 700 nm, as expected from the donor–acceptor–donor triad design.[47]

Using the polymer as an electron donor and PCBM as an electron acceptor, Wong and coworkers fabricated and optimized the performance of a photovoltaic cell equipped with aluminium as an electron collector and poly(3,4-ethylene-dioxythiophene)/poly(styrene sulfonate)-coated indium tin oxide as a hole collector.[47] The optimized photovoltaic cell outperforms a poly(3-hexylthiophene) (P3HT)/PCBM photovoltaic cell fabricated under similar fabrication conditions (Figure 4.12). Also, the optimized cell remains the most efficient platinum-containing polymer/PCBM photovoltaic cell to date, with a PCE of 4.93%.[47] Nonetheless, using the same platinum-containing polymer (Figure 4.12) as a donor and PCBM as an acceptor, Schanze and coworkers[41] as well as Jen and coworkers,[50] obtained less than 2% power conversion efficiency. As the material properties and fabrication conditions are critical to the performance of cells, it is reasonable to assume that the discrepancy observed by different research groups is due to differences in the optimization process and fabrication conditions. Wong and coworkers demonstrated, with compelling evidence, that the superior performance of their optimized photovoltaic cell is due to balanced charge transport and strong absorption of solar radiation.[47]

Another approach to enhance the efficiency of these cells involves controlling the concentration of the electron-donating or -accepting groups in the polymer backbone to tune the band gap energy to absorb light throughout the solar spectrum. For instance, by increasing the number of thiophene units in a monomer unit of a platinum-containing polymer (Figure 4.13), Wong and coworkers lowered the optical band gap, as evidenced by the decrease in the onset of absorption with increasing thiophene units (Figure 4.13).[51] The photocurrent generated with this polymer is not responsive to changes in the number of thiophene units when only electron-donating thiophene units are in the polymer framework, but responsive when alternating electron-donating thiophene and electron-accepting bis(oligothienyl)-2,2-bithiazole are present (Figure 4.13).[51] In the latter scenario, increasing the number of thiophene units increases the PCE as well as the electron and hole mobilities (Figure 4.13).[51] The bis(oligothienyl)-2,2-bithiazole-containing polymer absorbs in the visible region (*ca.* 452–481 nm), and emits from the singlet excited state, due to its small band gap. Also, the charge transfer excited state effectively promotes photoinduced charge separation.[51]

Replacing thiophene units with a stronger electron-donating unit, such as (ethylenedioxy)thiophene, raises the HOMO levels, decreasing the band gap from 2.07 to 1.93 eV as well as slowing down the hole mobility from $(1.4 \pm 0.3) \times 10^{-7}$ to $(1.1 \pm 0.2) \times 10^{-8}$ $cm^2\ V^{-1}\ s^{-1}$.[41] Despite the small band gap of the (ethylenedioxy)thiophene analog, its PCE (0.81%) is inferior to that (1.45%) of the thiophene analog, due to an increase in the short-circuit current of the former compared with the latter.[41] Perhaps aggregation of the

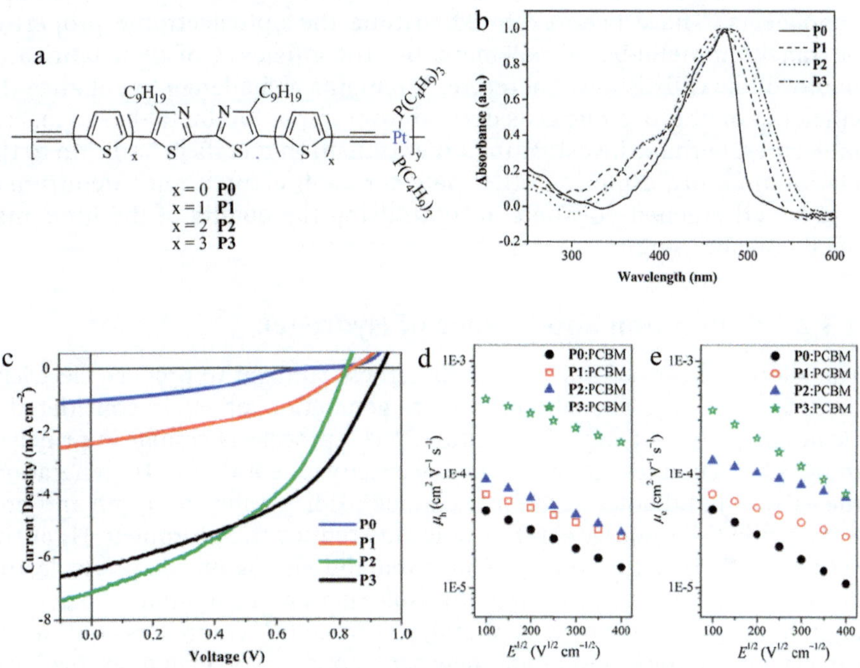

Figure 4.13 (a) Schematic of a platinum-containing polymer with varying numbers of donor thiophene units. Effect of number of thiophene units on (b) the optical band gap revealed by the onset of absorption, (c) current density, (d) hole mobility, and (e) electron mobility. Reprinted with permission from ref. 51. Copyright (2007) American Chemical Society.

(ethylenedioxy)thiophene analog deteriorates the quality of the film, increasing the short-circuit current, and leading to light scattering, as evidenced by the broadened absorption spectrum of the (ethylenedioxy)thiophene polymer/PCBM film.[41] This result unequivocally underlines the importance of processing conditions (for instance, on film quality) on the performance of photovoltaic cells.

The rigidity of the polymer backbone impacts efficiency, with an increase in stiffness to enhance electronic coupling between the electron-donating and -acceptor units in the polymer backbone. For instance, Jen *et al.* tuned the hole mobility in a platinum-containing polymer from 6.1×10^{-5} cm^2 V^{-1} s^{-1} to 1.5×10^{-3} cm^2 V^{-1} s^{-1} by replacing a thiophene unit with a more rigid thieno[3,2-*b*]thiophene unit in the backbone.[50] Also, the switch to the stiff thieno[3,2-*b*]thiophene enhances the PCE. The alkyl side chains that enable better packing of the polymer chain also enhance hole mobility to increase the PCE.[50]

Various approaches, including those by conjugating a zinc(II) porphyrinate complex to a platinum-containing polymer,[52] or cyclometalated platinum complex to a π-conjugated polymer,[53] as well as by constructing porphyrin-derived metal–organic frameworks,[54] or bis(dipyrrinato)zinc(II) nanosheets,[55] have been explored to tune the optoelectronic properties of transition-metal-based polymers, but the efficiency of their fabricated photovoltaic cells is low. Therefore, it remains a challenge to enhance the efficiency of photovoltaic cells derived from transition-metal-based macromolecules. Perhaps investigating other transition metals in addition to the platinum family, exploring other polymer architectures, such dendrimers and hyperbranched polymers, or optimizing the quality of the films may offer a breakthrough.

4.3.2 Generation and Storage of Hydrogen

The alluring dream of scientists is to harness sunlight to produce electricity or hydrogen (H_2). The photochemical generation of H_2 is considered a clean energy technology of the future.[56] The process is similar in principle to the photochemical generation of electricity except that in H_2 generation, the valence band holes at the photoanode oxidize water to oxygen and protons, while the photogenerated electrons reduce the protons to H_2 at the cathode.[57,58] Since the work of Fujishima and Honda on the photoelectrochemical generation of H_2 using ultraviolet light and a titanium dioxide catalyst, a process also known as photolysis of water,[59] researchers continue to expand on the idea, developing new semi-conductors, including low band gap polymers to improve the conversion efficiency.

A glaring challenge in the photolysis of water is to ensure the activity and stability of the semi-conductor in an aqueous medium[56,60] and to tune the spectral window of the semi-conductor into the visible region of the electromagnetic spectrum.[57] Incorporating redox-active transition metals into macromolecular frameworks is an excellent strategy to ensure stability, since macromolecules are more chemically robust than small molecules. Polyoxometalates (POMs), polyanionic clusters of three or more transition metals and oxygen atoms, can be integrated into insoluble porous coordination polymers known as metal–organic frameworks. For instance, molybdenum clusters mediate several natural redox processes such as nitrogen fixation[61,62] and are explored as catalysts in the electrochemical generation of H_2 from water. As an example, two octamolybdate-derived metal–organic frameworks (MOFs), $Cu_3[Mo_8O_{26}][H_2O]_2[OH]_2[L1]_4$ and $Ag_4[Mo_8O_{26}][L2]_{2.5}[H_2O]$ (where L1 is 4*H*-4-amino-1,2,4-triazole, and L2 is 3,5-dimethyl-4-amino-4*H*-1,2,4-triazole) (Figure 4.14) showed excellent electrocatalytic activity and stability in acidic media, generating H_2 from aqueous sulfuric acid.[60] The MOF provides stability, as revealed by its thermograms (Figure 4.14), whereas the molybdate cluster, $[Mo_8O_{26}]_4$, enables the catalytic conversion of water to H_2.

MOFs are robust platforms to protect labile but highly efficient catalysts. The labile [FeFe] active site of hydrogenase enzyme is covalently attached to a thermally and chemically stable Zr(IV)-based UiO-66 MOF to construct a robust [FeFe]-containing MOF catalyst (Figure 4.15), which generates H_2 in the presence of a ruthenium polypyridyl photosensitizer

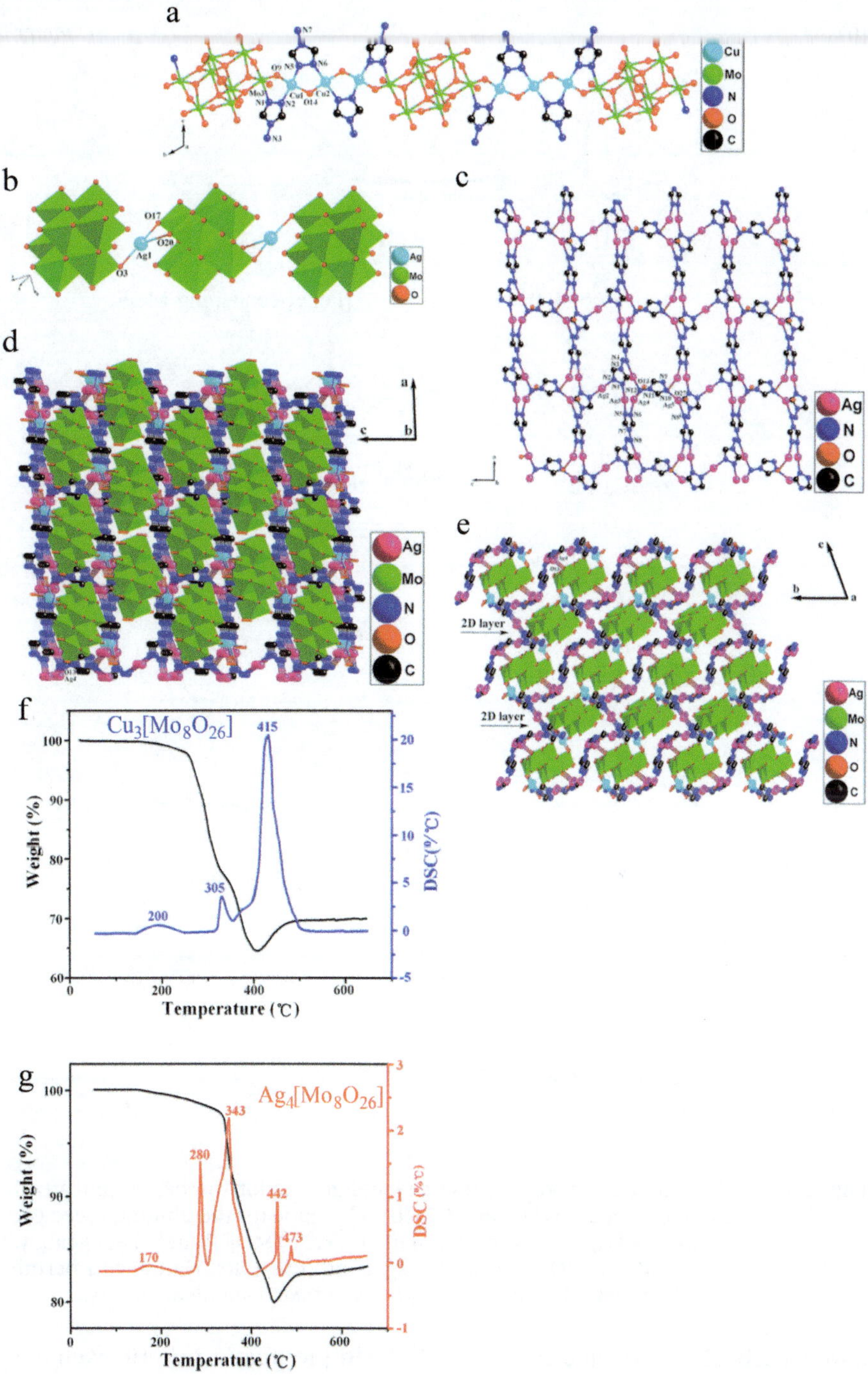

Figure 4.14 (a) 1D structure of $Cu_3[Mo_8O_{26}]$; (b) 1D structure of $Ag_4[Mo_8O_{26}]$; (c) 2D MOF layer of $Ag_4[Mo_8O_{26}]$; (d) 1D structure in 2D MOF layer of $Ag_4[Mo_8O_{26}]$; (e) 2D MOF layer linked by 1D chain in 3D structure of $Ag_4[Mo_8O_{26}]$. TGA and DSC of (f) $Cu_3[Mo_8O_{26}]$ and (g) $Ag_4[Mo_8O_{26}]$ MOFs. Reprinted with permission from ref. 60. Copyright (2013) American Chemical Society.

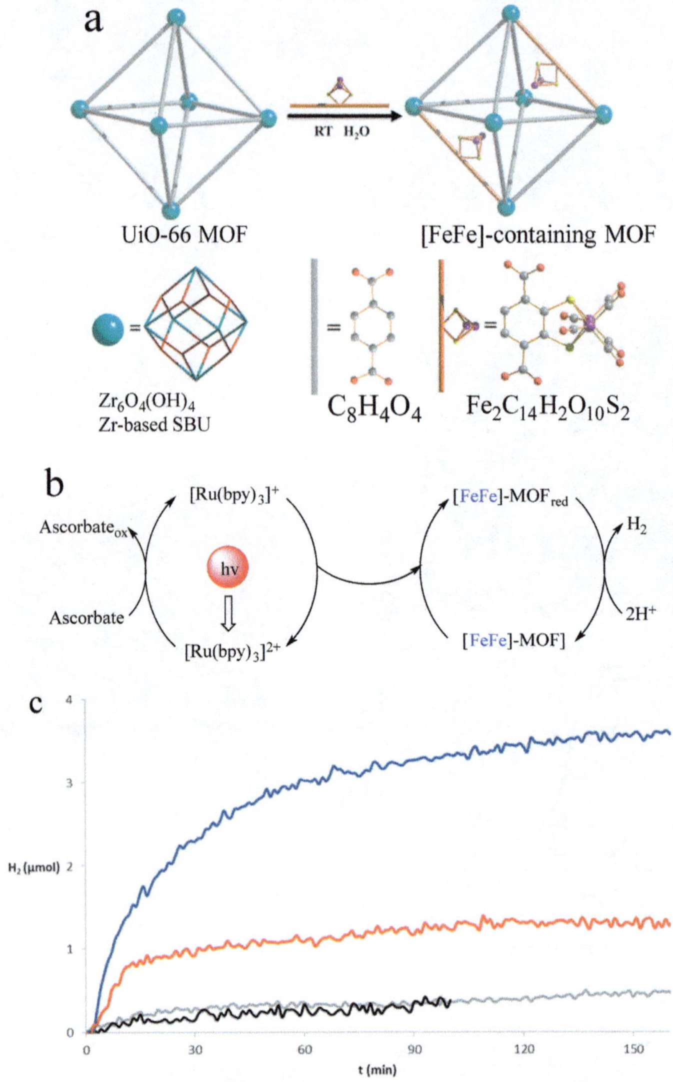

Figure 4.15 (a) Schematic of post-synthetic exchange reaction for the covalent attachment of [FeFe] to UiO-66 MOF; (b) schematic for the photoinduced generation of H_2; (c) H_2 generation using [FeFe]-MOF (blue), [FeFe]-catalyst (red); UiO-66 MOF (black); background (gray). Reprinted with permission from ref. 17. Copyright (2013) American Chemical Society.

and ascorbate as the electron donor.[17] A simple post-synthetic exchange reaction that covalently attaches the active site to the MOF ensures the integrity of the catalyst in the assembled macromolecule. For instance, Fourier-transform infrared spectroscopy and Fe K-edge extended X-ray absorption fine structure spectroscopy provided convincing evidence to conclude that the catalyst is intact within the MOF. The [FeFe]-containing

MOF catalyzes the formation of H_2 when suspended in an acidic solution containing a ruthenium polypyridyl photosensitizer and ascorbate electron donor (Figure 4.15). The [FeFe]-containing MOF outperforms other systems that consist of the isolated catalyst, UiO-66 MOF, or both in generating H_2 (Figure 4.15).[17]

The robustness, well-defined porosity, and high surface areas of MOFs continue to attract researchers to explore their potential in heterogeneous catalysis. A well-implemented synthesis allows the photosensitizer to be integrated into MOFs to increase the efficiency of H_2 generation. This approach ensures fast electron injection from the photosensitizers to the catalytic site. Lin *et al.* explored this idea, incorporating a catalytic polyoxometalate into a ruthenium polypyridyl photosensitizer-derived MOF.[63] The concentration of the catalyst is varied to obtain a series of catalyst–MOF macromolecular assemblies, a strategy that allows control over the photophysical and catalytic properties of assembled macromolecules. For instance, increasing the concentration of the catalytic polyoxometalate decreases the lifetime of the ruthenium complex phosphorescence, indicating a quenching phenomenon, which the authors attributed to electron transfer from the excited states of the complex to the catalytic polyoxometalate. Using methanol as a sacrificial electron donor, Lin and coworkers demonstrated that these catalytic-MOFs generate H_2 from an acidic solution upon white light irradiation. The concentration of the catalytic polyoxometalate influences the turnover number, *viz.*, increasing the concentration would decrease the turnover number. These assembled catalytic-MOFs retain their structures after use, but slightly lose activity after three cycles. In contrast, an isolated catalytic polyoxometalate and ruthenium polypyridyl complex pair did not catalyze the generation of H_2.[63]

The different transition metal oxyanions in polyoxometalates can act in synergy to enhance the catalytic activity on the formation of H_2. For instance, by using a nickel-containing polyoxometalate, $[Ni_4(H_2O)_2(PW_9O_{34})_2]_{10}$, instead of a Wells–Dawson type polyoxometalate, $[P_2W_{18}O_{62}]_6$, and an iridium-derived photosensitizer in place of its ruthenium analog, the turnover number increases from 79 to 1476.[64] The mechanism of the photo-induced generation of H_2 involves multielectron transfer from the photosensitizer to the catalyst. On exposure to visible light, the phosphorescent ruthenium or iridium polypyridyl complex is excited to the singlet charge transfer state, then yields the triplet charge transfer state *via* intersystem crossing before returning to the ground state. The sacrificial electron donor or the catalytic polyoxometalate quenches the photosensitizer excited state phosphorescence. Phosphorescence quenching experiments indicated that the catalyst more effectively quenches the emission than the electron donor, implying a more efficient electron transfer from the photosensitizer to the catalyst.[64]

To enable hydrogen-based energy technologies for practical applications, we must store hydrogen safely and conveniently. Current technologies allow hydrogen to be stored as gas in high pressure tanks, as liquid under cryogenic conditions, and as adsorbed or bound hydrogen in a solid support

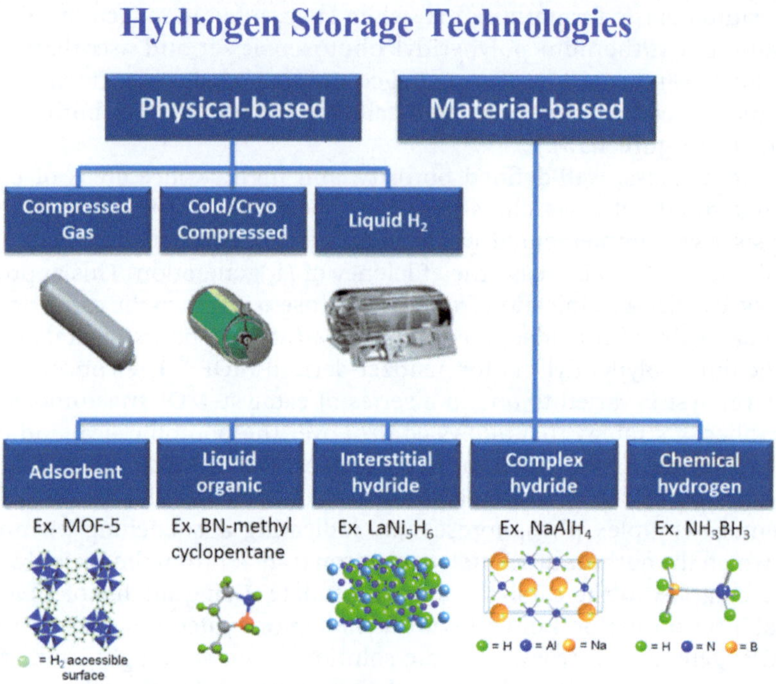

Figure 4.16 Hydrogen storage technologies. Source: United State Department of Energy, Office of Energy Efficiency and Renewable Energy (https://energy.gov/eere/fuelcells/hydrogen-storage).

(Figure 4.16).[65] Materials scientists are exploring various chemistries to adsorb or bind hydrogen onto solid supports as the technology can be controlled to meet the expectation of end users and regulatory bodies. Success in this regard means that hydrogen must be reversibly stored, the release controlled, and refilling of the storage vessel rapid. In this respect, it must be acknowledged that no impactful breakthroughs have been reported and the challenge remains daunting.[10] However, the rigorous effort and intense interest of various researchers to store hydrogen have resulted in many encouraging discoveries that are advancing the field.

The porous and high surface area of MOFs, for instance, offer the possibility for high hydrogen uptake. Indeed, various MOFs have been reported as platforms for the storage of hydrogen.[13,15,66–70] Yaghi *et al.* reported the use of MOF-5 composed of $Zn_4(BDC)_3$ (BDC = 1,4-benzenedicarboxylate) in hydrogen storage, demonstrating that hydrogen molecules bind to zinc and the BDC ligand in MOF-5.[15] At room temperature and 20 bar, MOF-5 absorbs 1.0% weight percent hydrogen. Also, Long and coworkers have explored MOFs to store hydrogen and showed that the hydrogen molecule binds to the metal center.[13,66–70] For instance, they constructed a chromium(II)-derived MOF (Figure 4.17) and explored its hydrogen

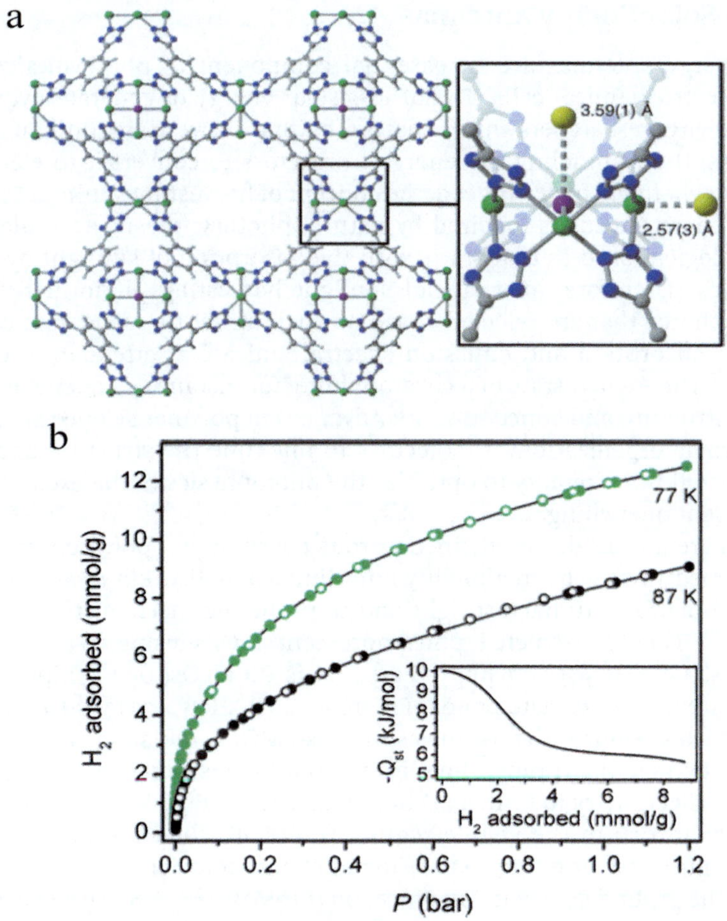

Figure 4.17 (a) Crystal structure of a chromium-containing MOF as revealed by neutron powder diffraction (dark green: Cr; purple: Cl; gray: C; blue: N; white: H). (b) Hydrogen adsorption isotherm (insert: isosteric heat of hydrogen adsorption). Reproduced from E. D. Bloch, W. L. Queen, M. R. Hudson *et al.*, Hydrogen Storage and Selective, Reversible O_2 Adsorption in a Metal–Organic Framework with Open Chromium(II) Sites, *Angewandte Chemie*,[13] John Wiley and Sons, © 2016 WILEY-VCH Verlag GmbH & Co. KGaA, Weinheim.

adsorption property.[13] The MOF strongly binds hydrogen as indicated by the −10.0 kJ mol^{-1} isosteric heat of hydrogen adsorption. The MOF binds hydrogen through its Cr^{2+} and μ_4-Cl^- ions, adsorbing 12 mmol g^{-1} (2.4%) of hydrogen at 77 K and 1 bar (Figure 4.17). So far, the results achieved are impressive but are yet to meet the United State Department of Energy hydrogen storage target of 6.5% for onboard hydrogen storage for light-duty fuel cell vehicles.[65]

4.3.3 Solar Energy Antennas

Solar energy antennas are an essential component of photovoltaics and photo-electrochemical cells. To harness solar energy to generate hydrogen or electricity, researchers mimic nature by first harvesting sunlight using antennas, then funneling the energy to reactors to convert it to electrical energy or chemical fuel.[71] Indeed, the concept of harvesting sunlight to cater for our energy needs is inspired by natural photosynthesis, an indispensable biological process that starts with the absorption of sunlight by chromophores. Therefore, an artificial sunlight harvesting platform requires chromophores that are selected based on their structure, band gap, photostability, absorption and emission spectra, and molar attenuation coefficient.[71,72] The excited state of a chromophore, for instance, is quenched by a certain structure and concentration. Advances in polymer science allow the microscopic organization of molecules to fine-tune the structure and precisely install functionality to optimize the photophysics of the excited state and prevent quenching.

MOFs are a class of well-defined porous coordination polymers that can be designed to target functionality and dimensionality. Lin *et al.* explored the MOF platform to harvest light and to probe the mechanism of energy transfer.[73] They constructed phosphorescent MOFs using a photoactive motif, $\{M[4,4'-(HO_2C)_2-bpy]_2bpy\}\{PF_6\}_2$, (M is Ru or Os; bpy is bipyridine), linked by zinc–carboxylate bonds (Figure 4.18).[73] X-Ray crystal structures of the Ru-MOFs confirm the synthesis of these MOFs, revealing that the carboxylate oxygen of the bipyridine ligand coordinates to the zinc metal in a tetrahedral geometry to form a 2D bilayer structure (Figure 4.18). The MOFs intensely absorb in the visible region, (Ru-MOF: 400–600 nm; Os-MOF: 400–650 nm), transitioning to the singlet charge transfer excited state and back to the ground state with an emission (Ru-MOF: λ_{max} ~620 nm; Os-MOF: λ_{max} ~720 nm).[73] Also, the Os-MOF exhibits a spin-allowed absorption band at 710 nm due to single–triplet intersystem crossing promoted by the strong spin–orbit coupling in Os^{3+}.

It is required that an antenna transports the energy stored in the excited state to a reactor for the photochemical transformations, thus it is imperative to understand the dynamics of energy transport in sunlight harvesting platforms. Using the Ru-MOF (Figure 4.18), Lin and coworkers probed the energy transport after sunlight capture by doping the Ru-MOF with varying concentrations of $\{Os[4,4'-(HO_2C)_2-bpy]_2bpy\}\{PF_6\}_2$ salt to obtain Ru/Os-MOFs.[73] Transient emission decay studies of the Ru/Os-MOF reveal fast and efficient energy transport between the metal centers. Notably, the Ru-MOF, Os-MOF, and Ru/Os-MOFs' decay fit a biexponential decay profile, and the lifetime of the excited state of Ru^{2+} decreases as the concentration of the osmium dopant in the MOF increases (Figure 4.18). It is apparent that the dopant quenches the emission of the Ru^{2+} complex, an indication of energy transfer from the excited state of Ru^{2+} to the dopant. Furthermore, the concurrent delay in emission from the excited state of osmium and decay of the emission

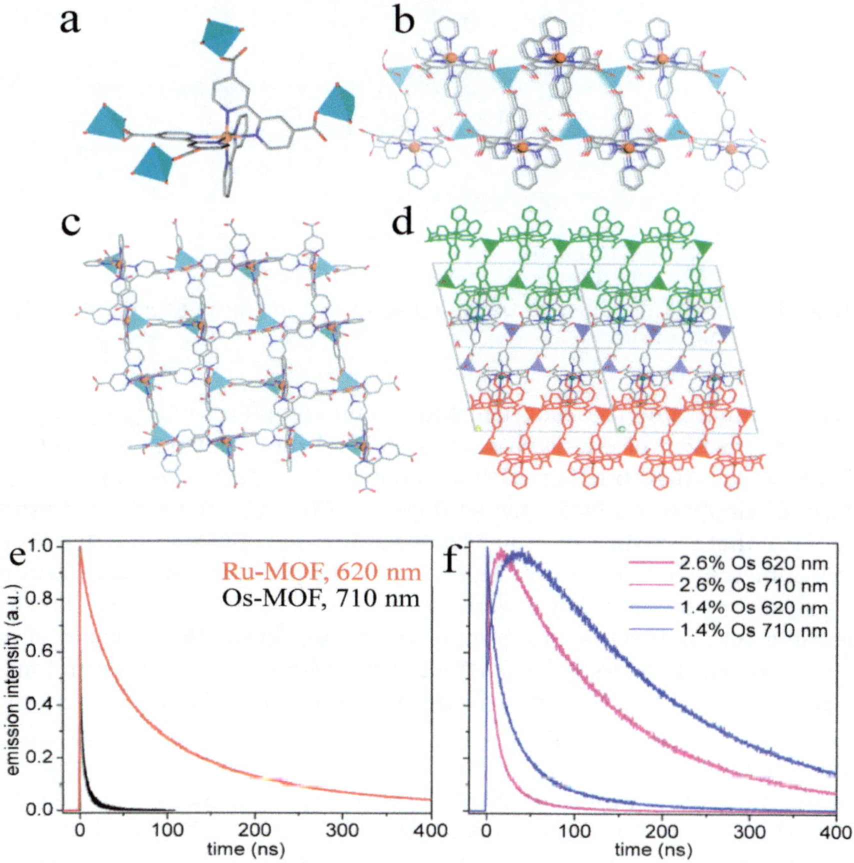

Figure 4.18 (a)–(d) 1D and 2D structures of Ru-MOF showing atom connectivity and bilayer packing. (e) Transient emission decays of Ru-MOF and Os-MOF. (f) Effect of osmium dopant on the decay profile of Ru-MOF. Reprinted with permission from ref. 73. Copyright (2010) American Chemical Society.

from the excited state of ruthenium highlights the dynamics of energy transfer within the MOF, specifically from ruthenium to an osmium charge trap or a ruthenium trap. The structure and dimensionality of the antenna affect the number of hops, and the hopping time it takes for an excited state to reach a trap. Lin and coworkers estimated that the excited state hops between layers (3D transport) in 25 ns and within a layer (2D transport) in 7 ns, with 15–55 hops between ruthenium to a ruthenium trap in a lifetime.[73]

Other transition metals, including platinum,[53,74] ruthenium,[75–78] cobalt,[79,80] zinc,[81–84] gadolinium,[85] europium,[85] and ytterbium,[85] have been incorporated into a polymer framework to harvest sunlight. Gadolinium(III)-, europium(III)- and ytterbium(III)-containing nanoscale coordination polymers (Figure 4.19), for instance, constructed *via* coordination-directed assembly

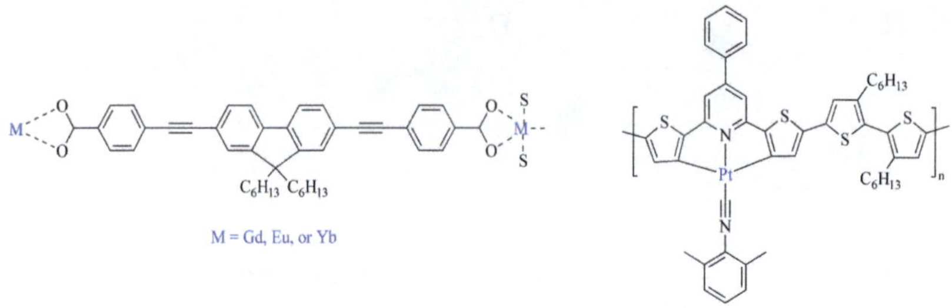

Figure 4.19 Some examples of transition-metal-containing polymers for sunlight harvesting.

of the π-conjugated dicarboxylate with metal complexes efficiently capture light and transfer the energy to an encapsulated guest molecule.[85] Also, Chan and coworkers incorporated a cyclometalated platinum complex into a π-conjugated thiophene-derived polymer (Figure 4.19) to obtain a light-harvesting platform that can channel triplet excitons.[53] In this platform, the thiophene backbone transfers the harvested energy to the platinum complex, forming a singlet excited state that generates a triplet excited state *via* intersystem crossing. A triplet exciton is subsequently formed during the transport of the triplet excited state along the polymer chain. Mimicking some natural macrocyclic antenna that harvest light from all directions, Kobuke and coworkers assembled zinc(II) porphyrin into pentameric or hexameric macrocycles through covalent and non-covalent linkages.[82] On a ^{1}H NMR timescale, the porphyrin units in the non-covalently assembled macrocycle rotate along the ethynyl axis even at temperatures as low as −50 °C in $CDCl_3$, an indication that the barrier to rotation along that axis is small in solution. The transitional dipoles are more aligned in the covalently-bonded hexameric macrocycle than its pentameric analog, as revealed from the shorter hopping time of the excitation energy in the former (12.8 ps) compared with the latter (21 ps).

4.4 Energy Storage Platforms: Molecular Batteries and Wires

To address today's energy crisis, policymakers and researchers strongly advocate the need to replace hydrocarbon-based energy sources with alternative energy derived from the sun, wind or falling water. These alternatives feature certain advantages over hydrocarbon-based sources, but their limitations are evident. For instance, whereas power generated from hydrocarbon-based sources can be regulated to meet demand, several of these alternative energy sources depend on irrepressible and intermittent natural sources, such as sunlight, the wind, or falling water, to generate power.[86] It is, therefore, difficult to match power outputs with demand. To realize the full potential of these technologies, we need a platform to store excess energy during peak power outputs, when production exceeds demand, and then release the energy when production drops below demand. Energy storage platforms are, therefore, an indispensable component of these alternative energy technologies. Also, the emergence and increasing use of mobile electronic and telecommunication devices, such as cell phones and laptops, highlight the need for energy storage platforms. This will ensure these devices function even without being connected to a power supply provided adequate energy is stored in their batteries.

The rechargeable lithium-ion battery currently dominates the energy storage industry.[10,86,87] Despite its excellent performance and commercial success, there is still need for storage platforms with longer discharging times, shorter recharging times, and increased power and energy density.[10,86,87] Advances in organometallic polymer science push the frontiers towards modification of the electrodes to enhance the performance of batteries. Wong *et al.* explored the excellent redox property of ferrocene and the electronic property of π-conjugated polymers to fabricate a cathode for a lithium-ion battery.[88] The redox reaction of ferrocene is fast and reversible and yields a relatively stable ferrocenium cation, while π-conjugated polymers feature extended delocalization of π-electrons along the polymer backbone, properties that are relevant to electronic materials. Against this background, Wong and coworkers aim to design cathode-active materials from a linear acetylenic polymer containing platinum in the backbone and redox-active ferrocene as a pendant (Figure 4.20).[88] The ferrocenyl moieties endow these acetylenic polymers with redox activity, as revealed by cyclic voltammograms, where reversible redox waves corresponding to the redox process of the ferrocene are found (Figure 4.20).[88] The anodic and cathodic currents decrease after oxidation of the ferrocene to the ferrocenium cation presumably because the polymer leaches from the electrode into the electrolyte (Figure 4.20). To enhance performance and ensure robustness, Wong and coworkers fabricated a ferrocene-containing acetylenic polymer–carbon composite electrode, which displayed better stability and electronic current compared with a thin layer of the polymer on the electrode. Also, the cyclic voltammograms of the electrode showed the cathodic and anodic currents that correspond to the charge and discharge processes. The composite electrodes feature impressive charge and discharge capacities, which depend on the structure of the polymer, with the lowest charge and discharge capacities being 15 mA h g^{-1} and 7 mA h g^{-1}, respectively (Figure 4.20). The authors attributed the low discharge capacity to leaching of the polymer from the electrode,[88] perhaps due to a change in solubility as the ferrocene is reduced to the polar ferrocenium cation. Remarkably, the charge and discharge reactions are stable after 50 cycles and display no n-type redox behaviour.

Molecular electronics has become a topical research field in our world of miniaturized gadgets. Molecular batteries and wires are an integral component of these electronics.[89–91] Whereas these batteries are stable redox-active electron reservoirs, the wires offer pathways to shuttle electrons between two

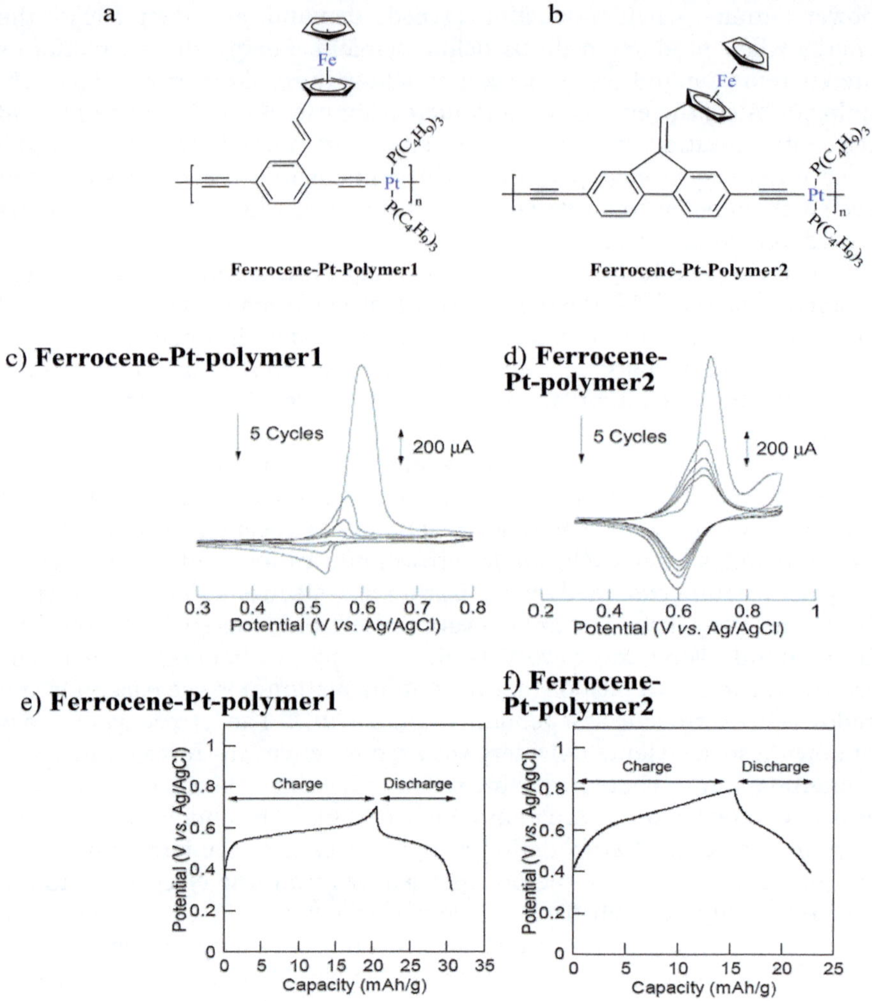

Figure 4.20 (a) and (b) Schematics of ferrocene/platinum-containing polymers; (c) and (d) cyclic voltammograms of polymer thin layer electrodes; (e) and (f) charging–discharging cycles of the composite electrode. Reprinted from *Journal of Organometallic Chemistry*, **812**, Z. Meng, K. Sato, T. Sukegawa *et al.*, Metallopolyyne polymers with ferrocenyl pendant ligands as cathode-active materials for organic battery application, 51–55.[88] Copyright (2016) with permission from Elsevier.

junctions. Conceptually, the batteries are energy storage systems because their robust reversible redox chemistry allows them to charge (reduce) and then discharge (oxidize).[92–96] Also, in an era faced with an energy crisis, molecular wires are an attractive platform to conserve energy, since their electron transport pathways may be more efficient and result in lesser energy losses than transport through space. Several transition-metal-containing macromolecules are redox-active and electron-rich to function as electron/energy reservoirs and channels. Indeed, many researchers incorporate transition metals into macromolecules to create these reservoirs and channels.[55,92–105] Noteworthy examples are macromolecules containing ferrocene or η^6-arene-η^5-cyclopentadienyliron(II) complexes, which are acknowledged for their robust redox activity.

Astruc *et al.*, for instance, constructed electron reservoirs by incorporating ferrocene (Fc), or pentamethylferrocene (Fc*) into the periphery of dendrimers.[93–95] Their synthetic approach allows the design of dendrimers from zeroth-generation (G_0) to seventh-generation (G_7) with 14 000 ferrocene units as peripheral units.[95] The redox-active ferrocenyl moieties are separated in space, precluding electrostatic interactions that will split their redox wave in a cyclic voltammogram. Thus, notwithstanding the large number of ferrocenes, the dendrimers feature a single reversible redox wave like molecular ferrocene (Figure 4.21).[93,95] Cyclic voltammetry of the low generation dendrimers in dichloromethane shows a diffusion-controlled reversible redox process, but the higher generations adsorb on the surface of the platinum electrode,[93] a phenomenon that can be exploited to modify electrodes with a redox-active layer. Also, cyclic voltammetry of smaller dendrimers in dimethylformamide shows evidence of chemically and electrochemically reversible redox chemistry controlled by diffusion, but shows proof of absorption with larger dendrimers.[93] While the chemical reversibility implies robustness of the ferrocene and ferrocenium cation, the electrochemical reversibility suggests that the electron transfer between the ferrocenyl units and the electrode is faster than the timescale of the cyclic voltammetry experiment. Perhaps electron hopping among the peripheral ferrocenes is efficient or molecular rotation of the dendrimer positions the ferrocenyl units in proximity with the electrode. Both mechanisms facilitate electron transfer within a period that is shorter than that of cyclic voltammetry.[92] The large number of ferrocene units implies these dendrimers are electron reservoirs and molecular batteries, as they can be charged or reduced with up to 14 000 electrons and rapidly discharged or oxidized, releasing these electrons.[92]

η^6-Arene-η^5-cyclopentadienyliron(II) complexes feature redox activity that is explored in the design of electron reservoir systems. Abd-El-Aziz *et al.* have developed a library of redox-active linear and star polymers as well as dendrimers (Figures 4.22–4.24), based on these complexes and demonstrated the reversibility of their redox chemistry at room and low temperatures.[104–114] For instance, through a sequential polymerization strategy, they obtained linear polymers with several redox-active iron centers within linear polymer frameworks (Figure 4.22). The redox-active η^6-arene-η^5-cyclopentadienyliron(II) moieties in these linear polymers are distinct from that of ferrocene, and show a single reversible redox wave at the negative potential, a property that suggests that the redox-active moieties are isolated (Figure 4.22).[104] With a star polymer, a single or split redox wave is observed, depending on the size of the polymer. For instance, a star polymer consisting of different generations of redox centers exhibits a split redox wave, in

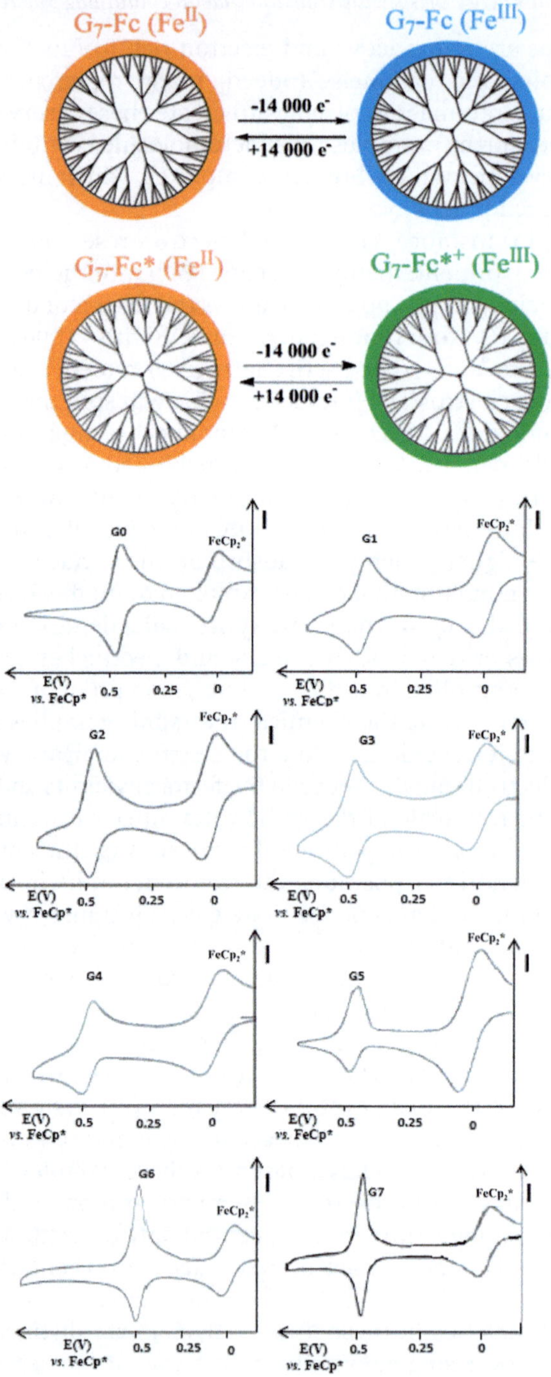

Figure 4.21 Schematic of charging and discharging of seventh-generation dendrimers with 14 000 ferrocene units. Cyclic voltammograms of zeroth-generation (G_0) to seventh-generation (G_7) dendrimers showing the single redox wave. Reprinted with permission from ref. 95. Copyright (2009) American Chemical Society.

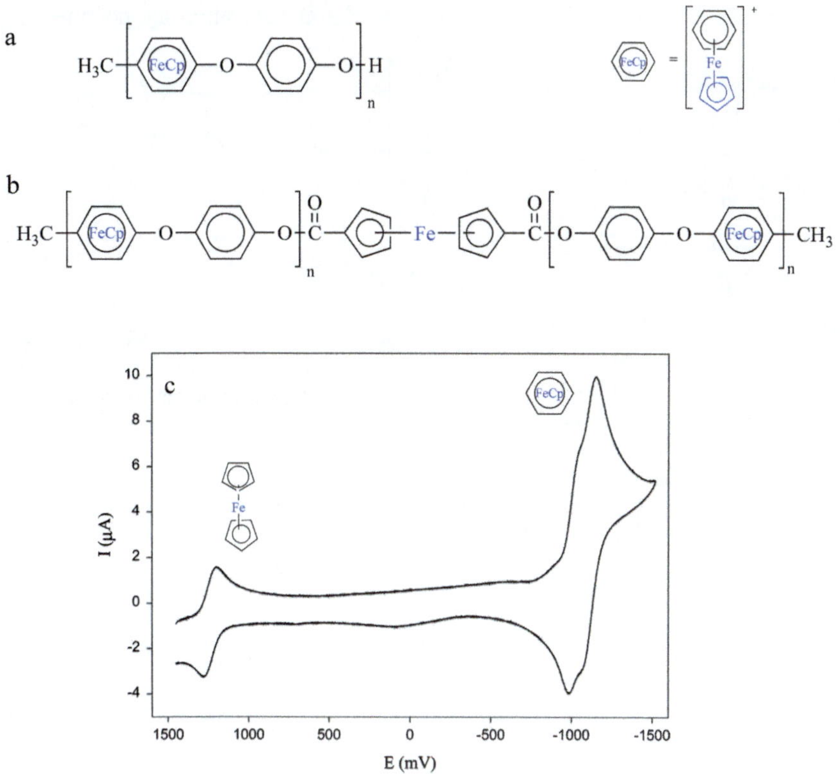

Figure 4.22 (a) and (b) Schematics of redox-active polymers; (c) cyclic voltammograms of a redox active polymer containing η^6-arene-η^5-cyclopentadienyliron(II) complexes and ferrocene. Experiment conducted at 0 °C. Reprinted with permission from ref. 104. Copyright (2005) American Chemical Society.

contrast with a polymer with a single generation (Figure 4.23).[106] Recently, Abd-El-Aziz and coworkers developed a synthetic strategy to incorporate up to 54 η^6-arene-η^5-cyclopentadienyliron(II) moieties in the branches of the dendrimers (Figure 4.24).[107,108,110,112] The chemical reversibility of the redox process is temperature dependent, with low temperatures ensuring chemical reversibility, but slowing down the electron transfer rate.[107,115] This finding suggests the possibility to explore these dendrimers as molecular batteries under low-temperature conditions. Abd-El-Aziz and coworkers also showed that the reduction potential of some of these star polymers and dendrimers is generation dependent.[108] Also, Astruc and coworkers used the η^6-arene-η^5-cyclopentadienyliron(II) complex to construct an electron-rich dendrimer-based molecular battery that contains 64 redox centers that are chemically and electrochemically robust within the cyclic voltammetry timescale.[94] Reduction (charging) of the dendrimer affords its 19-electron analog, which releases the electrons to reduce many fullerene molecules to

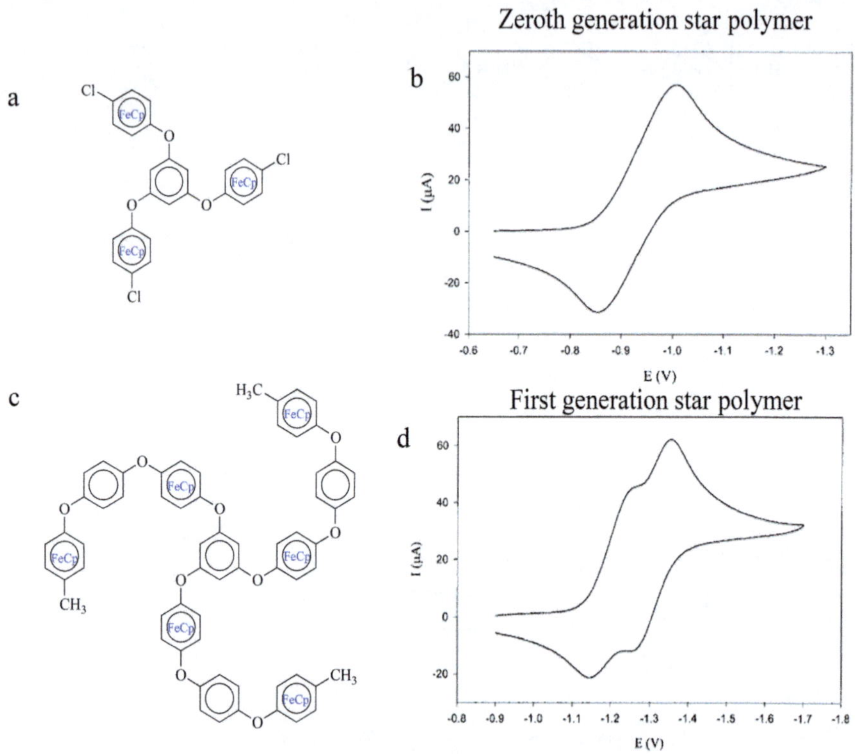

Figure 4.23 Schematics of (a) G_0 and (c) G_1 redox-active star polymers. (b) and (d) Cyclic voltammograms of star polymers. Reproduced from A. S. Abd-El-Aziz, E. K. Todd and T. H. Afifi, Star-Shaped Polyaromatic Ethers Coordinated to Redox-Active Cyclopentadienyliron Moieties, *Macromolecular Rapid Communications*,[106] John Wiley and Sons, © 2002 WILEY-VCH Verlag GmbH, Weinheim, Fed. Rep. of Germany.

the radical anion, confirming its potential as an electron reservoir or molecular battery.[94]

As mentioned previously, molecular wires ensure efficient transfer of electronic current or charges between junctions, such as electrodes. Thus, they are critical to molecular electronics and are, conceptually, useful energy-conserving technologies. From the circuitry perspective of molecular electronics, a molecular wire should shuttle charges over a relatively long distance, usually in the range of 1–100 nm.[90,116] The effort of various researchers has provided evidence to confirm that charge transport over 1–100 nm in a single molecule is feasible and is driven over short distances by tunneling, flickering resonance or superexchange charge transfer, and over long distances by charge hopping.[90,116,117] Advancement in the field is contingent on enhancing charge transport in these molecular interconnects, which are primarily based on single π-conjugated organic molecules.

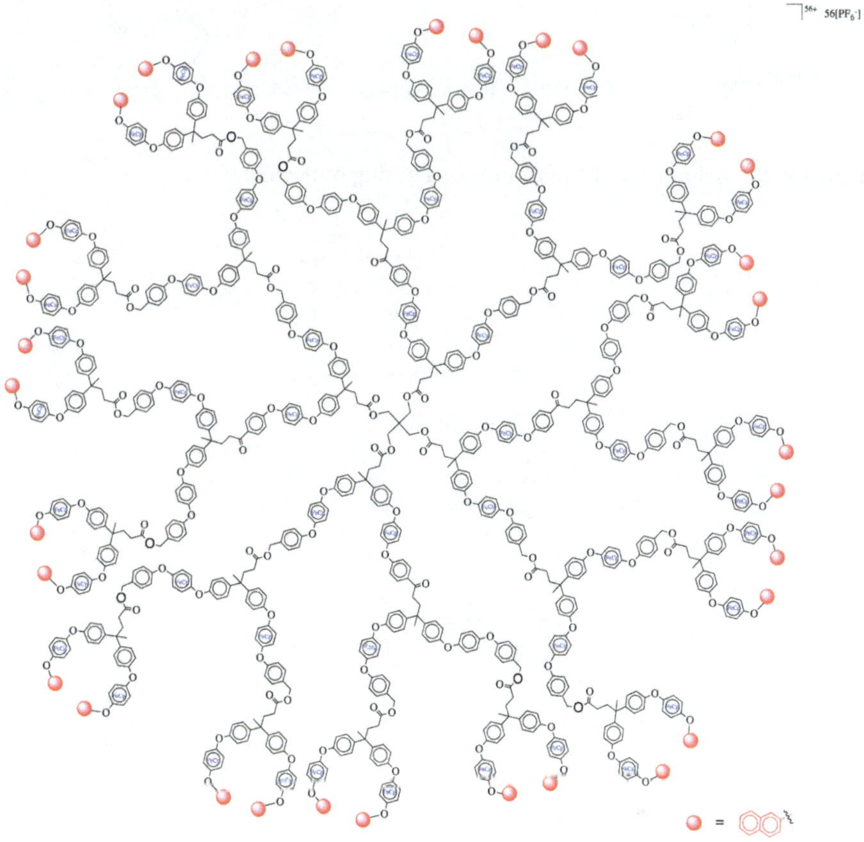

Figure 4.24 Schematic of a second-generation redox-active dendrimer with 56 redox-active iron centers.

A seminal strategy explores incorporating transition metals within π-conjugated organic systems to access new optoelectronic properties. It is assumed that the d-orbitals of the transition metals can overlap with the p-orbitals of the alkynyl carbon atoms to extend the delocalization of the π-electrons and enhance intrachain charge transport.[37] It is also possible to tune the molecular orbital levels relative to the Fermi levels of the electrodes by inserting transition metals within the backbone of a π-conjugated organic polymer.[118] Indeed, this strategy is pursued to enhance and understand charge transport in molecular wires. For instance, to gain an understanding of the mechanism of exciton and negative polaron (electron) transport in single molecules, Schanze *et al.* constructed a series of monodispersed platinum-containing acetylide oligomers end-capped with exciton and electron traps (Figure 4.25), and probed intrachain triplet exciton and electron transport.[116] Fast timescale pump-probe spectroscopy and pulse radiolysis indicate that charge carriers rapidly traverse over

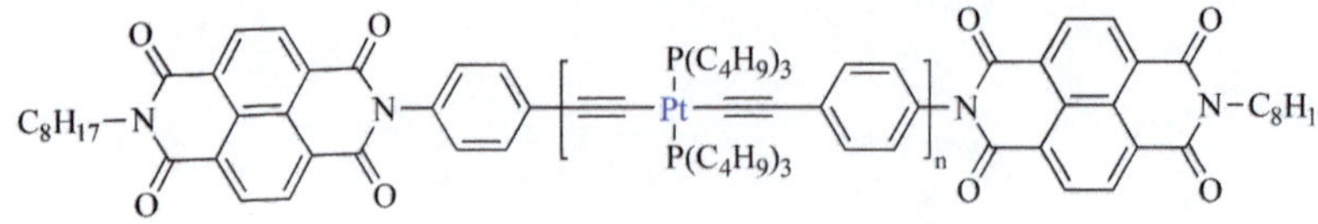

Figure 4.25 Schematic of a platinum-containing molecular wire.

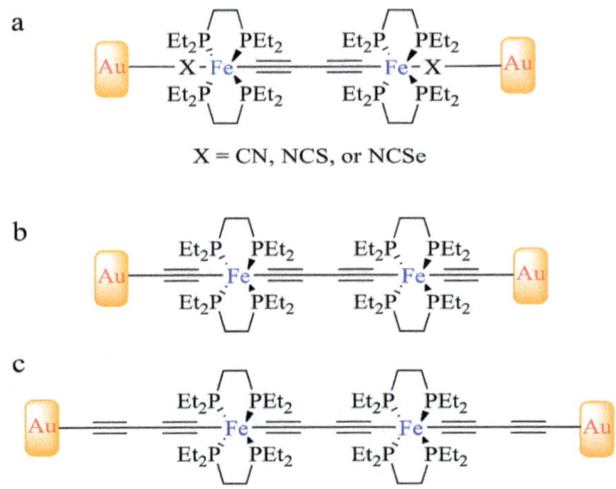

Figure 4.26 Schematic of iron-containing molecular wires (a) coordinated *via* a metal–ligand interaction to an Au electrode (b) and (c) covalently attached to Au electrodes.

3–5 nm of the platinum–acetylide chain to the trap in less than 200 ps. The intrachain charge transport proceeds *via* incoherent charge hopping, with electrons (10 ps per hop) hopping faster than excitons (~27 ps per hop).[116]

Also, in a transition-metal-containing molecular wire, charges are better transported in a covalently π-conjugated system than in a metal–ligand coordinated system. For instance, Lörtscher *et al.* investigated the charge transport in a series of transition-metal-containing oligomeric molecular wires, $X(RR')_2FeC_4Fe(RR')_2X$ (RR′ is $Et_2PCH_2CH_2PEt_2$; and X is CN, NCS, NCSe, C_4SnMe_3, or C_2SnMe_3) (Figure 4.26).[118] In this study, the metal center in some wires is covalently connected to the electrode, or coordinatively bonded *via* metal–ligand interactions (Figure 4.26) to understand how bond type affects transport. Although the CN ligand results in an unstable connection, the others give reproducible stable connections with the room-temperature coupling strength increasing in the order: Fe-NCS-electrode < Fe–NCSe-electrode < Fe–C_4-electrode < Fe–C_2-electrode. Also, under ultrahigh vacuum and variable temperature conditions, the conductance (8.9×10^{-3} G_0) measured with the wire derived from the covalently π-conjugated C_4 spacer (Fe–C_4-electrode) is three orders of magnitude higher than that (8.1×10^{-7} G_0) obtained with the coordinatively-bonded NCS spacer

(Fe–NCS-electrode). Perhaps, the overlap of the metal d-orbitals with the π-orbitals ensures a more extended intrachain delocalization of electrons than the chelating effect, electron donation, or back-donation present in metal–ligand interactions.

Although charge transport through coordination bonds is shown to be inferior to covalent π-conjugated systems, the coordinative interaction of metals with ligands remains a well-explored facile synthetic route to assemble molecular wires. Nishihara *et al.* constructed several molecular wires using the coordination reaction of 2,2′:6′,2″-terpyridine with transition metals. They obtained a series of novel redox-active metal complexes, which were linked *via* π-conjugated organic systems to obtain linear or branched oligomeric wires with an impressive charge transport property (Figure 4.27).[100–102,119] The cyclic voltammograms of the assembled oligomeric wires show the characteristic Fe^{3+}/Fe^{2+} redox couple. Peak current and separation in peak-to-peak potential increase with the number of metal-complexation cycles (n), which corresponds to n and $2n + 1$ numbers of redox centers in the linear and branched wire, respectively.[119] With branched oligomeric wires, higher metal-complexation cycles result in steric repulsion forcing the wire to grow only in one dimension.[101] The distinctive current *vs.* time plots of the wires eliminate through-space diffusional electron transfer dynamics, supporting a through-bond mechanism that involves sequential hopping of electrons over a long distance between adjacent redox sites.[101] The attenuation constant, defined by long-range electron transport over the wire, is independent of the nature of the terpyridine that anchors the wire to the gold electrode and the terpyridine that caps the wire, but is dependent on the nature of the bridging terpyridine ligand. For instance, the attenuation constant is lower for wires with a bridging bis(terpyridine)cobalt complex (0.002–0.004 $Å^{-1}$) than with a bis(terpyridine)iron complex (0.008–0.07 $Å^{-1}$), suggesting better electron transport property with the cobalt-based wire.[119] The nature of the anchoring terpyridine and terminal terpyridine ligands affects the zero-distance electron transfer rate constant, a property that describes the absolute rate constant for electron transfer between the electrode and the terminal redox center.[100,102] As an example, changing the terminal redox center from ferrocene to a cyclometalated ruthenium complex increases the zero-distance electron transfer rate constant.[100] Presumably, the higher reorganization energy of the ferrocene compared with the ruthenium complex explains this trend.[100]

4.5 Energy-efficient Technologies: Light-emitting Macromolecules

Meeting the global energy demand remains a challenge but presents opportunities to efficiently harness, distribute, and utilize energy. Light is indispensable to our survival and comfort; accordingly, generating light is

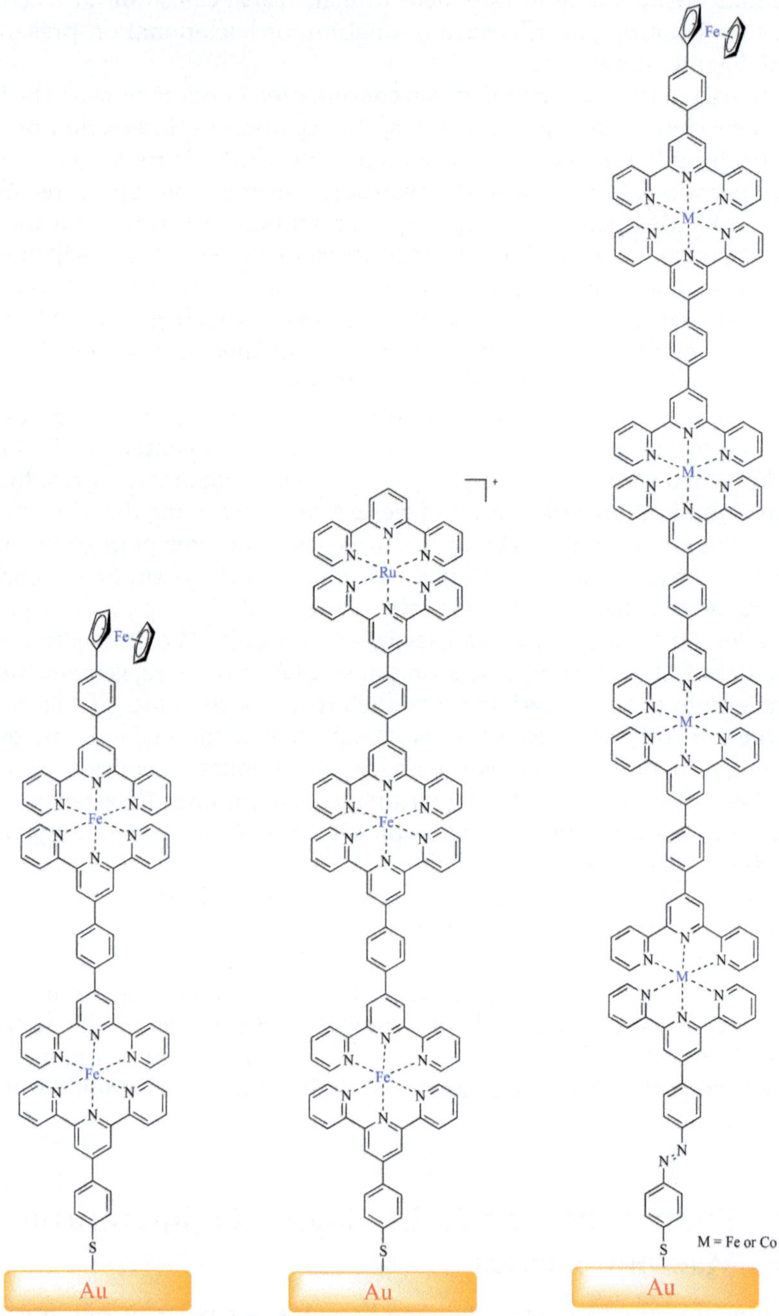

Figure 4.27 Examples of some coordination polymer molecular wires.

a major end use that drains the global energy pipeline. The development of energy-efficient lighting technologies offers a lot of potential to conserve energy, lower utility bills, and reduce greenhouse gas emission.[120–123] Indeed, the advent of energy-efficient lighting technologies has significantly reduced the amount of energy required for lighting in residential and commercial buildings. For instance, in the United States, the adoption of energy-efficient lighting technologies remarkably reduced the percentage of electrical energy required for lighting in the commercial sector from 38% in 2003 to 11% in 2016.[122,123] Among the energy-efficient lighting technologies, light-emitting diodes or devices (LED) are highly efficient and environmentally-friendly, producing light that meets consumer expectation without generating unwanted heat as in incandescent light bulbs, or constituting an environmental threat, as in fluorescent bulbs.[120,124] White polymer light-emitting devices are a type of LED that emits white light from semi-conducting polymers, and have attracted attention due to their low cost and light weight, and their potential for solid-state lighting, as well as easy and economical fabrication procedures. These promising attributes are challenged by the color instability, and long-term operation instability of some of the state-of-the-art white polymer light-emitting devices.[124]

A strategy to access better performing white polymer light-emitting devices involves covalently attaching to a polymer framework moieties that emit the three primary colors (red, green, and blue) in the right proportion.[14,125] This approach ensures better control over color bias than when the polymer is doped or blended with emitting moieties. Recently, Lu *et al.* covalently attached lanthanide complexes (europium(III) (Eu^{3+}) or terbium(III) (Tb^{3+})) to styrene monomers, which were copolymerized with methyl methacrylate monomer (complex–monomer ratio: 1:50, 1:100, or 1:200) to construct color tunable and white-light-emitting polymers (Figure 4.28).[126] The idea is to combine the red and green emissions of Eu^{3+} and Tb^{3+}, respectively, with the blue emission of the polymer backbone at the molecular level and fine-tune these emissions to realize pure white light. These polymers absorb as well as emit in the solid state (Figure 4.28).[126] Polymerization alters the energy level of the chelating organic ligand, decreasing the decay lifetime of the Eu^{3+}-containing polymer and increasing that of the Tb^{3+}-containing polymer, relative to their respective monomers. To fine-tune the emission to realize white light, Lu and coworkers constructed a heterometallic polymer containing Eu^{3+} and Tb^{3+} complexes, which emits from the ligand at 400–475 nm as well as from Eu^{3+} and Tb^{3+} (Figure 4.29).[126] By varying the excitation wavelength or the ratio of Eu^{3+}–Tb^{3+} from 2:1, 5:1, to 7:1, the relative emission intensity is tuned to a white light. For instance, excitation of the polymer containing a 2:1 ratio of Eu^{3+} and Tb^{3+} at 250–280 nm results in a predominantly green light emission from the Tb^{3+}, whereas excitation at 290 nm increases the intensity of the blue light from the ligand, resulting in a cyan emission.[126] The

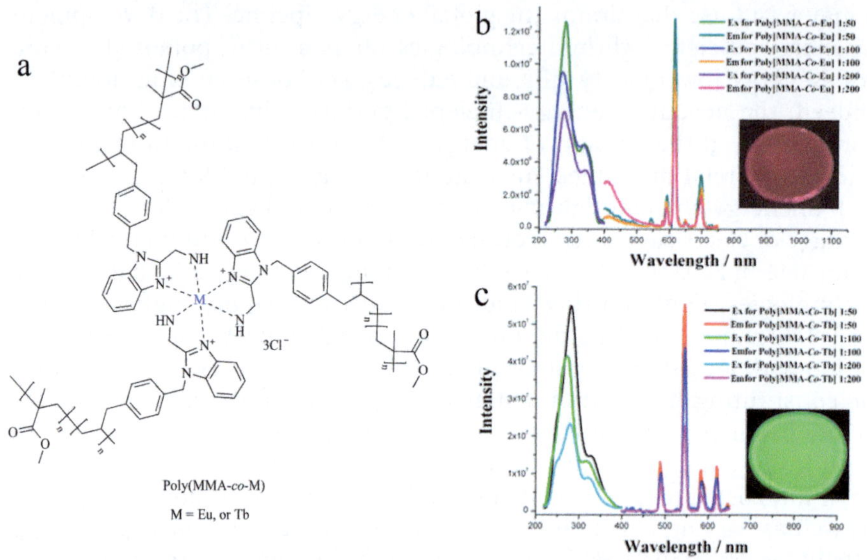

Figure 4.28 (a) Schematic of white-light-emitting europium- or terbium-containing polymer. Solid state emission and excitation spectra of europium- (b) and terbium- (c) containing polymer (inset: polymer film showing solid state emission). Reproduced from ref. 126 with permission from The Royal Society of Chemistry.

polymer emits yellow light or pure white light when the ratio of Eu^{3+}–Tb^{3+} is increased to 5 : 1 or 7 : 1, respectively (Figure 4.29).[126]

Lu and coworkers also constructed a red- or greenish-yellow-light-emitting polymer by copolymerizing a red-light-emitting Eu^{3+} complex-containing monomer and a blue-green-light-emitting Zn^{2+}–Gd^{3+} complex-containing monomer with methyl methacrylate monomer.[125] A 2 : 1 stoichiometric ratio of the Eu^{3+} to the Zn^{2+}–Gd^{3+} complex gives a red-light emission while a ratio of 1 : 2 results in a greenish-yellow light.[125] Wong *et al.* designed a series of light-emitting platinum(II)-polyyne conjugated polymers (Figure 4.30) that exhibit phosphorescence due to the presence of platinum.[127] Wong and coworkers explored these polymers as phosphorescent dopants rather than a neat emissive layer in polymer light-emitting devices, because of aggregation-induced self-quenching in the solid state. The performance of the light-emitting device fabricated with the phosphor dopant is concentration dependent (Figure 4.30). As an example, external quantum efficiency decreases when the dopant concentration increases above 1%, due to concentration-caused quenching. Also, Wong and coworkers demonstrated the potential of these polymers as single-dopant white light sources, exploiting the concurrent fluorescent emission at ~485–490 nm from the 2,9-dimethyl-4,7-diphenyl-1,10-phenanthroline (BCP) substrate in the device, and phosphorescent emission at ~555 nm from the polymer

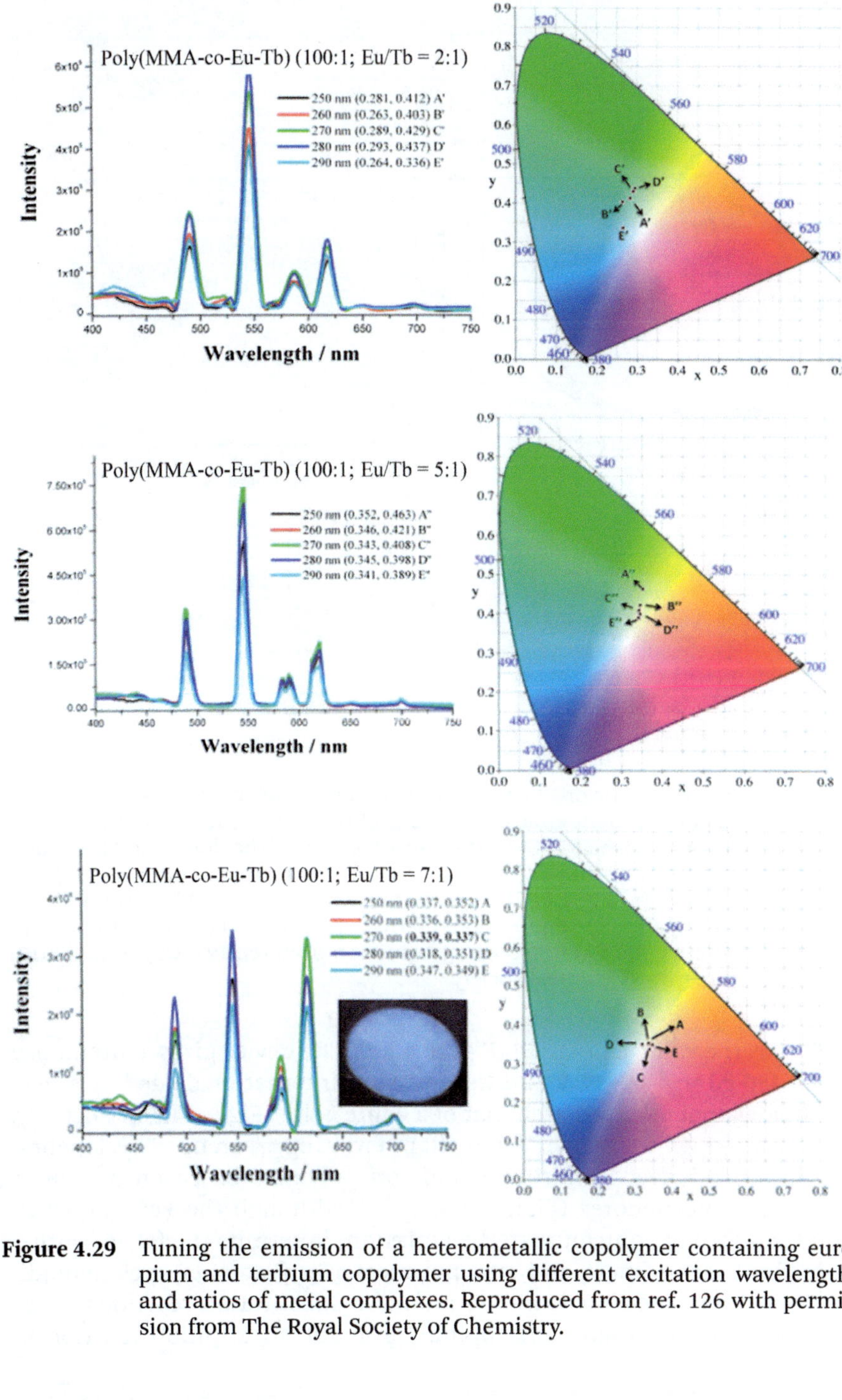

Figure 4.29 Tuning the emission of a heterometallic copolymer containing europium and terbium copolymer using different excitation wavelengths and ratios of metal complexes. Reproduced from ref. 126 with permission from The Royal Society of Chemistry.

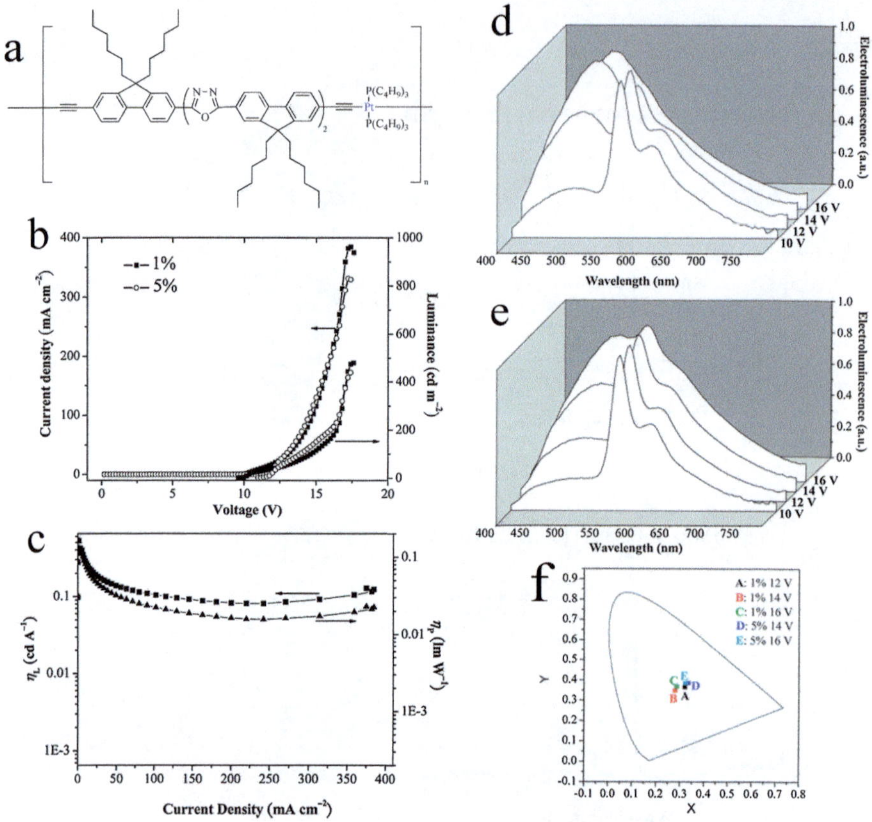

Figure 4.30 (a) Schematic of a light-emitting platinum polymer used as a dopant in a polymer light-emitting device (PLED). (b) Dependence of the current density of the PLED on the concentration of the dopant. (c) Electroluminescence (EL) efficiencies of the PLED fabricated with 1% dopant. (d) and (e) EL of PLED dependence on DC voltage (1% (d) and 5% (e) of dopant). (f) Commission Internationale de l'Eclairage coordinates of white light. Reprinted with permission from ref. 127. Copyright (2010) American Chemical Society.

dopant, as the voltage is varied (Figure 4.30). The device gives a luminance of 440–480 cd m^{-2} at ~17 V and Commission Internationale de l'Eclairage coordinates that closely match that of a white color (Figure 4.30). Recently, Abd-El-Aziz *et al.* constructed blue-light-emitting η^6-arene-η^5-cyclopentadienyliron(II)-derived dendrimers and star polymers based on pyrene or anthracene fluorophores (Figure 4.31).[108,109] Although the Fe^{2+} sandwich complex does not contribute to the emission, it permits the facile synthesis of these dendrimers and star polymers *via* η^6-arene-η^5-cyclopentadienyliron(II)-mediated nucleophilic aromatic substitution reactions. They also explored this simple nucleophilic aromatic substitution reaction to

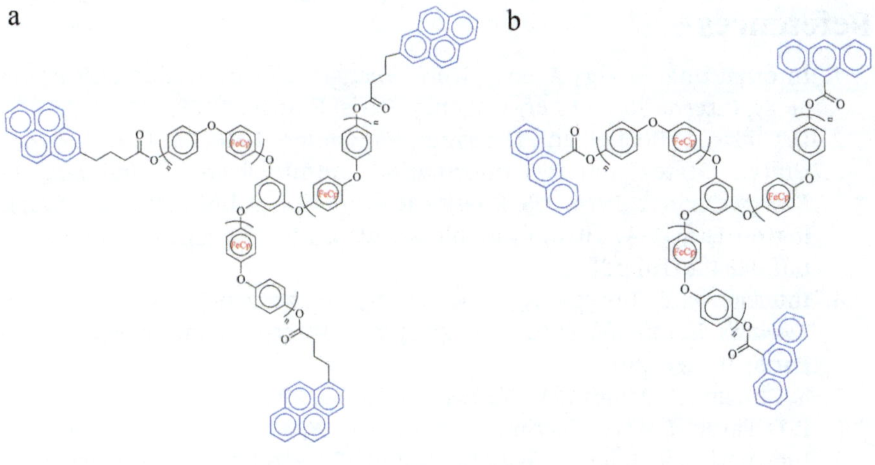

Figure 4.31 Schematics of blue-light-emitting (a) pyrene- and (b) anthracene-containing dendrimers designed *via* a η^6-arene-η^5-cyclopentadienyliron(II)-mediated nucleophilic aromatic substitution reaction.

construct tetraphenylethylene-containing organometallic dendrimers that emit in the solid state.[128]

4.6 Summary and Outlook

The bulk of the global energy supply is derived from fossil fuel, which is unsustainable and poses a considerable threat to our environment. Given the vital role of energy in our society, policymakers, scientists, and engineers are working together to find alternatives that are sustainable and clean. Materials scientists are exploring the rich chemistry of the transition metals to develop new energy technologies. The redox activity, as well as optical and electronic properties of these metals, offer many opportunities to efficiently transport electrons or energy. For instance, transition metals are conjugated into polymers to harvest sunlight, transport excitons, and promote charge separation and transfer in photovoltaic cells. Also, several research groups have incorporated transition metals in polymers to construct polymer light-emitting devices with the goal of designing energy-efficient platforms. Although most of these developments are still far from commercialization, the results are impactful and demonstrate the interest of materials scientists in finding an alternative to fossil fuel and ensuring efficient management of the current energy pipeline. To push the frontiers, we need a greater understanding of the chemistry of the transition metals, especially their redox, optical, and electronic properties in the excited state. Such fundamental understanding will be helpful in harvesting sunlight and transporting photo-generated charges to reactors for chemical transformation to enhance the efficiency of photovoltaic and photoelectrochemical cells.

References

1. International Energy Agency, *World Energy Outlook 2016: Executive Summary*, International Energy Agency, Paris, France, 2016.
2. D. Gielen, F. Boshell and D. Saygin, *Nat. Mater.*, 2016, **15**, 117.
3. United States Energy Information Administration, *International Energy Outlook 2016*, US Energy Information Administration, Washington DC, USA, 2016, available at: https://www.eia.gov/outlooks/ieo/pdf/0484(2016).pdf.
4. International Energy Agency, *Energy Technology Perspective 2016: Towards Sustainable Urban Energy System*, International Energy Agency, Paris, France, 2017.
5. S. Chu and A. Majumdar, *Nature*, 2012, **488**, 294.
6. P. D. Farah, *J. World Energy Law Bus.*, 2015, **8**, 34.
7. International Energy Agency, *Energy Technology Perspectives 2012: Pathways to a Clean Energy System*, International Energy Agency, Paris, France, 2012.
8. R. O. Keohane and D. G. Victor, *Nat. Clim. Change*, 2016, **6**, 570.
9. United Nations, *Paris Agreement*, United Nations, Paris, France, 2015.
10. M. Dresselhaus and I. Thomas, *Nature*, 2001, **414**, 332.
11. S. Schlömer, T. Bruckner, L. Fulton, E. Hertwich, A. McKinnon, D. Perczyk, J. Roy, R. Schaeffer, R. Sims, P. Smith and R. Wiser, Annex III: technology-specific cost and performance parameter, in *Climate Change 2014: Mitigation of Climate Change. Contribution of Working Group III to the Fifth Assessment Report of the Intergovernment Panel on Climate Change*, ed. O. Edenhofer, R. Pichs-Madruga, Y. Sokona, E. Farahani, S. Kadner, K. Seyboth, A. Adler, I. Baum, S. Brunner, P. Eickemeier, B. Kriemann, J. Savolainen, S. Schlömer, C. von Stechow, T. Zwickel and J. C. Minx, Cambridge University Press, Cambridge, United Kingdom and New York, NY, USA, 2014.
12. L. Schlapbach and A. Züttel, *Nature*, 2001, **414**, 353.
13. E. D. Bloch, W. L. Queen, M. R. Hudson, J. A. Mason, D. J. Xiao, L. J. Murray, R. Flacau, C. M. Brown and J. R. Long, *Angew. Chem., Int. Ed.*, 2016, **55**, 8605.
14. J. Xiang, C.-L. Ho and W.-Y. Wong, *Polym. Chem.*, 2015, **6**, 6905.
15. N. L. Rosi, J. Eckert, M. Eddaoudi, D. T. Vodak, J. Kim, M. O'Keeffe and O. M. Yaghi, *Science*, 2003, **300**, 1127.
16. J. Qin, D. Du, W. Guan, X. Bo, Y. Li, L. Guo, Z. Su, Y. Wang, Y. Lan and H. Zhou, *J. Am. Chem. Soc.*, 2015, **137**, 7169.
17. S. Pullen, H. Fei, A. Orthaber, S. M. Cohen and S. Ott, *J. Am. Chem. Soc.*, 2013, **135**, 16997.
18. M. K. Debe, *Nature*, 2012, **486**, 43.
19. C. Zhao, P. Gai, R. Song, Y. Chen, J. Zhang and J. Zhu, *Chem. Soc. Rev.*, 2017, **46**, 1545.
20. V. Soukharev, N. Mano and A. Heller, *J. Am. Chem. Soc.*, 2004, **126**, 8368.
21. R. Szamocki, V. Flexer, L. Levin, F. Forchiasin and E. Calvo, *Electrochim. Acta*, 2009, **54**, 1970.

22. N. Mano, F. Mao and A. Heller, *J. Am. Chem. Soc.*, 2002, **124**, 12962.
23. N. Mano, F. Mao, W. Shin, T. Chen and A. Heller, *Chem. Commun.*, 2003, 518.
24. N. Mano, F. Mao and A. Heller, *Chem. Commun.*, 2004, 2116.
25. N. Mano, J. L. Fernandez, Y. Kim, W. Shin, A. J. Bard and A. Heller, *J. Am. Chem. Soc.*, 2003, **125**, 15290.
26. F. Mao, N. Mano and A. Heller, *J. Am. Chem. Soc.*, 2003, **125**, 4951.
27. P. Scodeller, R. Carballo, R. Szamocki, L. Levin, F. Forchiassin and E. J. Calvo, *J. Am. Chem. Soc.*, 2010, **132**, 11132.
28. N. P. Godman, J. L. DeLuca, S. R. McCollum, D. W. Schmidtke and D. T. Glatzhofer, *Langmuir*, 2016, **32**, 3541.
29. J. Lee, H. Ahn, I. Choi, M. Boese and M. J. Park, *Macromolecules*, 2012, **45**, 3121.
30. M. T. Meredith, D. Kao, D. Hickey, D. W. Schmidtke and D. T. Glatzhofer, *J. Electrochem. Soc.*, 2011, **158**, B166.
31. C. Bunte, O. Prucker, T. König and J. Rühe, *Langmuir*, 2010, **26**, 6019.
32. United States Energy Information Administration, *Annual Energy Outlook 2017 with Projections to 2050*, United States Energy Information Administration, Washington DC, USA, 2017.
33. United States Energy Information Administration, *Monthly Energy Review: May 2017*, United States Energy Information Administration, Washington DC, USA, 2017.
34. United States Department of Energy, *The SunShot Initiative's 2030 Goal: 3 cent per Kilowatt Hour for Solar Electricity*, United States Department of Energy, Washington, DC, USA, 2016.
35. Z. Shahan, *Low Solar Prices Scaring Companies Away from Solar Auctions*, CleanTechnica, July 27 2016, https://cleantechnica.com/2016/07/27/low-solar-prices-scaring-companies-away-solar-auctions/, accessed: June 21, 2017.
36. K. Yoshikawa, H. Kawasaki, W. Yoshida, T. Irie, K. Konishi, K. Nakano, T. Uto, D. Adachi, M. Kanematsu, H. Uzu and K. Yamamoto, *Nat. Energy*, 2017, **2**, 17032.
37. Y. Liu, S. Wang, Y. Tao and W. Huang, *Chin. Chem. Lett.*, 2016, **27**, 1250.
38. W.-Y. Wong and C.-L. Ho, *Acc. Chem. Res.*, 2010, **43**, 1246.
39. M. Li, K. Gao, X. Wan, Q. Zhang, B. Kan, R. Xia, F. Liu, X. Yang, H. Feng, W. Ni, Y. Wang, J. Peng, H. Zhang, Z. Liang, H. Yip, X. Peng, Y. Cao and Y. Chen, *Nat. Photonics*, 2017, **11**, 85.
40. S. Goswami, M. K. Gish, J. Wang, R. W. Winkel, J. M. Papanikolas and K. S. Schanze, *ACS Appl. Mater. Interfaces*, 2015, **7**, 26828.
41. J. Mei, K. Ogawa, Y. Kim, N. C. Heston, D. J. Arenas, Z. Nasrollahi, T. D. McCarley, D. B. Tanner, J. R. Reynolds and K. S. Schanze, *ACS Appl. Mater. Interfaces*, 2009, **1**, 150.
42. F. Guo, Y. Kim, J. R. Reynolds and K. S. Schanze, *Chem. Commun.*, 2006, 1887.
43. F. Guo, K. Ogawa, Y. Kim, E. O. Danilov, F. N. Castellano, J. R. Reynolds and K. S. Schanze, *Phys. Chem. Chem. Phys.*, 2007, **9**, 2724.

44. G. L. Schulz and S. Holdcroft, *Chem. Mater.*, 2008, **20**, 5351.
45. A. Köhler, H. Wittmann, R. Friend, M. Khan and J. Lewis, *Synth. Met.*, 1996, **77**, 147.
46. A. Köhler, H. Wittmann, R. Friend, M. Khan and J. Lewis, *Synth. Met.*, 1994, **67**, 245.
47. W.-Y. Wong, X.-Z. Wang, Z. He, A. B. Djurišić, C.-T. Yip, K.-Y. Cheung, H. Wang, C. S. K. Mak and W.-K. Chan, *Nat. Mater.*, 2007, **6**, 521.
48. J. S. Wilson, N. Chawdhury, M. R. Al-Mandhary, M. Younus, M. S. Khan, P. R. Raithby, A. Köhler and R. H. Friend, *J. Am. Chem. Soc.*, 2001, **123**, 9412.
49. N. Chawdhury, A. Köhler, R. H. Friend, W.-Y. Wong, J. Lewis, M. Younus, P. R. Raithby, T. Corcoran, M. Al-Mandhary and M. S. Khan, *J. Chem. Phys.*, 1999, **110**, 4963.
50. N. S. Baek, S. K. Hau, H. Yip, O. Acton, K. Chen and A. K. Jen, *Chem. Mater.*, 2008, **20**, 5734.
51. W.-Y. Wong, X.-Z. Wang, Z. He, K.-K. Chan, A. B. Djurišić, K.-Y. Cheung, C.-T. Yip, A. M.-C. Ng, Y. Y. Xi, C. S. K. Mak and W.-K. Chan, *J. Am. Chem. Soc.*, 2007, **129**, 14372.
52. H. Zhan, S. Lamare, A. Ng, T. Kenny, H. Guernon, W.-K. Chan, A. B. Djurišić, P. D. Harvey and W.-Y. Wong, *Macromolecules*, 2011, **44**, 5155.
53. L. Du, W. Xiong, S. Cheng, H. Shi, W. K. Chan and D. L. Phillips, *J. Phys. Chem. Lett.*, 2017, **8**, 2475.
54. E. D. Spoerke, L. J. Small, M. E. Foster, J. Wheeler, A. M. Ullman, V. Stavila, M. Rodriguez and M. D. Allendorf, *J. Phys. Chem. C*, 2017, **121**, 4816.
55. R. Sakamoto, K. Hoshiko, Q. Liu, T. Yagi, T. Nagayama, S. Kusaka, M. Tsuchiya, Y. Kitagawa, W.-Y. Wong and H. Nishihara, *Nat. Commun.*, 2015, **6**, 6713.
56. S. Shima, O. Pilak, S. Vogt, M. Schick, M. S. Stagni, W. Meyer-Klaucke, E. Warkentin, R. K. Thauer and U. Ermler, *Science*, 2008, **321**, 572.
57. M. Grätzel, *Nature*, 2001, **414**, 338.
58. Q. Huang, Z. Ye and X. Xiao, *J. Mater. Chem. A*, 2015, **3**, 15824.
59. A. Fujishima and K. Honda, *Nature*, 1972, **238**, 37.
60. Y. Gong, T. Wu, P. G. Jiang, J. H. Lin and Y. X. Yang, *Inorg. Chem.*, 2013, **52**, 777.
61. J. Liu, M. S. Kelley, W. Wu, A. Banerjee, A. P. Douvalis, J. Wu, Y. Zhang, G. C. Schatz and M. G. Kanatzidis, *Proc. Natl. Acad. Sci. U. S. A.*, 2016, **113**, 5530.
62. J. G. Rebelein, Y. Hu and M. W. Ribbe, *Angew. Chem., Int. Ed.*, 2014, **53**, 11543.
63. Z. Zhang, T. Zhang, C. Wang, Z. Lin, L. Long and W. Lin, *J. Am. Chem. Soc.*, 2015, **137**, 3197.
64. X. Kong, Z. Lin, Z. Zhang, T. Zhang and W. Lin, *Angew. Chem., Int. Ed.*, 2016, **128**, 6521.
65. U. S. DRIVE Partnership, *Target Explanation Document: Onboard Hydrogen Storage for Light-duty Fuel Cell Vehicle*, U. S. DRIVE Partnership, Washington DC, USA, 2017.

66. S. S. Kaye, A. Dailly, O. M. Yaghi and J. R. Long, *J. Am. Chem. Soc.*, 2007, **129**, 14176.
67. M. Dinca, A. Dailly, Y. Liu, C. M. Brown, D. A. Neumann and J. R. Long, *J. Am. Chem. Soc.*, 2006, **128**, 16876.
68. M. Dincă and J. R. Long, *Angew. Chem., Int. Ed.*, 2008, **47**, 6766.
69. L. Pan, M. B. Sander, X. Huang, J. Li, M. Smith, E. Bittner, B. Bockrath and J. K. Johnson, *J. Am. Chem. Soc.*, 2004, **126**, 1308.
70. M. Dincă, A. F. Yu and J. R. Long, *J. Am. Chem. Soc.*, 2006, **128**, 8904.
71. G. D. Scholes, G. R. Fleming, A. Olaya-Castro and R. Van Grondelle, *Nat. Chem.*, 2011, **3**, 763.
72. B. P. Krueger, G. D. Scholes and G. R. Fleming, *J. Phys. Chem. B*, 1998, **102**, 5378.
73. C. A. Kent, B. P. Mehl, L. Ma, J. M. Papanikolas, T. J. Meyer and W. Lin, *J. Am. Chem. Soc.*, 2010, **132**, 12767.
74. Z. Chen, H. Hsu, M. Arca and K. S. Schanze, *J. Phys. Chem. B*, 2015, **119**, 7198.
75. C. N. Fleming, K. A. Maxwell, J. M. DeSimone, T. J. Meyer and J. M. Papanikolas, *J. Am. Chem. Soc.*, 2001, **123**, 10336.
76. Y. Sun, Z. Chen, E. Puodziukynaite, D. M. Jenkins, J. R. Reynolds and K. S. Schanze, *Macromolecules*, 2012, **45**, 2632.
77. P. Wang, S. M. Zakeeruddin, J. E. Moser, M. K. Nazeeruddin, T. Sekiguchi and M. Grätzel, *Nat. Mater.*, 2003, **2**, 402.
78. D. Kuang, S. Ito, B. Wenger, C. Klein, J. Moser, R. Humphry-Baker, S. M. Zakeeruddin and M. Grätzel, *J. Am. Chem. Soc.*, 2006, **128**, 4146.
79. D. Y. Lee, I. Lim, C. Y. Shin, S. A. Patil, W. Lee, N. K. Shrestha, J. K. Lee and S. Han, *J. Mater. Chem. A*, 2015, **3**, 22669.
80. W. Xiang, W. Huang, U. Bach and L. Spiccia, *Chem. Commun.*, 2013, **49**, 8997.
81. H. Imahori, *J. Phys. Chem. B*, 2004, **108**, 6130.
82. F. Hajjaj, Z. S. Yoon, M. Yoon, J. Park, A. Satake, D. Kim and Y. Kobuke, *J. Am. Chem. Soc.*, 2006, **128**, 4612.
83. M. Choi, T. Aida, T. Yamazaki and I. Yamazaki, *Angew. Chem.*, 2001, **113**, 3294.
84. M. Choi, T. Aida, T. Yamazaki and I. Yamazaki, *Chem.–Eur. J.*, 2002, **8**, 2667.
85. X. Zhang, M. A. Ballem, M. Ahrén, A. Suska, P. Bergman and K. Uvdal, *J. Am. Chem. Soc.*, 2010, **132**, 10391.
86. D. Larcher and J. Tarascon, *Nat. Chem.*, 2015, **7**, 19.
87. S. Chu, Y. Cui and N. Liu, *Nat. Mater.*, 2017, **16**, 16.
88. Z. Meng, K. Sato, T. Sukegawa, K. Oyaizu, C.-L. Ho, J. Xiang, Y. Feng, Y. H. Lo, H. Nishide and W.-Y. Wong, *J. Organomet. Chem.*, 2016, **812**, 51.
89. M. Ratner, *Nat. Nanotechnol.*, 2013, **8**, 378.
90. S. Ho Choi, B. Kim and C. D. Frisbie, *Science*, 2008, **320**, 1482.
91. N. Robertson and C. A. McGowan, *Chem. Soc. Rev.*, 2003, **32**, 96.
92. D. Astruc, *Nat. Chem.*, 2012, **4**, 255.
93. S. Nlate, J. Ruiz, V. Sartor, R. Navarro, J. Blais and D. Astruc, *Chem.–Eur. J.*, 2000, **6**, 2544.

94. J. Ruiz, C. Pradet, F. Varret and D. Astruc, *Chem. Commun.*, 2002, 1108.
95. C. Ornelas, J. Ruiz, C. Belin and D. Astruc, *J. Am. Chem. Soc.*, 2009, **131**, 590.
96. D. Astruc, *Chem.–Asian J.*, 2011, **6**, 1679.
97. J. R. Aranzaes, C. Belin and D. Astruc, *Angew. Chem., Int. Ed.*, 2006, **45**, 132.
98. C. Ornelas, J. R. Aranzaes, E. Cloutet, S. Alves and D. Astruc, *Angew. Chem.*, 2007, **119**, 890.
99. D. Astruc, M. Daniel and J. Ruiz, *Chem. Commun.*, 2004, 2637.
100. S. Katagiri, R. Sakamoto, H. Maeda, Y. Nishimori, T. Kurita and H. Nishihara, *Chem.–Eur. J.*, 2013, **19**, 5088.
101. Y. Nishimori, K. Kanaizuka, M. Murata and H. Nishihara, *Chem.–Asian J.*, 2007, **2**, 367.
102. T. Kurita, Y. Nishimori, F. Toshimitsu, S. Muratsugu, S. Kume and H. Nishihara, *J. Am. Chem. Soc.*, 2010, **132**, 4524.
103. H. Maeda, R. Sakamoto and H. Nishihara, *Coord. Chem. Rev.*, 2017, **346**, 139.
104. A. S. Abd-El-Aziz, E. K. Todd, R. M. Okasha, P. O. Shipman and T. E. Wood, *Macromolecules*, 2005, **38**, 9411.
105. A. S. Abd-El-Aziz, C. R. de Denus, E. K. Todd and S. A. Bernardin, *Macromolecules*, 2000, **33**, 5000.
106. A. S. Abd-El-Aziz, E. K. Todd and T. H. Afifi, *Macromol. Rapid Commun.*, 2002, **23**, 113.
107. A. S. Abd-El-Aziz, C. Agatemor, N. Etkin and R. Bissessur, *Macromol. Chem. Phys.*, 2015, **216**, 369.
108. A. S. Abd-El-Aziz, A. A. Abdelghani, B. D. Wagner and E. M. Abdelrehim, *Polym. Chem.*, 2016, **7**, 3277.
109. A. S. Abd-El-Aziz, A. A. Abdelghani, B. D. Wagner, J. K. Pearson and M. K. Awad, *Polymer*, 2016, **98**, 210.
110. A. S. Abd-El-Aziz, A. A. Abdelghani, S. K. El-Sadany, D. P. Overy and R. G. Kerr, *Eur. Polym. J.*, 2016, **82**, 307.
111. A. S. Abd-El-Aziz, A. A. Abdelghani, J. K. Pearson, M. K. Awad, D. P. Overy and R. G. Kerr, *Macromol. Chem. Phys.*, 2016, **217**, 987.
112. A. S. Abd-El-Aziz, C. Agatemor, N. Etkin, D. P. Overy, M. Lanteigne, K. McQuillan and R. G. Kerr, *Biomacromolecules*, 2015, **16**, 3694.
113. A. S. Abd-El-Aziz, C. R. de Denus, M. J. Zaworotko and L. R. MacGillivray, *J. Chem. Soc., Dalton Trans.*, 1995, 3375.
114. A. S. Abd-El-Aziz and C. R. de Denus, *J. Chem. Soc., Chem. Commun.*, 1994, 663.
115. C. R. de Denus, P. Baker, J. Toner, S. McKevitt, E. K. Todd and A. S. Abd-El-Aziz, *Macromol. Symp.*, 2003, **196**, 113.
116. J. M. Keller, K. D. Glusac, E. O. Danilov, S. McIlroy, P. Sreearuothai, A. R. Cook, H. Jiang, J. R. Miller and K. S. Schanze, *J. Am. Chem. Soc.*, 2011, **133**, 11289.
117. C. Franco, P. M. Burrezo, V. Lloveras, R. Caballero, I. Alcón, S. T. Bromley, M. Mas-Torrent, F. Langa, J. T. López Navarrete, C. Rovira, J. Casado and J. Veciana, *J. Am. Chem. Soc.*, 2017, **139**, 686.

118. F. Schwarz, G. Kastlunger, F. Lissel, H. Riel, K. Venkatesan, H. Berke, R. Stadler and E. Lörtscher, *Nano Lett.*, 2014, **14**, 5932.
119. Y. Nishimori, K. Kanaizuka, T. Kurita, T. Nagatsu, Y. Segawa, F. Toshimitsu, S. Muratsugu, M. Utsuno, S. Kume, M. Murata and H. Nishihara, *Chem.–Asian J.*, 2009, **4**, 1361.
120. N. T. Kalyani and S. Dhoble, *Renewable Sustainable Energy Rev.*, 2012, **16**, 2696.
121. Natural Resources Canada, *Energy-efficient Products*, July 20 2016, http://www.nrcan.gc.ca/energy/products/12509, accessed: June 22, 2017.
122. United States Energy Information Administration, *How Much Electricity Is Used for Lighting in the United States?* May 23 2017, https://www.eia.gov/tools/faqs/faq.php?id=99&t=3, accessed: June 19, 2017.
123. United States Energy Information Administration, *Trends in Lighting in Commercial Building*, May 17 2017, https://www.eia.gov/consumption/commercial/reports/2012/lighting/, accessed: June 19, 2017.
124. L. Ying, C.-L. Ho, H. Wu, Y. Cao and W.-Y. Wong, *Adv. Mater.*, 2014, **26**, 2459.
125. L. Liu, H. Li, P. Su, Z. Zhang, G. Fu, B. Li and X. Lü, *J. Mater. Chem. C*, 2017, **5**, 4780.
126. W. Feng, S. Yin, M. Pan, H. Wang, Y. Fan, X. Lü and C. Su, *J. Mater. Chem. C*, 2017, **5**, 1742.
127. T. Goudreault, Z. He, Y. Guo, C.-L. Ho, H. Zhan, Q. Wang, K. Y.-F. Ho, K.-L. Wong, D. Fortin, B. Yao, Z. Xie, L. Wang, W.-M. Kwok, P. D. Harvey and W.-Y. Wong, *Macromolecules*, 2010, **43**, 7936.
128. A. S. Abd El Aziz, C. Agatemor, N. Etkin and B. Wagner, *Macromol. Rapid Commun.*, 2016, **37**, 1235.

CHAPTER 5

Biomedical Applications of Transition-metal-containing Macromolecules

5.1 Introduction

Transition metals are increasingly explored to prevent, diagnose, and treat diseases. The application of transition metals in disease management precedes modern science and medicine, dating back to ancient civilizations. For instance, a 1600 BC Egyptian medical treatise, the Edwin Smith Papyrus, describes the use of a copper salt as an astringent.[1,2] The biomedical applications of metals among ancient civilizations probably originated from serendipitous discoveries and evaluations of available materials in injury and disease management. Interest in the biomedical applications of metals is increasing and is motivated by breakthroughs such as the clinical outcome of certain platinum compounds, specifically cisplatin (*cis*-$[PtCl_2(NH_3)_2]$) in cancer therapy.[3,4] Recent advances in science provide sophisticated synthesis toolkits to rationally design and synthesize metal-based therapeutic and diagnostic agents with the desired function in mind. Multidisciplinary research that integrates breakthroughs and fundamental knowledge from synthetic inorganic chemistry and macromolecular science with biology opens new frontiers where metals are precisely installed into a macromolecular framework to tackle various healthcare problems.

This chapter examines the biomedical applications (Figure 5.1) of transition-metal-containing macromolecules as biosensing and bioimaging probes, therapeutic agents, and drug delivery platforms. These applications are possible due to the unique properties of metals, specifically their

Polymer Chemistry Series No. 27
Macromolecules Incorporating Transition Metals: Tackling Global Challenges
By Alaa S. Abd-El-Aziz, Christian Agatemor and Wai-Yeung Wong

Published by the Royal Society of Chemistry, www.rsc.org

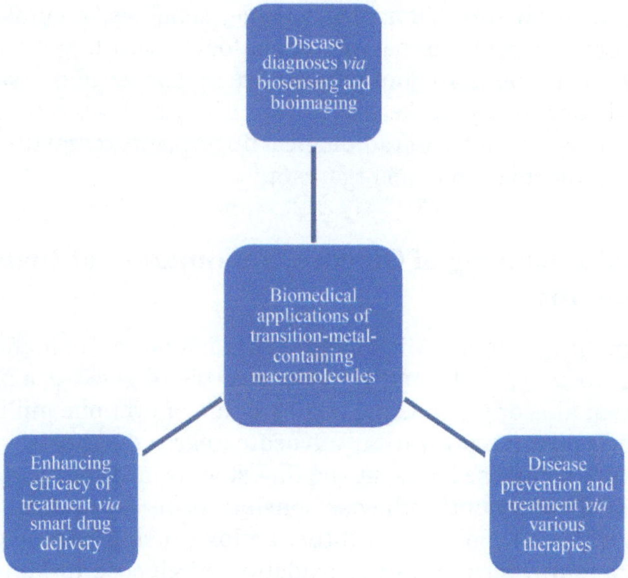

Figure 5.1 Biomedical applications of transition-metal-based macromolecules.

photoactivity and redox properties, as well as coordination chemistry. The first section of this chapter focuses on biosensing and bioimaging applications, highlighting how the photoactivity and redox properties of metals translate biochemical processes and interactions into anatomical and functional information to diagnose diseases. The next section is on the topical applications of transition-metal-based macromolecules in chemotherapy, radiotherapy, and photodynamic therapy, as well as in the fight against drug-resistant infections. We will also examine how the redox property or coordination chemistry of transition metals enables the design of smart drug delivery systems.

5.2 Transition-metal-containing Macromolecules in Diagnosis: Bioimaging and Biosensing Applications

Medical diagnosis, a central component of clinical procedures, furnishes information on the presence and progress of diseases *via* laboratory testing or imaging of the biology of the living organisms. Advances in science and technology allow medical professionals to extract crucial information from complex biological components and interactions using sophisticated or easy-to-use sensing or imaging modalities. The complexity of biological systems hampers the effectiveness of some of these modalities, exemplified by limited sensitivity and specificity, which are challenges to the scientific community. A fundamental understanding of macromolecular science and transition metal chemistry enables reliable, sensitive, accurate, and convenient biosensing and bioimaging technologies with target specificity. In biosensing, the redox-active and photoactive responses of transition metals identify and confirm biomarkers of various diseases,[5,6] whereas in bioimaging, photoactive, magnetic, or radiological fingerprints reveal the anatomical or functional information of an organism.[7]

5.2.1 Redox Sensing of Glucose, a Biomarker of Diabetes Mellitus

Diabetes mellitus, a metabolic disorder that is due to the ineffective regulation of glucose, typically results in elevated blood glucose, a pathological condition that kills or worsens the quality of life of over one million people.[8] The World Health Organization advocates early diagnosis, particularly screening blood glucose levels, as the first step towards reducing the threat of diabetes.[8] Consequently, glucose sensing forms an integral element in the management of diabetes mellitus. Redox sensing of glucose remains the gold standard, relying on the oxidation of glucose to gluconolactone by glucose oxidase, glucose-1-dehydrogenase, or hexokinase enzymes (Figure 5.2), a biochemical redox reaction that can be mediated by redox-active molecules.[9–13] Ferrocene, an iconic organometallic molecule, is a well-explored mediator in the redox sensing of glucose. This is due to the chemical and electrochemical reversibility of its redox process as well as

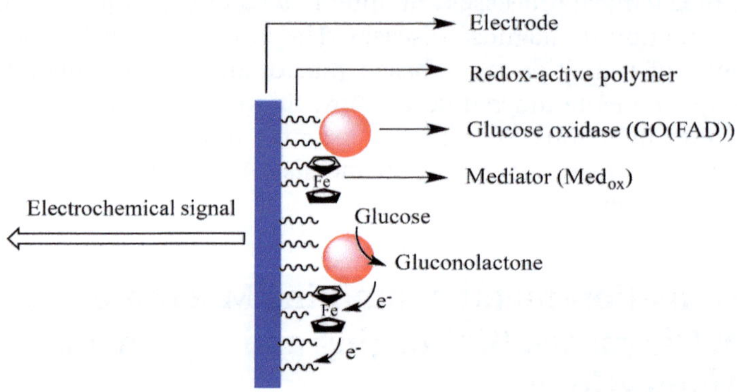

$$\text{Glucose} + \text{GO(FAD)} \longrightarrow \text{GO(FADH}_2\text{)} + \text{Gluconolactone}$$

$$\text{GO(FADH}_2\text{)} + 2\text{Med}_{ox} \longrightarrow \text{GO(FAD)} + 2\text{H}^+ + 2\text{Med}_{red}$$

$$2\text{Med}_{red} \longrightarrow 2\text{Med}_{ox} + 2e^- \quad \text{(at the electrode)}$$

Figure 5.2 General scheme for the redox sensing mechanism of glucose on an electrode surface. GO(FAD), oxidized glucose oxidase; and GO($FADH_2$), reduced glucose oxidase.

the independence of its redox response on pH and oxygen.[9] Accordingly, ferrocene remains at the forefront in the design of glucose biosensors. A conventional ferrocene glucose biosensor includes an electrode on which the mediator and enzyme are immobilized (Figure 5.2). The mediator shuttles electrons between the enzyme and the electrode, and the total electron transfer, which is related to glucose concentration,[9] is transduced to an electrochemical signal. Conjugating the mediator to a macromolecular framework ensures effective immobilization of the enzymes on the electrode, leading to an efficient transfer of electrons between the enzyme and the electrode, and eventually enhanced sensitivity.[11] Also, covalently attaching the mediator to a macromolecule prevents leakage, a major problem with small-molecule mediators.[11]

To achieve efficient electron transfer between the enzyme and electrode, the mediator ferrocene is covalently bonded to the surface of the enzyme, glucose oxidase.[10] The length of the linker between redox-active sites of ferrocene and the enzyme tunes the rate of electron transfer.[10] Long flexible linkers more efficiently transport electrons since such linkers easily allow the active sites to make contacts. In other approaches, ferrocene and the enzyme were covalently bonded to a polymer immobilized on the electrode surface,[11,12] or a ferrocene-containing polymer was blended with the enzyme and graphite to form a paste, which was fabricated into an electrode.[13] For instance, Şenel *et al.* constructed a glucose biosensor *via* covalently attaching glucose oxidase onto a ferrocene copolymer, poly(glycidyl methacrylate-*co*-vinyl ferrocene)[11] or polypyrrole,[12] immobilized on a glassy carbon electrode. These electrodes detect glucose, evidenced by an increase in the anodic current and a decrease in the cathodic current (Figure 5.3).[11] In the presence of glucose, the coenzyme FAD reduces to $FADH_2$, which reduces ferrocenium cations to ferrocene. The oxidation of ferrocene at the applied potential increases the anodic current while the reduction of the ferrocenium cation by $FADH_2$, as well as the electron transfer between the ferrocenium cation and ferrocene, decreases the cathodic current.[11] The biosensor linearly and rapidly responds to changes in glucose concentration, attaining a steady-state current in less than five seconds, but this response is influenced by pH and temperature (Figure 5.3).[11] In a real application using human plasma samples, the biosensor gives results that agree reasonably well with those obtained from a commercial biochemical analyzer, exhibits an acceptable level of selectivity for glucose, and is stable over a period of 60 days.[11]

5.2.2 Redox Sensing of Biologically Relevant Molecules

The redox behavior of some transition-metal-containing molecules, such as ferrocene, changes on interacting with different molecules including adenosine mono-, di-, and triphosphate, and DNA.[14–16] For instance, the ferrocenium cation modulates its redox properties when its cationic iron center electrostatically interacts with an anionic molecule or when a functional

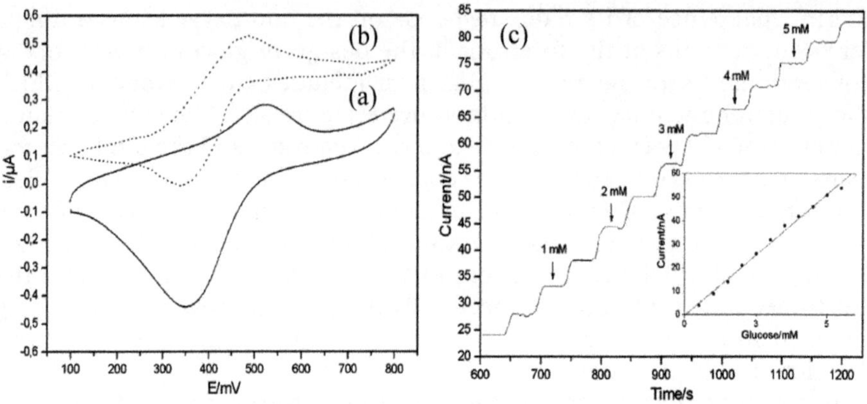

Figure 5.3 Cyclic voltammograms of a poly(glycidyl methacrylate-*co*-vinyl ferrocene)/glucose oxidase electrode in the (a) absence and (b) presence of glucose. (c) Amperometric response of a poly(glycidyl methacrylate-*co*-vinyl ferrocene)/glucose oxidase electrode to sequential injection of glucose and calibration curve (see inset). Reproduced from M. Şenel and M. F. Abasıyanık, Construction of a Novel Glucose Biosensor Based on Covalent Immobilization of Glucose Oxidase on Poly(glycidyl methacrylate-*co*-vinylferrocene), *Electroanalysis*,[11] John Wiley and Sons, © 2010 WILEY-VCH Verlag GmbH & Co. KGaA, Weinheim.

group on its cyclopentadienyl ligand chelates or hydrogen-bonds with another molecule.[17] Such interaction-induced changes in redox behavior are transduced into measurable electrochemical signals, allowing ferrocene or its derivatives to function as sensitive receptors for the qualitative or quantitative sensing of other molecules. Astruc *et al.* explored the use of a ferrocenyl dendrimer (Figure 5.4) as a redox sensor for many molecules of biological relevance.[18–20] They also showed that conjugating ferrocene to a dendrimer scaffold increases the strength of these interactions between the ferrocene receptor and the analyte, as indicated by the difference in potential (ΔE^o) between the free and bound forms of the dendrimer, as well as the apparent association constant (K^+).[20] This topology effect increases with dendrimer generation, a positive dendritic effect that results in better molecular recognition at higher dendrimer generation.[1]

The electrostatic interaction of the ferrocenium cation with anionic molecules, such as phosphate anions, and the concurrent perturbation of the ferrocenium cation's redox property facilitates redox sensing of adenosine-5′-triphosphate (ATP^{2-}) or phosphate anion (H_2PO^{4-}). Astruc and coworkers demonstrated that dendrimers with peripheral ferrocene recognize ATP^{2-} or H_2PO^{4-} by the appearance of a new redox wave at a less positive potential.[18] They reasoned that the anions donate electrons to the iron centers, making it more susceptible to oxidation. Phosphate anion recognition is not limited to ferrocene, because the iron cluster $[\{CpFe(\mu_3\text{-}CO)\}_4]$, (where Cp is cyclopentadienyliron ligand) (Figure 5.5) exhibits this ability when covalently attached to the periphery of a dendrimer.[19]

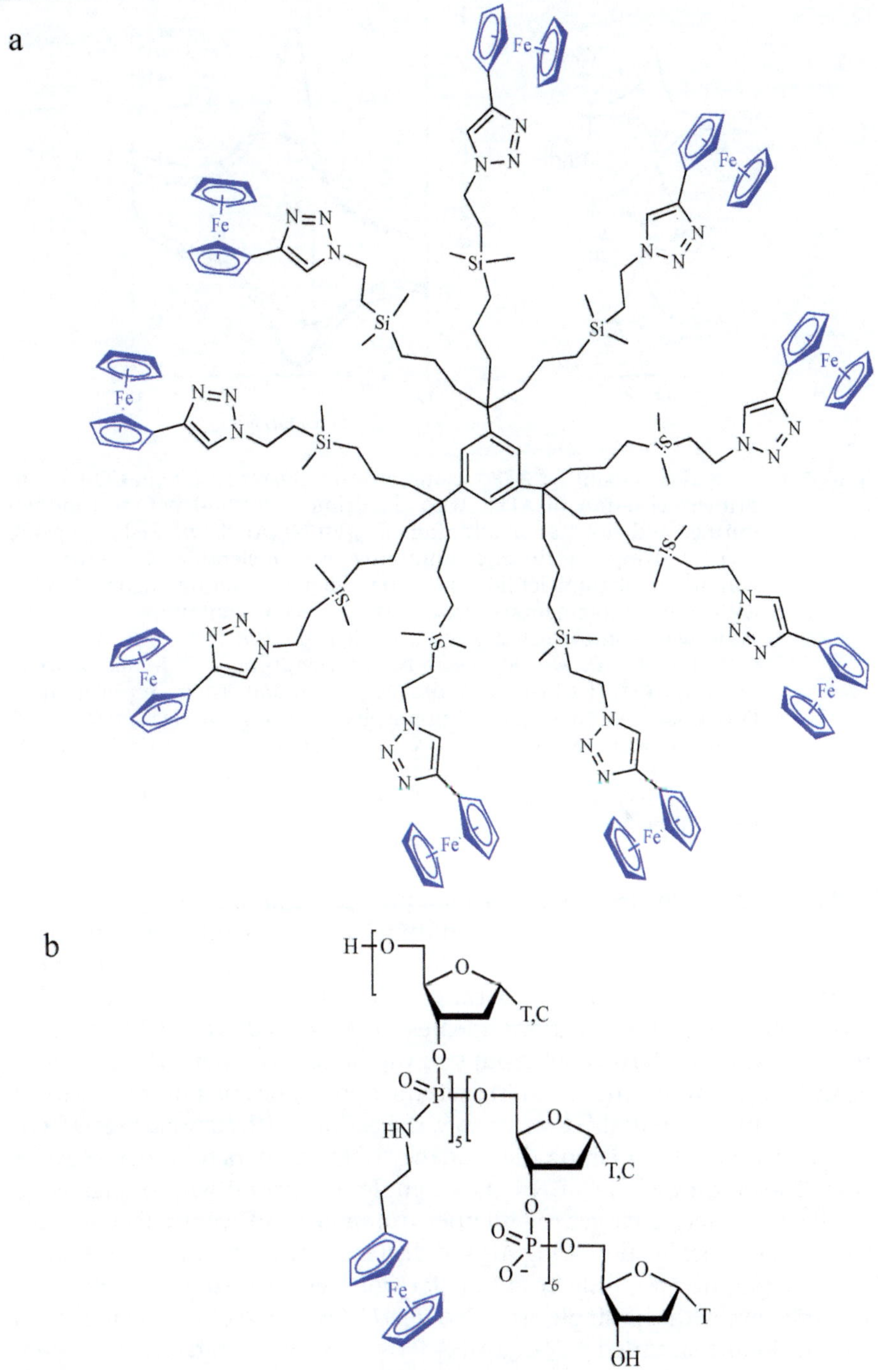

Figure 5.4 (a) Ferrocene-containing dendrimer for redox sensing of biologically relevant molecules such as adenosine-5′-triphosphate. (b) Ferrocene-containing oligonucleotide for redox sensing of its complementary DNA.

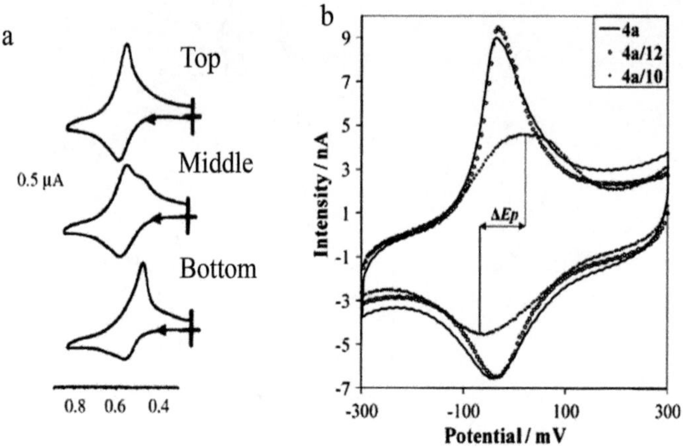

Figure 5.5 (a) Redox sensing of ATP^{2-} using an iron cluster $[\{CpFe(\mu_3\text{-}CO)\}_4]$ dendrimer. Titration of ATP^{2-} with dendrimer (bottom) before, (middle) during, and (top) after addition of $(n\text{Bu}_4\text{N})_2(\text{ATP})$. (b) Redox sensing of DNA using a ferrocene-containing oligonucleotide. **4a**, ferrocene-containing oligonucleotide, **4a/12** ferrocene-containing oligonucleotide with non-complementary DNA; **4a/10** ferrocene-containing oligonucleotide with complementary DNA. (a) Reproduced from J. R. Aranzaes, C. Belin and D. Astruc, Assembly of Dendrimers with Redox-Active $[\{CpFe(\mu_3\text{-}CO)\}_4]$ Clusters at the Periphery and Their Application to Oxo-Anion and Adenosine-5′-Triphosphate Sensing, *Angewandte Chemie*,[19] John Wiley and Sons, © 2006 WILEY-VCH Verlag GmbH & Co. KGaA, Weinheim. (b) Reproduced from ref. 21 with permission from The Royal Society of Chemistry and the Centre National de la Recherche Scientifique.

Like ferrocene, the iron cluster senses phosphate anions, as evidenced by the appearance of a new CV wave upon titration of the dendrimer with ATP^{2-} (Figure 5.5).[19] Also, the interaction of functional groups on the Cp ligand of ferrocene permits redox sensing of cations[18] and DNAs.[21] As an example, chelation of transition metal cations such as Cu^{+}, Cu^{2+}, Pd^{2+}, or Pt^{2+} by triazole groups linking the ferrocenyl groups to the dendrimer, perturbs the redox response of the iron center, resulting in the appearance of a new CV wave at a more positive potential.[18] Presumably, the cations withdraw electrons from the iron centers, diminishing the tendency of the centers to undergo oxidation.[18] The binding of the β-DNA to its complementary α-DNA, conjugated to a ferrocene moiety, changes the microenvironment and eventually the redox property, specifically increasing ΔE and decreasing the anodic current of the ferrocene (Figure 5.5).[21] Electron transfer from ferrocene to the electrode is more efficient with the single-stranded α-DNA-ferrocene bioconjugates than with the double-stranded β-DNA/α-DNA-ferrocene bioconjugates, where conformation changes in the ferrocene may hamper electron transfer.[21] This responsive redox behavior could be exploited in the detection of DNA.

5.2.3 Optical Bioimaging

Optical bioimaging, an ensemble of imaging modalities, relies on electromagnetic radiation in the ultraviolet, visible, and infrared regions to visualize biological structures. The intent is to diagnose diseases or understand the cellular and molecular mechanisms in physiology or pathology. These imaging modalities rely on ballistic or diffusive light as well as the photoacoustic effect of light, which thus exclude classical microscopy.[22] Optical imaging techniques are attractive because they are non-invasive, sensitive, offer high spatial resolution, are well-suited for soft tissue imaging, and offer the possibility to differentiate diseased and healthy tissues *via* differential accumulation of endogenous or exogenous contrast agents in these tissues.[23] Nonetheless, high loading of contrast agent may be required to realize objectives if the target biomolecule is expressed at low concentration. Unfortunately, at high loading, self-quenching may occur, limiting the sensitivity. Also, the absorption and emission spectra of most fluorescent dyes employed as the contrast agents tend to overlap, due to their characteristic small Stokes shift, further stressing the need for an advancement in the field.[24]

Fundamental knowledge of inorganic chemistry can be exploited to design contrast agents with a large Stokes shift, long luminescence lifetime, and that absorb and emit above the tissue-damaging UV region. Agents that emit from the triplet excited state have a long lifetime as emission from this state is forbidden, and have large Stokes shifts since singlet-to-triplet state conversion dissipates energy.[25,26] Transition metals facilitate singlet-to-triplet conversion *via* intersystem crossing due to spin–orbit coupling. Accordingly, transition metal complexes with a predominantly metallic character in the excited state orbitals emit from the triplet state with large Stokes shifts and long luminescence lifetimes as intersystem crossing is promoted.[26] For instance, polypyridyl complexes of ruthenium absorb *via* an allowed singlet metal-to-ligand charge transfer mechanism, and convert to the triplet state *via* intersystem crossing, finally emitting with a larger Stokes shift and long lifetime.[26] Lin *et al.* reasoned that incorporating a polypyridyl complex of ruthenium into nanoparticles allows access to contrast agents that emit even at high loading, due to their large Stokes shift.[24]

Exploiting the excellent photophysical property of the ruthenium polypyridyl complex and the coordination chemistry of zirconium, Lin and coworkers constructed a nanoscale coordination polymer (NCP) that emits at 57.4% contrast agent loading (Scheme 5.1).[24] The NCP nanoparticles were coated with silica to control the release of the ruthenium complex, pegylated to prevent aggregation and increase aqueous dispersibility, and functionalized with a sigma-receptor binding ligand, anisamide, to target H460 human non-small-cell lung cancer cells (Scheme 5.1).[24] These surface modifications tune the photophysical properties, the aggregation behavior, ruthenium

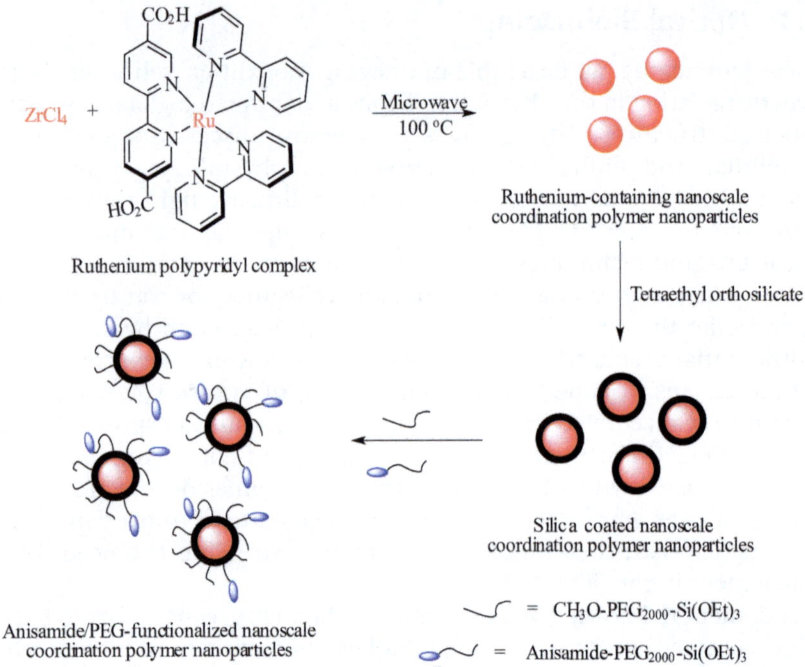

Scheme 5.1 Synthesis of ruthenium nanoscale coordination polymer nanoparticles for optical imaging.

release profile, and the ruthenium loading from 55 to 27%.[24] *In vitro*, the surface-modified nanoparticles successfully target and accumulate in H460 human non-small-cell lung cancer cells, enabling the visualization of these cells with confocal laser fluorescence microscopy (Figure 5.6).[24] Also, the nanoparticles were compatible with the cells, inducing no remarkable cell death after 24 hours (Figure 5.6).[24]

5.2.4 Other Bioimaging Modalities

The chemistry of transition-metal-containing macromolecules advances other imaging modalities including X-ray computed tomography (CT),[27] positron emission tomography (PET),[28] and magnetic resonance imaging (MRI).[2] These imaging modalities are useful technologies to diagnose and monitor diseases; for instance, X-ray CT provides three-dimensional spatial resolution of biological structures,[30] whereas MRI and PET have depth penetration superior to optical imaging modalities.[28,30] Transition metals with their high atomic number attenuate X-rays and radioactivity, so are used as contrast agents in X-ray CT and PET. For instance, a Hf-containing nanoscale metal–organic framework provides enhanced attenuation of X-ray CT signals for *in vivo* imaging of the spleens and livers of mice (Figure 5.7).[2]

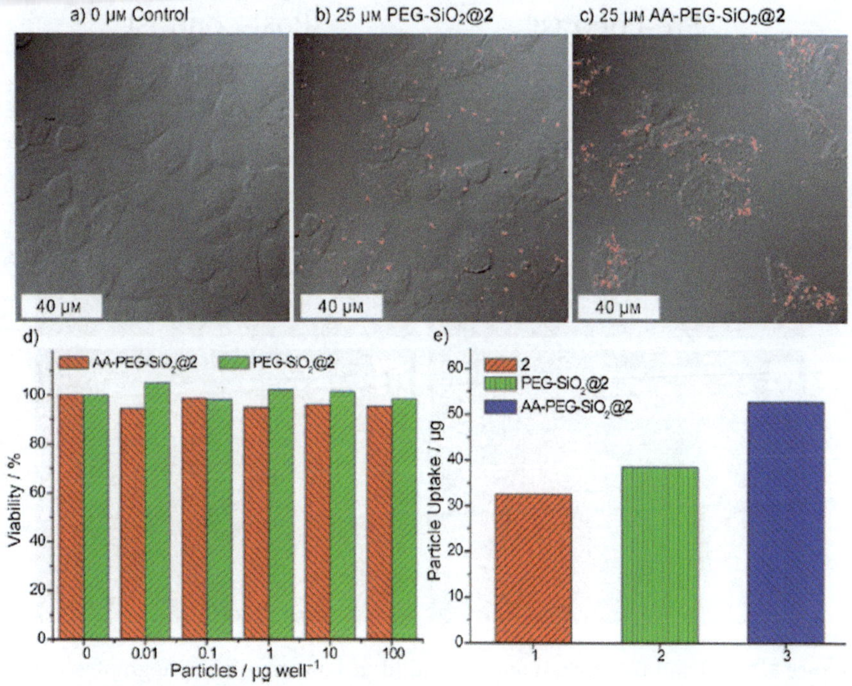

Figure 5.6 Optical imaging of H460 cells imaged (a) without nanoparticles (b) PEG-functionalized NCP nanoparticles (PEG-SiO_2@**2**), (c) anisamide/PEG-functionalized (AA-PEG-SiO_2@**2**) NCP nanoparticles using confocal laser fluorescence microscopy. (d) Viability assay for H460 cells using different amounts of PEG-SiO_2@**2** and AA-PEG-SiO_2@**2**; (e) uptake of non-functionalized NCP (**2**), PEG-SiO_2@**2**, AA-PEG-SiO_2@**2** by H460 cells. Reproduced from D. Liu, R. C. Huxford and W. Lin, Phosphorescent Nanoscale Coordination Polymers as Contrast Agents for Optical Imaging, *Angewandte Chemie*,[24] John Wiley and Sons, © 2011 WILEY-VCH Verlag GmbH & Co. KGaA, Weinheim.

5.2.5 Imaging-guided Therapy

Imaging-guided therapy is an emerging approach that integrates bioimaging with a therapeutic modality to improve efficacy. The approach provides post-treatment spatiotemporal information of the diseased cells or tissues, as well as the administered therapeutics. Imaging modalities including optical imaging, photoacoustic imaging, MRI, and X-ray CT have been integrated with chemotherapy or photodynamic therapy to realize effective treatments of cancer.[29–33] For instance, a triple-modality NCP constructed for imaging-guided therapy exploits the magnetic resonance response of gadolinium(III) (Gd^{3+}), the optical property of ruthenium polypyridyl complex, and the X-ray attenuation property of ytterbium(III) (Yb^{3+}) to improve the efficacy of photodynamic therapy (Figure 5.8).[30] This triple modality platform exhibits the characteristic red fluorescence of the ruthenium complex. Also, the

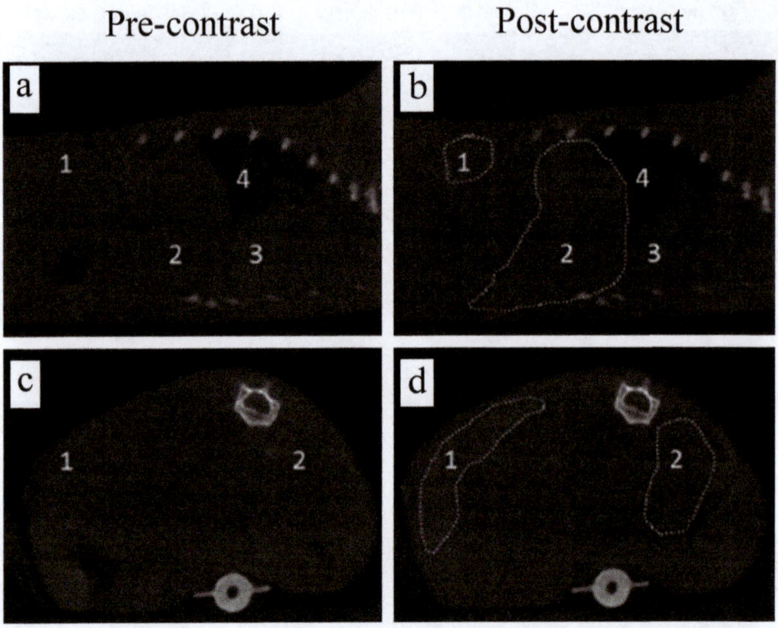

Figure 5.7 (a) and (b) Sagittal and (c) and (d) axial computer tomography imaging of the spleen (1) liver (2) of mice using a Hf-containing metal–organic framework. The heart (3) and lungs (4) are not imaged. Reproduced from ref. 27 with permission from The Royal Society of Chemistry.

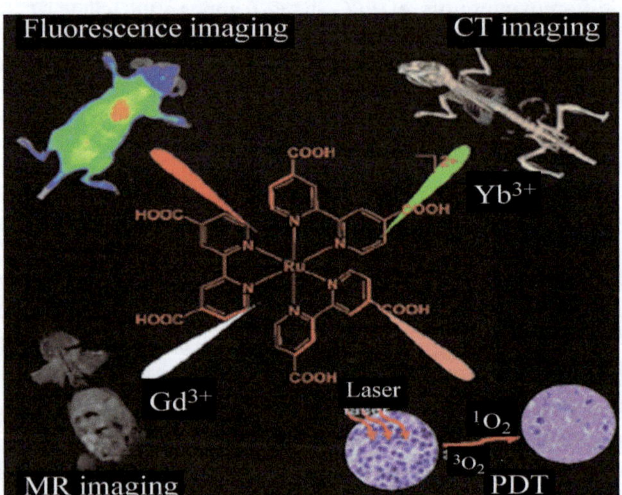

Figure 5.8 Triple-modality nanoscale coordination polymer for optical (fluorescence imaging), magnetic (MR), and computer tomography (CT) imaging-guided photodynamic therapy (PDT). Reproduced from ref. 30 with permission from The Royal Society of Chemistry.

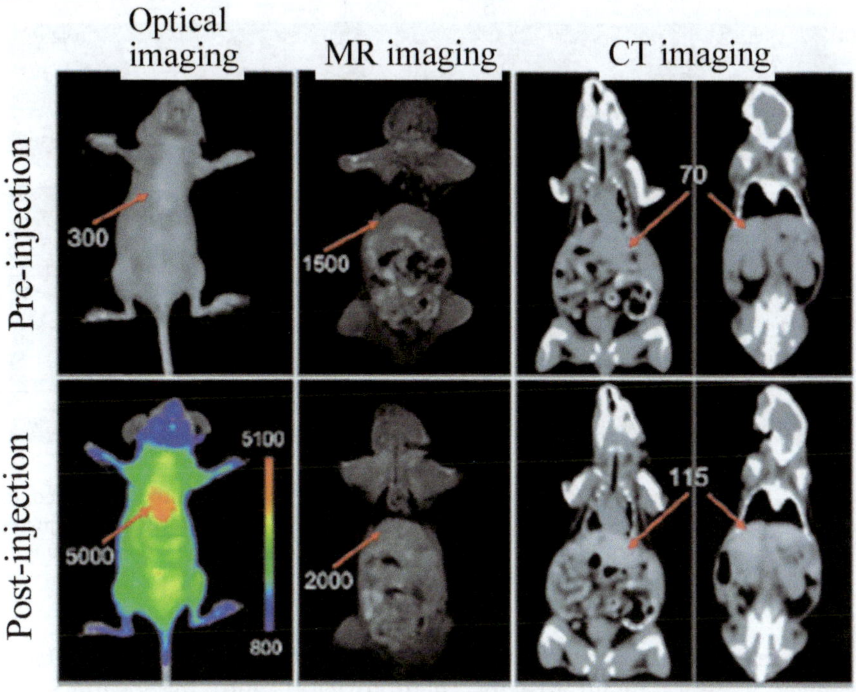

Figure 5.9 Optical (fluorescence), magnetic resonance, and X-ray computer tomography imaging of a mouse liver (red arrow) after injection with triple-modality NCP nanoparticles. Reproduced from ref. 30 with permission from the Royal Society of Chemistry.

magnetic resonance signals of Gd^{3+} and X-ray computed tomography contrast of Yb^{3+} are evidenced in the NCP. *In vivo* experiments with mice confirm the potential of this triple-modality imaging platform. For instance, the high sensitivity of the optical imaging, 3D spatial resolution of the CT, and depth penetration of the MR enable clear visualization of the liver, kidney and abdominal aorta (Figure 5.9).[30] Also, this platform allows the selective imaging of cancer cells in HepG2 mice xenografts, suggesting the potential of this triple-modality imaging platform in diagnosis and imaging-guided therapy (Figure 5.10).[30]

5.3 Transition-metal-containing Macromolecular Therapeutics

5.3.1 Chemotherapeutics

We previously mentioned the clinical successes of platinum(II) complexes including cisplatin, carboplatin, and oxaliplatin, in treating various types of cancers, as examples of the therapeutic potential of transition metal

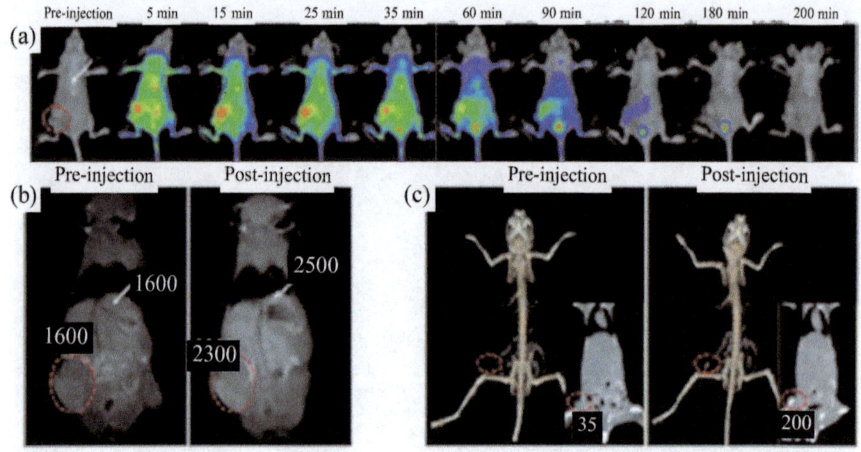

Figure 5.10 (a) Time-dependent optical (fluorescence); (b) magnetic resonance; (c) X-ray CT imaging of the cancer cells of HepG2 mice xenografts injected with triple-modality NCP nanoparticles. Reproduced from ref. 30 with permission from The Royal Society of Chemistry.

complexes. These iconic anticancer drugs cause unpleasant side effects, namely nausea, nephrotoxicity, neurotoxicity, and myelosuppressive effects.[34–41] Moreover, most transition metals coordinate to various endogenous ligands[1] as exemplified by the binding of platinum in cisplatin and oxaliplatin to DNA to form adducts,[37,42] a life-threatening process that limits the therapeutic potential of these metals. Academic and industry researchers are exploring many strategies to mitigate these side effects while harnessing the therapeutic potential. The development of transition-metal-containing macromolecular drugs is a part of the ongoing efforts towards more effective and less life-threatening metal-based drugs.

Transition-metal-containing macromolecules constructed with therapeutic activity are no longer mere hypothetical conjectures or interesting laboratory curiosities but now real solutions to some of the most feared healthcare problems. Through a prodrug approach, a biologically inactive metal complex is integrated into a macromolecular framework to protect, transport, and release a pharmacologically active metal complex with target specificity. This prodrug approach minimizes undesirable systemic toxicity, increasing the therapeutic window.[43–45] Indeed, macromolecular prodrugs are explored to mitigate the systemic toxicity of platinum-based anticancer therapeutics. For instance, Access Pharmaceuticals is collaborating with other researchers to develop an anticancer polymer-based prodrug, ProLindac™, which is currently undergoing clinical trials.[34–37] Their approach involves coordinating a biologically inactive diaminocyclohexyl platinum complex to a *N*-(2-hydroxypropyl)methacrylamide copolymer, through the pH-sensitive amidomalonato ligand (Figure 5.11).[34–36]

The design objectives of ProLindac™ include prolonged plasma half-life, inactivity during systemic circulation, and selectivity towards cancer

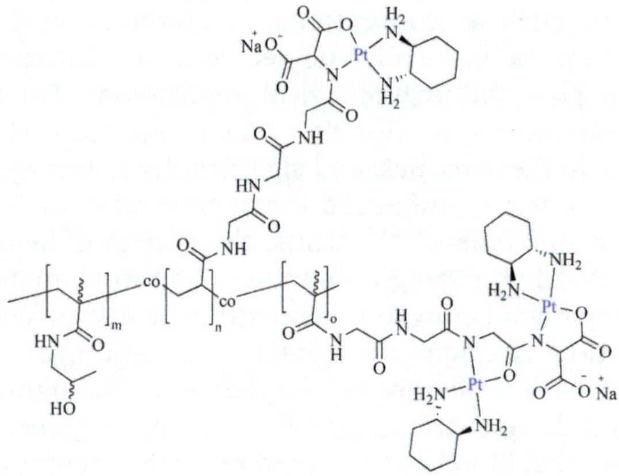

Figure 5.11 Schematic of ProLindac™.

cells, with the goal of an enhanced therapeutic window.[34–37] The prodrug, currently in phase II clinical trials, is a linear random copolymer characterized by a weight-averaged molecular weight of 25 000 Da and ~10% platinum by weight.[34,46] The prodrug transforms into a pharmacologically active diaminocyclohexyl platinum complex in the acidic microenvironment of the cancer cells.[37] The pH of the microenvironment controls the release of the prodrug and its transformation, eventually tuning the therapeutic efficacy and systemic toxicity. As an example, within 24 hours, ProLindac™ releases 3.5 ± 0.1% of free platinum species in a pH 7.4 phosphate buffered saline containing a physiological concentration of chloride but 23.6 ± 0.1% in a pH 5.4 phosphate/citrate buffered saline.[34] As expected, the results imply ProLindac™ will be inactive and less toxic at pH 7.4, which is the pH level of the bloodstream and extracellular fluid of normal tissues, but will release and transform into the active platinum complex in acidic cancer tissues.

Preclinical assessment of the efficacy and pharmacokinetics of ProLindac™ in mice with B16F10 melanoma, as well as with M5076 or 2008 ovarian carcinomas, reveals a promising therapeutic window, specifically an anticancer activity and a safety profile that are superior to that of the small-molecule commercially-available anticancer oxaliplatin.[34] For instance, oxaliplatin is ~20-fold more toxic to healthy cells and is also less effective in delaying the growth of cancer cells than ProLindac™.[34] In this mouse model, ProLindac™ outperforms cisplatin in treating cisplatin-sensitive and cisplatin-resistant variants of the M5076 ovarian carcinoma.[34] Phase I/II clinical trials in 35 patients confirm the superior therapeutic window of ProLindac™ over cisplatin and carboplatin in treating recurrent ovarian cancer.[36,46] Compared with these drugs, ProLindac™ features minimal nephrotoxicity and myelosuppressive effects with none of the patients experiencing neurotoxicity.[36]

Nanomaterials are increasingly explored in modern medicine, as their properties are well-suited for biomedical applications, differing from those of bulk materials such as conventional small-molecule drugs.[47–50] Also, these properties can be rationally engineered at the nanoscale dimension to broaden their potential in biomedical applications. For instance, anticancer nanotherapeutics can selectively target cancerous tissues, preferentially accumulate in these tissues, and significantly reduce systemic toxicity with the ultimate effect of enhanced therapeutic windows compared with small-molecule counterparts.[48,51,52] Thus, the emerging field of nanomedicine is rapidly evolving, relying on advances and fundamentals in diverse fields including synthetic inorganic chemistry to realize its goal.

The coordination of transition metals to organic ligands enables the design of NCPs with wide-ranging implications for nanomedicine.[53–55] Through rational design strategies, NCPs can be engineered to prolong their lifetime in the bloodstream, preferentially accumulate in cancer tissue, reduce systemic toxicity, and eventually improve the therapeutic window. Lin *et al.* explored the coordination chemistry of transition metals to develop NCPs as anticancer chemotherapeutics.[56–63] As an example, Lin and coworkers coordinated terbium(III) (Tb^{3+}) to carboxylate ligands to construct an NCP that incorporates an anticancer prodrug, *cis,cis,trans*-diamminedichloro-disuccinato platinum(IV) (Scheme 5.2).[63] The reductive microenvironment of cancerous tissues triggers the release of the prodrug, which is transformed to a pharmacologically active platinum(II) (Pt^{2+}) complex (Scheme 5.2).[63] As the NCP nanoparticles dissociate in excess water, it was critical to stabilize them by coating with a silica layer ~2 nm or ~7 nm thick. The thickness of the silica layer controls the half-life ($t_{1/2}$) of dissolution of the polymer and, therefore, the release profile of

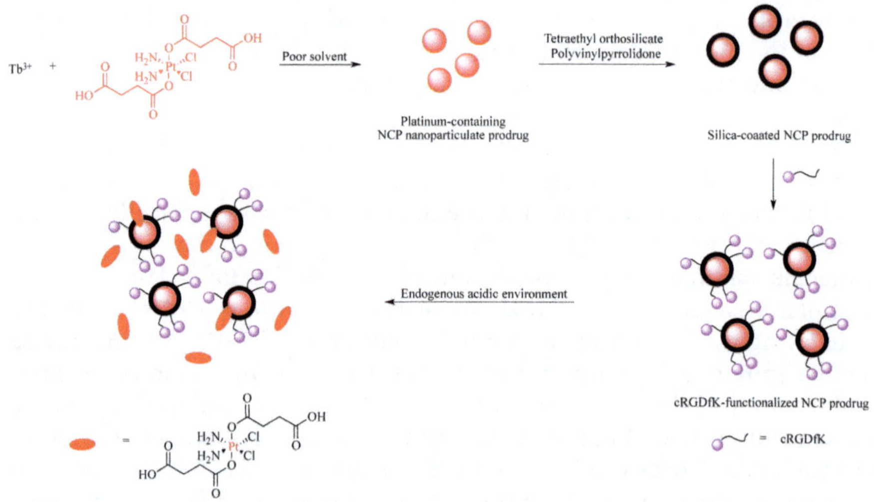

Scheme 5.2 Schematic of the assembly of a platinum-containing polymer prodrug and the release of the drug in the endogenous acid environment of cancer cells.

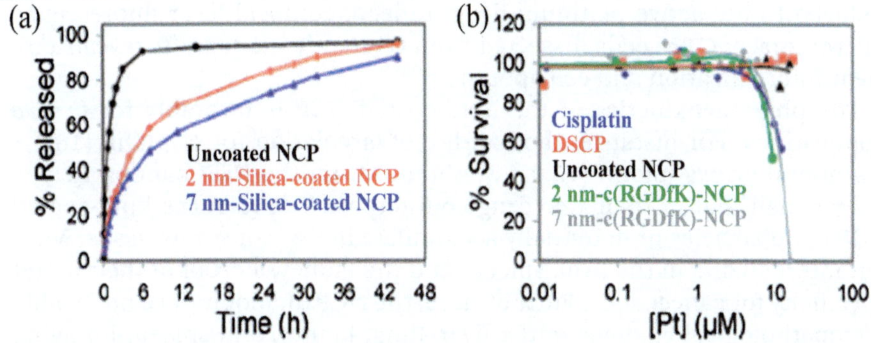

Figure 5.12 (a) Time-dependent release profile of platinum from an NCP prodrug. (b) *In vitro* anticancer activity of prodrugs and cisplatin against an angiogenic human colon carcinoma cell line (HT-29). DSCP is *cis,cis,trans*-diamminedichlorodisuccinato platinum(IV). Adapted with permission from ref. 63. Copyright (2008) American Chemical Society.

platinum (Figure 5.12).[63] For instance, uncoated NCP nanoparticles, and ~2 nm- or ~7 nm-thick silica-coated polymers have a $t_{1/2}$ of ~1, ~5.5, or ~9 hours, respectively. To target angiogenic cancer cells, such as HT-29, the NCP nanoparticles were tagged with silyl-derived cyclo(RGDfK), a peptide that strongly binds to the upregulated αvβ3 integrin in angiogenic cancer cells. This strategy ensures internalization of the NCP in HT-29 cells *via* receptor-mediated endocytosis. Internalization is pivotal to the therapeutic function of the prodrug as endogenous reducing agents, such as glutathione, are needed to reduce the released inactive platinum(IV) (Pt^{4+}) complex to an active Pt^{2+} complex. An *in vitro* cytotoxicity assay gives evidence of the anticancer efficacy of the cyclo(RGDfK)-tagged ~2 nm- or ~7 nm-thick silica-coated NCP prodrug as IC_{50} values of 9.7 or 11.9 μM, which are lower than that of cisplatin (13.0 μM), were obtained (Figure 5.12).[63]

For *in vivo* applications, it is important that nanotherapeutics are relatively stable, biocompatible, and exhibit favorable pharmacokinetics. As metals form stronger coordination bonds with phosphonates than with carboxylates,[64] Lin *et al.* explored the more stable metal–phosphonate coordination bonds to obtain stable anticancer nanotherapeutics. In this approach, Zn^{2+} was coordinated to a phosphonate ligand to obtain NCP nanoparticles with cisplatin or oxaliplatin prodrugs in the backbone/core and a lipid outer coating.[60] The lipid coating prolongs the half-life in the bloodstream, and tunes the release kinetics of the platinum prodrug, specifically slowing down the release and preventing burst release. As an example, in a cell culture medium at 37 °C, the lipid-coated NCP releases less than 10% of the prodrug compared with over 50% release by the uncoated NCP nanoparticles. *In vitro*, murine colon adenocarcinoma 26 (CT26) cells internalize the NCP nanoparticles into their cytoplasm *via* a pathway that includes energy-dependent receptor-mediated endocytosis.[60] Once inside the cytoplasm, the nanoparticles induce cell apoptosis as the prodrug is reduced to the active platinum drug. Indeed, confocal laser fluorescence microscopy of CT26 cells dissected from CT26 mice xenografts reveals efficient internalization and cell apoptosis.[60]

The pharmacokinetics of the lipid-coated NCP is favorable for *in vivo* applications. For instance, the half-life of circulation for cisplatin (16.4 ± 2.9 hours) or oxaliplatin (12.0 ± 3.9 hours) from the NCP nanoparticle is longer than that of their free drug counterpart.[60] *In vivo*, the lipid-coated NCP nanoparticles preferentially accumulate in the cancerous tissue, without internalizing in the liver, spleen, and the kidney, a proof of their target specificity for cancer cells. Furthermore, the NCP did not induce noticeable histopathological changes in the liver, lung, kidney, and spleen, implying a lack of systemic toxicity. Whole body fluorescence imaging of CT26 mice xenografts injected with chlorin e6-doped lipid-coated NCP reveals preferential accumulation of the nanoparticles in the cancerous tissue due to the enhanced permeability and retention (EPR) effects (Figure 5.13).[60] Lin and coworkers screened the lipid-coated NCP for *in vivo* anticancer activity using mice xenografts transplanted with CT26, H460 human non-small cell lung cancer, or AsPC-1 human pancreatic cancer cells.[60] The anticancer activity of the NCP is superior to that of the free drugs. For instance, the oxaliplatin-containing NCP nanoparticles more effectively inhibit the growth of cancer cells in mice transplanted with AsPC-1 cancer cells compared with free oxaliplatin (Figure 5.13).[60]

The threat of drug resistance stresses an urgent need for efficacious therapies. Combination chemotherapy is an emerging therapeutic modality, which combines multiple chemotherapeutic agents to act in synergy and *via* distinct mechanisms of action to destroy resistant pathways, genes, and cell replication cycles.[65–67] Compared with single-agent therapy, combination chemotherapy enhances the efficacy of treatment and reduces systemic toxicity, leading to an increase in the therapeutic window *via* the synergistic effect. The United States Food and Drug Administration approval of gemcitabine hydrochloride and carboplatin to treat advanced recurring ovarian cancer, and gemcitabine hydrochloride and cisplatin to treat locally advanced or metastatic non-small cell lung cancer attests to the efficacy and prospect of the strategy.[62]

The coordination chemistry of transition metals enables the combination of multiple drugs at the molecular level within a macromolecule. Lin *et al.* used this chemistry to combine oxaliplatin or carboplatin with gemcitabine within an NCP framework to provoke a synergistic effect, and selectively target cancer cells.[61,62] The oxaliplatin and gemcitabine act in synergy to enhance the anticancer activity of the NCP nanoparticles against AsPc-1 and BxPc-3 cancer cells.[61] The IC_{50} values of the oxaliplatin-gemcitabine-containing NCP, which are several folds lower than those of free oxaliplatin, free gemcitabine, or mono-therapeutic NCP,[61] gives evidence of the superiority of combination therapy over mono-drug therapy. Also, the synergistic effect is demonstrated *in vivo* by the superior efficacy of the oxaliplatin-gemcitabine-containing NCP over mono-drug therapy to suppress the growth of AsPc-1 and BxPc-3 cancer cells transplanted in mice (Figure 5.14).[61] The pharmacokinetics of

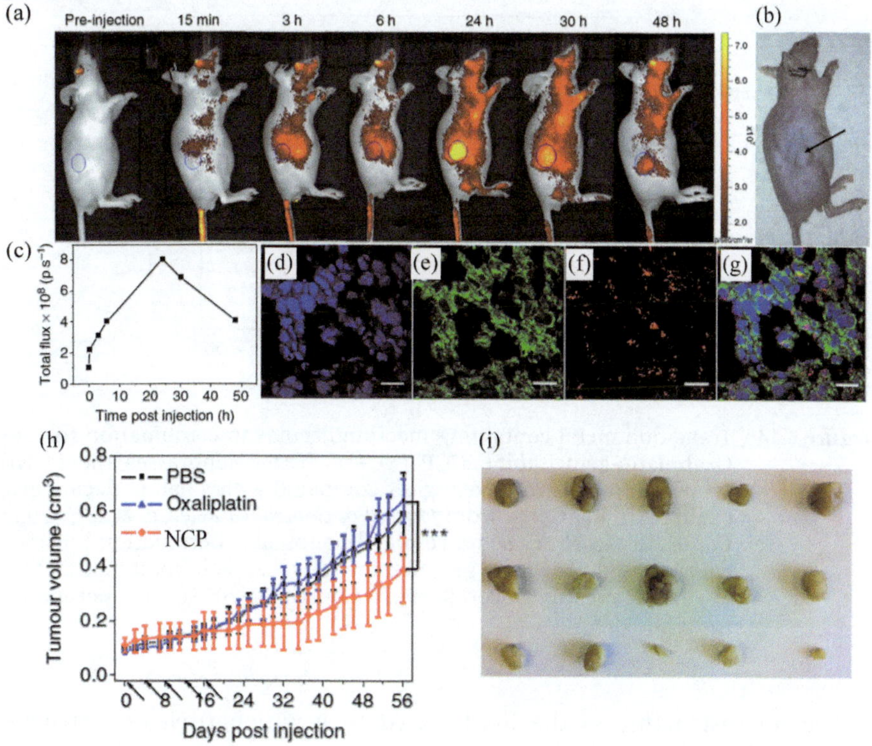

Figure 5.13 (a) Time-dependent whole-body fluorescence images reveal preferential accumulation of NCPs in the cancerous tissue of a CT26-tumor-bearing mouse. (b) Digital camera image indicates tumor position. (c) Time-dependent fluorescence emission intensity of tumor position. (d–g) Confocal laser fluorescence microscopy confirms apoptosis of CT26 cancer cells (channels: blue, DAPI nuclear stain; green, apoptosis of CT26 cells by TUNEL assay; red, chlorin e6 particle). (h) *In vivo* inhibition of AsPC-1 cancer cells. (i) Digital camera image indicates inhibition of AsPC-1 cancer cells by NCP (from top to bottom: PBS, oxaliplatin, NCP). Reprinted by permission from Macmillan Publishers Ltd: [*Nature Communications*] (ref. 60), copyright (2014).

the oxaliplatin-gemcitabine-containing NCP are excellent. Typical of nanotherapeutics, the NCP features a long blood circulation half-life for platinum (10.1 ± 3.3 hours), which is over 1000-fold longer than that of free oxaliplatin, as well as for gemcitabine (8.0 ± 2.3 hours), after intravenous injection.[61] Also, the NCP preferentially accumulates in the cancerous tissues with an uptake of 8.8 ± 2.0% ID per g (percentage injected dose per gram), which is six times higher than that of free oxaliplatin, which rapidly cleared from the bloodstream.[61] The superior ability of the NCP to accumulate in the cancer cells, as well as the synergy between the oxaliplatin and gemcitabine, results in a high degree of cancer cell apoptosis.

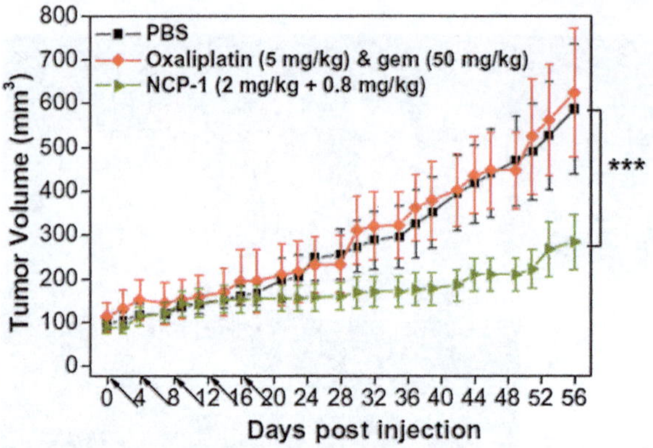

Figure 5.14 Transition-metal-containing macromolecules in combination therapy. Oxaliplatin-gemcitabine NCP (NCP-1) better suppresses the *in vivo* growth of AsPc-1 mice xenograft compared with small-molecule drug oxaliplatin. Reprinted from *Journal of Controlled Release*, **201**, C. Poon, C. He, D. Liu *et al.*, Self-assembled nanoscale coordination polymers carrying oxaliplatin and gemcitabine for synergistic combination therapy of pancreatic cancer, 90–99.[61] Copyright (2015), with permission from Elsevier.

The microstructure of the lipid-coated NCP nanoparticles constructed by Lin and coworkers can be engineered to incorporate therapeutic agents within the lipid shell. Such nanoengineering provides a route to combination chemotherapy to tackle drug resistance as the core incorporates a therapeutic agent. Lin and coworkers pioneered this concept, incorporating a pool of small interfering RNA (siRNA) within the lipid layer and a platinum prodrug within the NCP framework to target cisplatin-resistant ovarian cancer cells (Figure 5.15).[56,58] The strategy is to employ siRNA to disrupt drug-resistant pathways or genes in resistant cells, and re-sensitize these cells towards treatment with cisplatin.[56,58] The therapeutic potential of siRNA is best realized if siRNA is protected from nuclease degradation, and is released from the endosome into the cytoplasm.[68] Lin and coworkers engineered the NCP to protect the siRNA and escape from the endosome, ensuring the target specificity of siRNA to cisplatin-resistant ovarian cancer cells to silent drug-resistant genes.[56,58] The electrostatic interaction of the cationic lipid layer of the NCP with the negatively charged endosome membrane disrupts the membrane integrity to trigger endosomal escape of the siRNA into the cytoplasm.[56] Alternatively, the carbon dioxide generated from the reduction of the prodrug to cisplatin (Figure 5.16) disrupts the endosomal membrane, releasing the siRNA into the cytoplasm.[58] Confocal laser scanning microscopy shows evidence of the successful escape of siRNA from endosomal entrapment into the cytoplasm (Figure 5.16).[58] This well-engineered NCP shields siRNA from nuclease degradation as revealed by time-dependent gel electrophoresis analysis of the siRNA-containing NCP in a serum (Figure 5.16).[58]

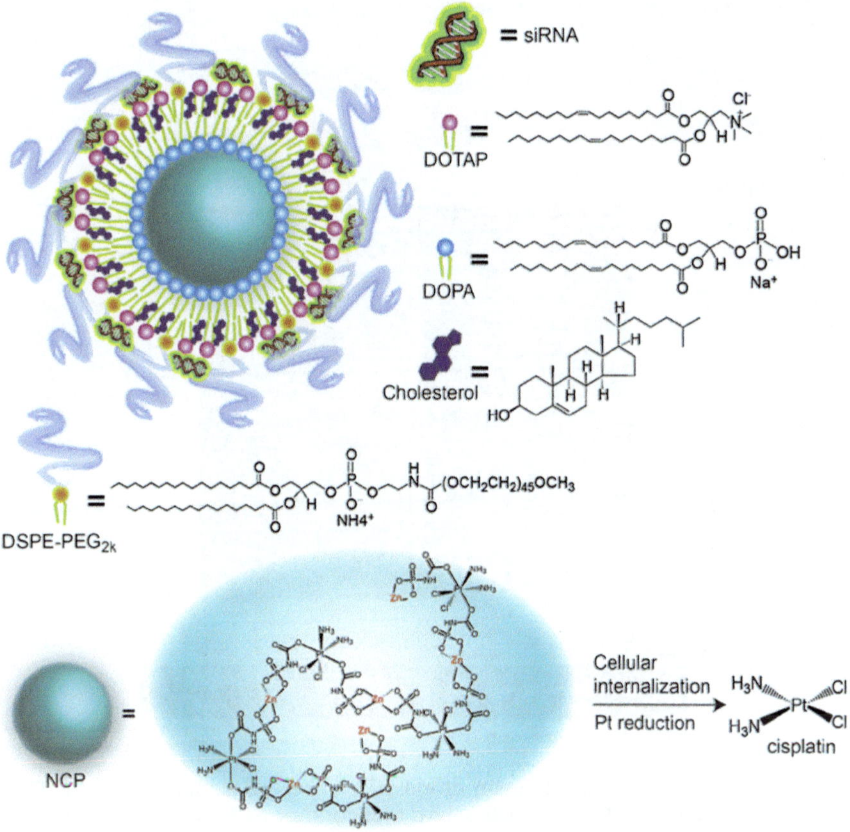

Figure 5.15 Schematic of the nanoscale coordination polymer nanoparticle carrying siRNA in the lipid layer and cisplatin in the polymer backbone and nanoparticle core. Release of cisplatin is triggered by the endogenous reducing agents in the cell. Reproduced from *Biomaterials*, **36**, C. He, D. Liu and W. Lin, Self-assembled nanoscale coordination polymers carrying siRNAs and cisplatin for effective treatment of resistant ovarian cancer, 124–133.[56] Copyright (2015), with permission from Elsevier.

The covalent binding of siRNA to the lipid layer of the NCP nanoparticles *via* disulfide linkage enables the simultaneous redox-controlled release of siRNA and cisplatin.[58] In phosphate buffered saline, release of siRNA is slow but significantly increases after a reducing agent, such as glutathione or cysteine, is added to the extracellular or intracellular environment.[58] These reducing agents also trigger the release of the cisplatin from the platinum prodrug. Undoubtedly, this redox-triggered release is possible inside the cell, which is rich in endogenous reducing agents, such as cysteine or glutathione. When administered *via* systemic injection to mice subcutaneously transplanted with cisplatin-resistant SKOV-3 ovarian cancer cells, the siRNA-containing NCP nanoparticles remarkably inhibit the growth of the cancer cells, contrasting the treatment with NCP, free cisplatin plus siRNA, free siRNA or NCP

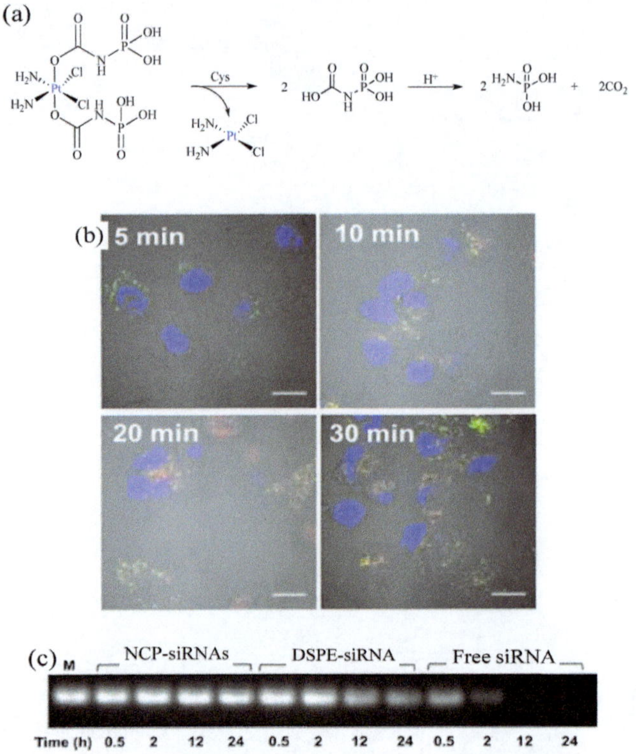

Figure 5.16 (a) A novel mechanism proposed for the endosomal escape of siRNA. Generation of CO_2 *via in vivo* reduction of NCP by endogenous reducing agents triggers the release of siRNA. (b) Confocal laser fluorescence microscopy confirms the time-dependent endosomal release of NCP-siRNA in SKOV-3 cells (red). Green, endosomes and nuclei; blue, DAPI. (c) Electrophoresis indicates the serum stability of siRNA in NCP–siRNA nanoparticles. Adapted with permission from ref. 58. Copyright (2016) American Chemical Society.

plus a control siRNA (Figure 5.17).[58] Terminal deoxynucleotidyl transferase (TdT) dUTP nick end labeling (TUNEL) assay shows evidence of cell apoptosis, specifically revealing that the combination therapy more effectively fragments the DNA of cancer cells. Also, the anticancer activity of this siRNA-containing NCP is demonstrated in mice intraperitoneally transplanted with cisplatin-resistant A2780/CDDP ovarian cancer cells. The NCP inhibits the growth of cancer cells in four out of six mice, as well as reduces the tumor weight and nodule count. Moreover, the combination therapy is well-tolerated, as body weight is maintained and immune response is lacking.[56,58]

5.3.2 Phototherapeutics

A dearth of safe therapeutics challenges the ongoing fight against many life-threatening diseases including cancer. Frequently employed anticancer therapies, such as radiotherapy, chemotherapy, immunotherapy, and

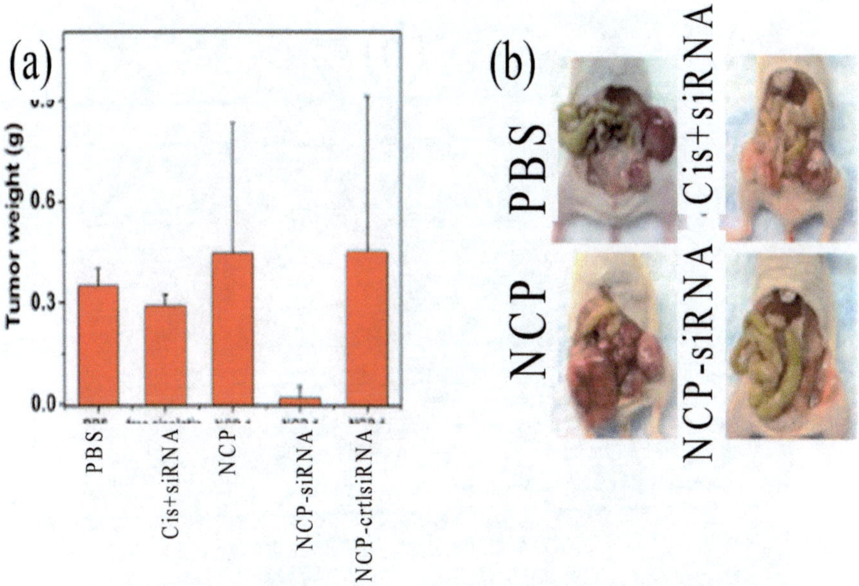

Figure 5.17 (a) and (b) Anticancer activity of NCP-siRNA nanoparticles compared with cisplatin plus siRNA, NCP nanoparticles, and NCP plus control siRNA. Adapted with permission from ref. 58. Copyright (2016) American Chemical Society.

surgery, exemplify this by side effects that are fatal or worsen the quality of life of patients. Driven by this challenge, scientists are exploring new therapeutic strategies, including phototherapy, a non-invasive yet effective light-triggered tumor-ablative therapy. Phototherapy is a proven curative therapy for advanced localized diseases (Figure 5.18).[69] The possibility to localize treatments and to repeatedly administer treatments with little or no side effects are two of its advantages. Two classes of phototherapy exist—photodynamic and photothermal therapies—both rely on the absorption of light by the therapeutic agent to eradicate diseased cells. In photothermal therapy, a light-absorbing agent converts the absorbed light energy into heat energy, which kills the target cells[70] whereas photodynamic therapy employs a light-absorbing photosensitizer to generate reactive oxygen species to kill target cells.[71]

To be effective, a photothermal agent or photosensitizer must satisfy certain requirements, including preferential accumulation in the target tissues and absorption of light above 700 nm as endogenous molecules absorb below this wavelength.[72] Many photothermal agents are organic molecules, but the presence of a metal atom is hypothesized to tune the optical properties of the agent due to the unique optical and electronic properties of metals. For instance, the presence of heavy metals in certain photosensitizers shifts the wavelength of maximum absorption (λ_{max}), reduces the difference in energy between the singlet and excited triplet states, alters the triplet state lifetime processes that influence intersystem crossing, and eventually affects the

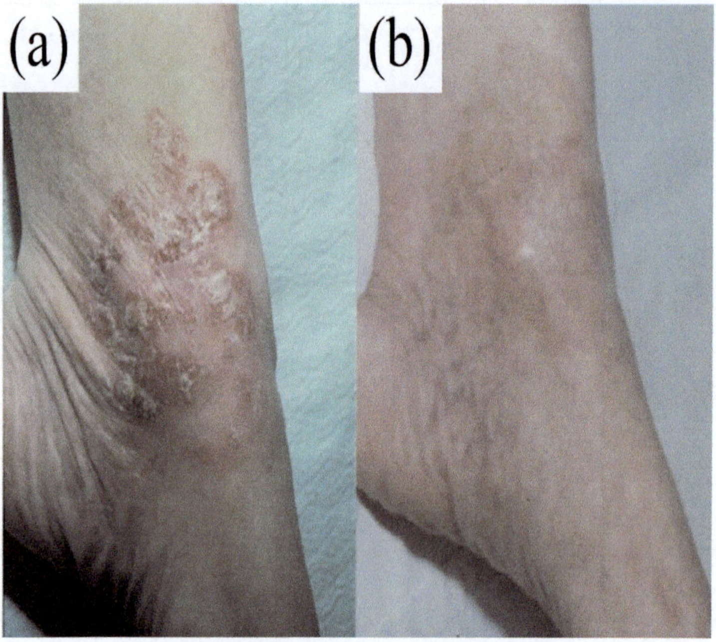

Figure 5.18 Proven efficacy of photodynamic therapy. (a) Patient with Bowen's disease before photodynamic therapy; (b) after photodynamic therapy. Reproduced from *The Lancet Oncology*, **5**, S. B. Brown, E. A. Brown and I. Walker, The present and future role of photodynamic therapy in cancer treatment, 497–508.[69] Copyright (2004), with permission from Elsevier.

generation of singlet oxygen yields.[72,73] Because photodynamic therapy efficiency depends on singlet oxygen yields, it is reasonable to assume that addition of a metal center to a photosensitizer affects the efficacy of the therapy. So far, empirical results are contradictory, with metalation impacting,[74,75] enhancing,[73] or having no effect on photosensitization.[74,75] Regardless, there are compelling results that demonstrate the favorable influence of metalation on photodynamic efficiency. For instance, the insertion of indium(III) into the photosensitizer, 3-(1′-hexyloxy)ethyl-3-devinyl-pyropheophorbide-cyanine dye, enhances photodynamic efficiency, specifically, lowering the therapeutic dose from 2.5 μmol kg^{-1} to 0.3 μmol kg^{-1} in mice transplanted with Colon26 tumors.[73] Also, the heavy atom effect is evident in hafnium–porphyrin nanoscale metal–organic frameworks, which demonstrated an enhanced photodynamic therapy efficiency.[76–78] The insertion of hafnium into 5,15-di(*p*-benzoate)porphyrin results in a red shift of all the porphyrin Q-bands as well as broadens the absorption band of the Soret band.[76] Also, the presence of hafnium doubles the efficiency of singlet oxygen generation due to the heavy metal effect that promotes intersystem crossing as indicated by a decrease in fluorescence lifetimes after metalation of the porphyrin. The photodynamic therapy efficacy of this hafnium–porphyrin metal–organic framework (Figure 5.19) is evident in *in vitro* and *in vivo* experiments with human

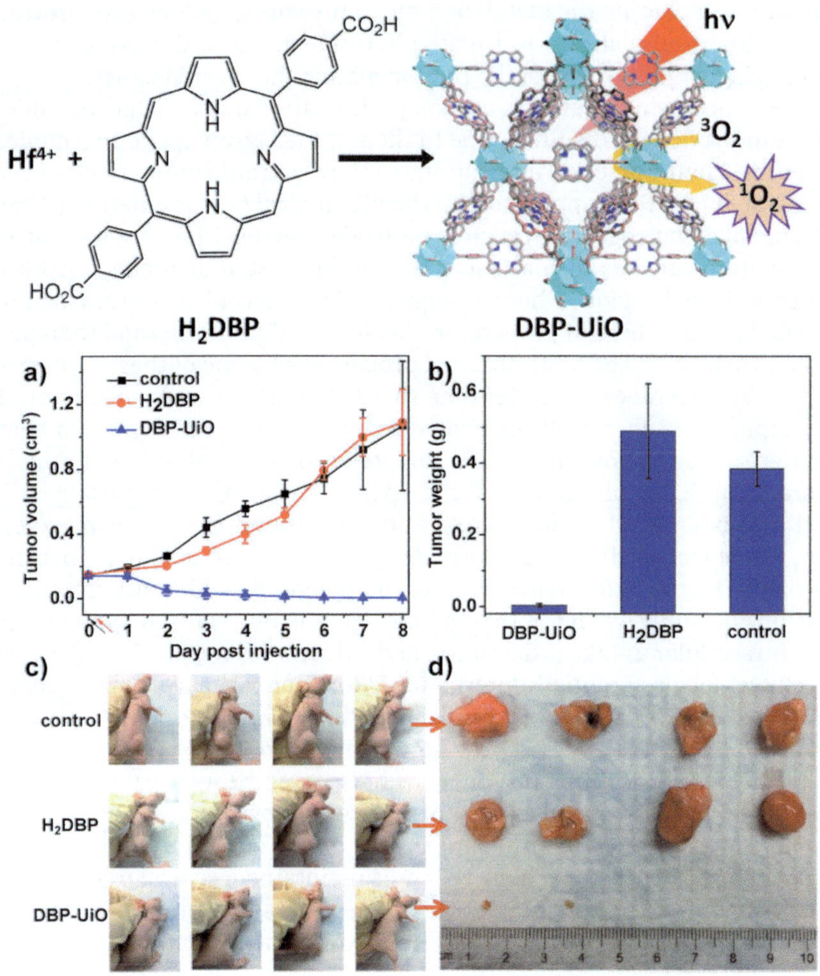

Figure 5.19 Schematic of a hafnium–porphyrin metal–organic framework (DBP–UiO) for photodynamic therapy. *In vivo* demonstration of the efficacy of photodynamic therapy of the hafnium–porphyrin metal–organic framework (a–d). Reprinted with permission from ref. 76. Copyright (2014) American Chemical Society.

head and neck cancer cells SQ20B (Figure 5.19).[76] As required for photodynamic therapy, the hafnium–porphyrin metal–organic framework selectively accumulates in cancer cells SQ20B transplanted into mice and completely eradicates the tumor in two out of four mice without causing skin damage.[76]

Incorporating the photosensitizer within a macromolecular framework prevents aggregation that, in some cases, quenches the excited states to decrease the efficient generation of singlet oxygen. For instance, constructing a dendrimer with the metalated photosensitizer as the core sterically confines the photosensitizer, isolating and preventing it from forming aggregates.[79–81] The steric confinement effect extends beyond the dendritic frameworks, including the metal–organic frameworks,[82] which also provides a macrostructural platform to isolate photosensitizers and prevent the formation of aggregates to boost the photodynamic therapy efficiency.

There is increasing interest in monolayer transition metal dichalcogenides, such as molybdenum disulfide, due to their optoelectronic, and mechanical properties. Their potential applications in the electronic industry is proven, and research into their application in the biomedical field is emerging. Transition metal dichalcogenides, such as molybdenum disulfide, titanium disulfide, titanium carbide, bismuth selenide, and tungsten disulfide, absorb in the near-infrared region, where endogenous biomolecules are transparent, a desirable photophysical property in the design of photothermal therapeutic platforms.[70,83–86] The photothermal properties of some of these transition metal dichalcogenides are better than those of graphene and gold nanorods as exemplified by chemically exfoliated molybdenum disulfide, which heats up water solution above 40 °C, the threshold for thermal ablation therapy (Figure 5.20).[86] The photothermal therapeutic potential of chemically exfoliated molybdenum disulfide (MoS_2) is demonstrated in an *in vitro* experiment, where HeLa cells treated with MoS_2 and irradiated at 800 nm become non-viable (Figure 5.20). Furthermore, the potential of photothermal therapy is obvious only when the therapeutic agent is internalized by the target cells. Intracellular uptake is demonstrated *in vitro* with murine breast cancer 4TI cells, which preferentially internalized MoS_2 nanosheets of smaller diameter as well as pegylated MoS_2 nanosheets.[84] *In vivo*, photothermal therapy with pegylated MoS_2 suppresses the growth of murine breast cancer 4TI cells transplanted into mice, and prolongs the lifespan of the mice compared with mice that received a control treatment.[84] Excitedly, evidence of life-threatening side effects are lacking in the mice after photothermal therapy with the pegylated MoS_2 because the body weight is maintained; critical organs, such as the heart, liver, spleen, lung, and kidney eliminated molybdenum; and mice remained healthy 40 days after administering treatment.[84]

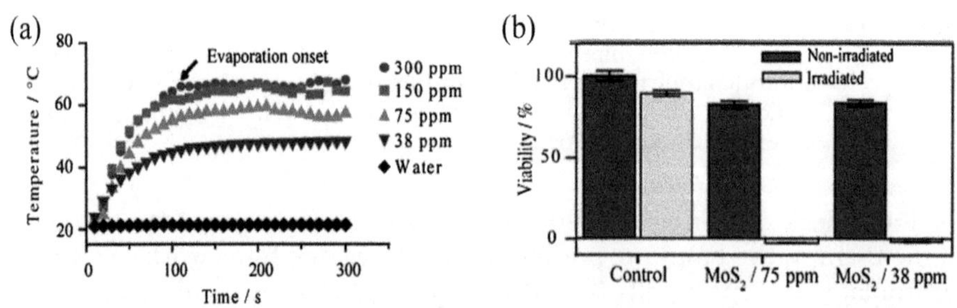

Figure 5.20 (a) Photothermal heating of water by different concentrations of chemically exfoliated molybdenum disulfide. (b) *In vitro* efficacy of photothermal therapy demonstrated with HeLa cells. Reproduced from S. C. Chou, B. Kaehr, J. Kim *et al.*, Chemically Exfoliated MoS_2 as Near-Infrared Photothermal Agents, *Angewandte Chemie*,[86] John Wiley and Sons, © 2013 WILEY-VCH Verlag GmbH & Co. KGaA, Weinheim.

5.3.3 Radiotherapy

Radiotherapy is an efficacious treatment modality administered to over 50% of cancer patients as a curative or palliative measure.[87,88] However, when applied as a single treatment modality, radiotherapy achieved the intended cure of many cancers but failed with others,[89] creating a gap that needs to be bridged. A challenge, therefore, is the design of technologies that broaden the scope of application, and mitigate the adverse effect of ionizing radiation. Absorption of radiation by endogenous tissues depends on the electron density in the microenvironment of the tissue. Therefore, it is reasoned that introduction of high-electron-density materials in the radiation pathway provides a strategy to improve the efficacy of the therapy. Indeed, nanoparticles of high-atomic-number elements, such as transition metals and rare-earth elements, absorb radiation to generate photo/Auger electrons, which eventually initiate the formation of reactive free radicals to sensitize cancer cells to radiotherapy.[90] A pegylated metal–organic framework constructed from hafnium, a high-atomic-number radiosensitizer, and tetrakis(4-carboxyphenyl)porphyrin, a photosensitizer, act in synergy to sensitize cancer cells to radiotherapy.[90] *In vitro* experiments demonstrated that pre-incubating 4T1 cells with the metal–organic framework enhances the efficacy of radiotherapy and increases damage to DNA, observations the authors attributed to the role of hafnium as a radiosensitizer.[90] Also, the presence of hafnium enhances the efficacy of radiotherapy in mice transplanted with 4T1 cells, although combined radiotherapy and photodynamic therapy are more efficacious.[90]

5.3.4 Antimicrobial Agents

Infectious diseases are among the leading causes of death globally,[91] and the emergence of drug-resistant strains of pathogens has hampered successful and sustainable treatment.[92] Drug resistance protracts illnesses, and increases mortality and healthcare costs. Indeed, globally, resistant infections are projected to cost an annual 10 million losses in human lives and a 100 trillion USD to the global economy by 2050.[93,94] A World Health Organization action plan challenges policymakers, academia, and the pharmaceutical industry to develop new technologies to prevent, diagnose, and treat resistance infections.[92] Towards this end, scientists are developing new antimicrobial agents *via* strategies that include incorporating transition metals into macromolecular frameworks with the assumption that the chemistry of the metal will, possibly, provide new mechanisms to destroy resistant pathways or genes.[1,95–102] This hypothesis is influenced by the established clinical outcome of cisplatin,[41] a platinum-containing drug, and the antimalarial activity of ferroquine, an iron-containing drug candidate under phase II clinical trials.[103] *In vivo* experiments with naïve and infected young mice demonstrated the efficacy of ferroquine against chloroquine-resistant strains of *Plasmodium falciparum*.[104] Mechanistic studies show evidence of increased reactive oxygen

$$Fe^{2+} + H_2O_2 \longrightarrow Fe^{3+} + OH^- + OH^\bullet \qquad \text{Fenton reaction}$$

$$Fe^{3+} + O_2^{\bullet -} \longrightarrow Fe^{2+} + O_2$$

$$O_2^{\bullet -} + H_2O_2 \qquad O_2 + OH^- + OH^\bullet \qquad \text{Haber-Weiss reaction}$$

Scheme 5.3 Schematic of the iron-mediated generation of reaction oxygen species.

species (ROS) in a subcellular probe of *P. falciparum* after treatment with ferroquine, indicating an oxidative stress-mediated antimalarial activity.[105] The generation of ROS is attributed to the redox activity of ferrocene. Indeed, it is argued that, under physiological conditions, some metals like iron generate ROS *via* the Fenton reaction (Scheme 5.3) and other metal-catalyzed free-radical chain reactions.[1] The mechanistic scenario driving the biological activity, specifically the antimicrobial function of metals, extends beyond the generation of ROS and includes other fatal biochemical processes like metal-mediated protein dysfunction, impaired membrane function, perturbed nutrient uptake, and genotoxicity.[1]

Abd-El-Aziz *et al.* developed strategies to incorporate iron into macromolecules including dendrimers and linear polymers (Figures 5.21 and 5.22), with the intent of accessing antimicrobial agents with activity against drug-resistant strains of bacteria.[106–110] Compared with small molecules, antimicrobial macromolecules are less volatile and more chemically robust, and therefore, possess longer lifetimes and better environmental compatibility, desirable properties in many applications. Also, the possibility to fabricate these macromolecules into sterile bandages, surgical gowns, medical devices, and implants as well as surface coatings for medical devices, hospital furniture, and shower walls to minimize microbial colonization, preventing the spread of infections, is attractive. The antimicrobial macromolecules constructed by Abd-El-Aziz and coworkers are highly effective against Gram-positive bacteria including methicillin-resistant *Staphylococcus aureus* (MRSA) and vancomycin-resistant *Enterococcus faecium* as revealed by their low minimum inhibitory concentrations.[106–110] They demonstrated that these iron-containing dendrimers catalyze the generation of free radicals as evidenced by electron spin resonance spectroscopy,[107] induce oxidative stress on MRSA as suggested by dichlorodihydrofluorescein oxidative stress assay, and disrupt cell membrane integrity (Figure 5.23).[106]

Various transition metals are being explored and incorporated into macromolecules, aiming for the "magic bullet" against drug-resistant strains. Cobalt, an essential metal and a component of vitamin B_{12}, is an emerging antimicrobial agent in the form of a cobaltocenium cation, a cobalt analog of ferrocene, and considered to be superior to ferrocene regarding hydrophilicity. In addition to the redox activity of the cobaltocenium cation, the presence of a cationic charge offers a mechanism of action against resistant strains. Tang *et al.* constructed macromolecular bioconjugates with potent activity against various resistant strains by exploiting the electrostatic interaction between the cationic charge on a cobaltocenium-containing

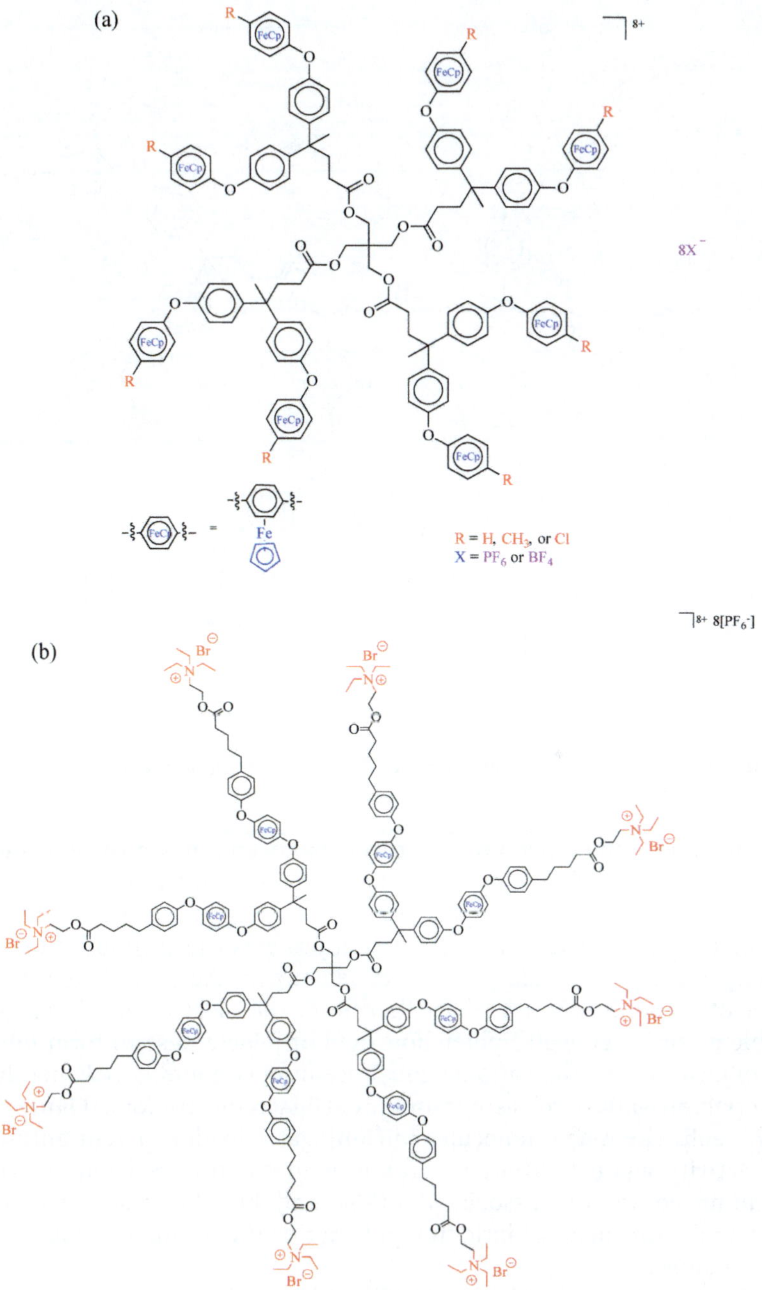

Figure 5.21 Schematic of antimicrobial dendrimers.

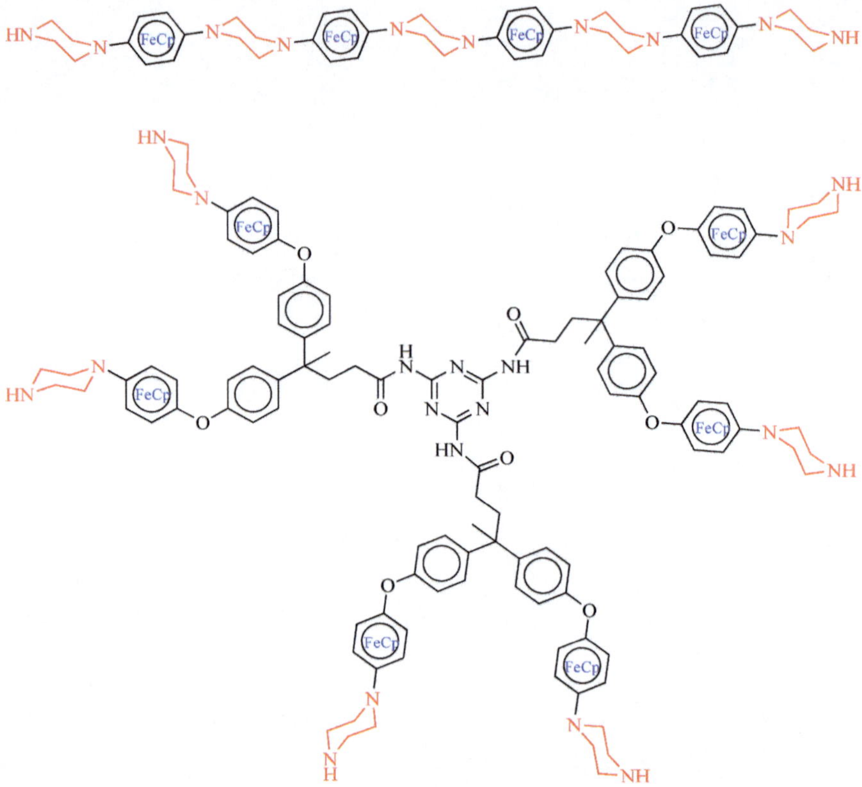

Figure 5.22 Schematic of an antimicrobial dendrimer and a linear polymer.

polymer, poly(2-(methacryloyloxy)ethyl cobaltoceniumcarboxylate), and the anionic charge on β-lactam antibiotics such as penicillin-G, amoxicillin, ampicillin, and cefazolin (Figure 5.24).[111,112] Tang and coworkers demonstrated that this electrostatic interaction protects the antibiotics from the activity of the β-lactamase enzyme, re-sensitizing the bacteria towards the effect of these antibiotics.[112] Furthermore, the electrostatic interactions enable anionic cell wall lipoteichoic acid or β-lactamase to form ion-pair bioconjugates with the cationic cobaltocenium-containing polymer, lysing bacterial cell walls or releasing complexed β-lactam antibiotic (Figure 5.24). As a result, the macromolecular bioconjugates exhibit potent antimicrobial activity against different strains of methicillin-resistant *S. aureus* including community-associated MRSA and healthcare-acquired MRSA compared with the antibiotic-free polymer or the antibiotic alone, which is less potent.[112]

Organometallic molecules are conjugated to antimicrobial agents such as peptides,[98,113–118] piperazine,[109] 2-mercaptobenzothiazole,[107] and quaternary ammonium groups to construct hybrid antimicrobial macromolecules with multiple mechanisms of action, which is an emerging strategy to combat

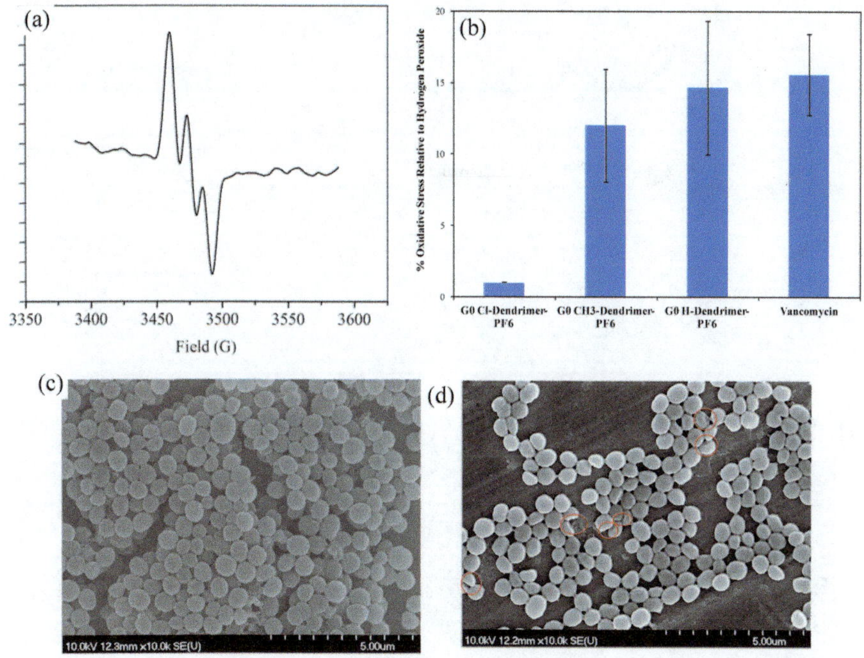

Figure 5.23 (a) Electron spin resonance spectrum of a nitroxyl radical generated by reaction of antimicrobial G_0 Cl-dendrimer-PF_6^- with a nitrosobenzene spin trap at room temperature. (b) Oxidative stress induced by antimicrobial G_0-CH_3-dendrimers-PF_6^- on methicillin-resistant *Staphylococcus aureus* (MRSA). (c) and (d) Antimicrobial G_0-CH_3-dendrimers-PF_6^- disrupt the cell membrane integrity of MRSA. MRSA cells (c) before incubation; (d) after incubation for 24 hours. (a) Reproduced from A. S. Abd-El-Aziz, C. Agatemor, N. Etkin *et al.*, Quaternized and Thiazole-Functionalized Free Radical-Generating Organometallic Dendrimers as Antimicrobial Platform against Multidrug-Resistant Microorganisms, *Macromolecular Bioscience*,[107] John Wiley and Sons, © 2017 WILEY-VCH Verlag GmbH & Co. KGaA, Weinheim. (b–d) Reprinted with permission from ref. 106. Copyright (2015) American Chemical Society.

resistant strains. Metzler-Nolte *et al.* synthesized metal-containing peptide bioconjugates in high yield and purity by covalently conjugating metallocene to peptides *via* an optimized solid-phase peptide synthesis methodology.[118] These bioconjugates are active against *Escherichia coli*, *Pseudomonas aeruginosa*, and *S. aureus*.[98,113–118] In a study aimed at gaining a fundamental understanding of the metal-specific activity of organometallic moieties on the antimicrobial activity of peptides, a cobaltocenium cation was conjugated to an arginine (H-WRWRW-NH_2) or tryptophan (H-RWRWRW-NH_2) peptide.[118] This strategy decreases activity against *E. coli*, *P. aeruginosa*, and *S. aureus*. In contrast, activity against *S. aureus* and *P. aeruginosa* remains the same when ferrocene is conjugated to the tryptophan peptide but is enhanced when it is conjugated to the

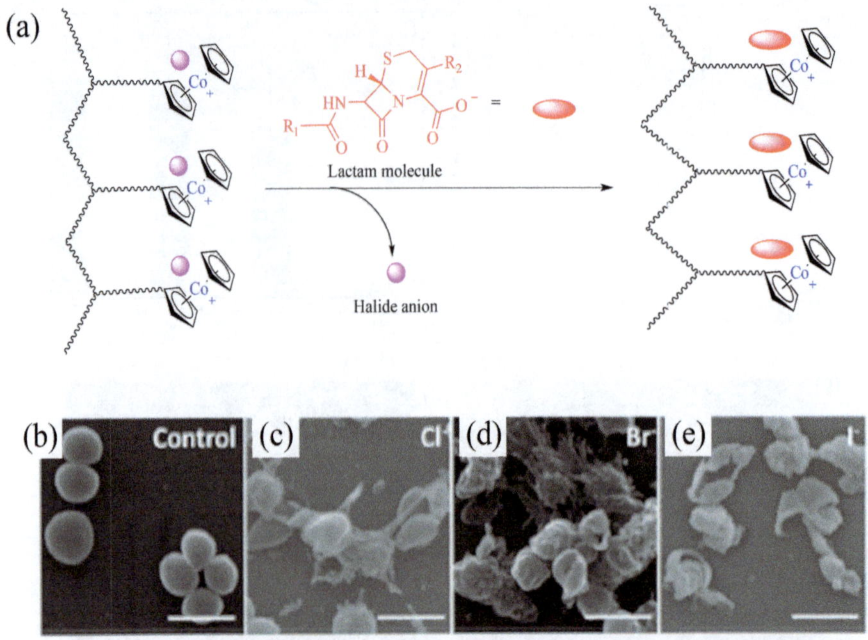

Figure 5.24 (a) Schematic of an antimicrobial cobaltocenium-containing polymer and its proposed mechanism of action. (b)–(e) Antimicrobial activity of the cobaltocenium-containing polymer against healthcare-acquired MRSA. SEM images of HA-MRSA cell before (b) incubation of different chloro-paired (c), bromo-paired (d), or iodo-paired (e) cobaltocenium-containing polymers. Reprinted with permission from ref. 112. Copyright (2014) American Chemical Society.

arginine peptide. To tune the antimicrobial activity of the WRWRW peptides against *E. coli*, *Acinetobacter baumannii*, *P. aeruginosa*, *Bacillus subtilis*, *S. aureus*, and methicillin-resistant *S. aureus*, Metzler-Nolte *et al.* replaced ferrocene with ruthenocene, a strategy that enhances the activity.[117] They proved that conjugating ruthenocene to the peptide tunes the killing kinetics of the peptide to better reduce the colony-forming units than the peptide alone.

5.4 Transition-metal-containing Macromolecules in Drug Delivery

The efficacy of drugs is directly dependent on how they are administered.[119–122] A good drug delivery system ensures that it is designed to protect the drug from undesirable metabolism and to deliver the drug to the target cells, tissues, or organs at the desired rate. Recent innovations in materials science have pushed the frontiers of drug delivery beyond the traditional matrix-based systems to smart hydrogels and nanocontainers. These disruptive innovations are aimed at endowing drug delivery systems with responsive behavior, enabling, *via* an external stimulus, the precise control of drug release, and targeting of specific body sites. The presence of metal centers in transition-metal-containing macromolecules imparts unique chemistries that could be triggered to control drug release and to target specific cells and tissues, allowing access to smart drug delivery technologies.

5.4.1 Redox-responsive Drug Delivery Systems

Metals undergo redox reactions, a ubiquitous chemistry that propels many life-dependent processes such as photosynthesis and cellular respiration,[123–128] and could, therefore, act as a trigger in redox-responsive drug delivery. The Fe^{3+}/Fe^{2+} redox couple, found in the ferrocenium cation/ferrocene redox couple, is at the forefront in the design of redox-responsive drug delivery technologies. Ferrocene is particularly attractive because its redox reaction is reversible and the product is stable. Importantly, the oxidation of ferrocene to the ferrocenium cation changes its solubility property, transforming the hydrophobic ferrocene to the hydrophilic ferrocenium cation as well as changing the predominantly hydrophobic effects to electrostatic interactions.

Towards smart delivery systems, Wang *et al.* constructed a ferrocene-containing amphiphilic block copolymer (Scheme 5.4) from hydrophobic poly(2-acryloyloxyethyl ferrocenecarboxylate) and hydrophilic poly(2-

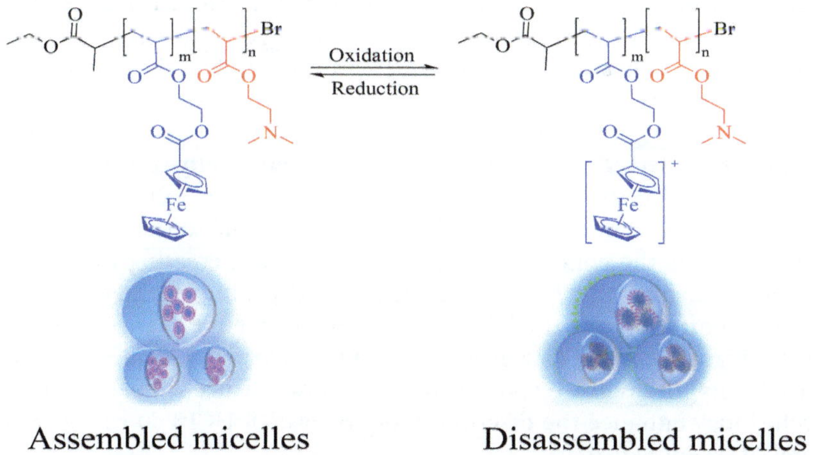

Scheme 5.4 Reversible redox-induced self-assembly and disassembly of poly(2-acryloyloxyethyl ferrocenecarboxylate)-*block*-poly(2-(dimethylamino) ethyl methacrylate) evidenced by the swelling and shrinking of the micelles. Reproduced from *Journal of Industrial and Engineering Chemistry*, **48**, Y. Wang, Y.-L. Luo, F. Xu *et al.*, Redox-responsive PAEFc-*b*-PDMAEMA amphiphilic block copolymer self-assembly micelles: Physicochemical properties and anticancer drug controlled release, 66–78.[129] Copyright (2017), with permission from the Korean Society of Industrial and Engineering Chemistry.

(dimethylamino) ethyl methacrylate).[129] The aromatic interactions between the ferrocenyl units drive the block copolymer to self-assemble, forming globular nanoscale micelles characterized by a poly(2-acryloyloxyethyl ferrocenecarboxylate) core and poly(2-(dimethylamino) ethyl methacrylate) shell.[129] The micelles exhibit reversible self-assembly/disassembly behavior that is tunable *via* control of the redox potential of the external environment. Specifically, addition of 1.8% m/v of $KMnO_4$, H_2O_2, NaClO or $FeCl_3$ to a micellar solution oxidizes the ferrocene moieties, $[Cp_2Fe^{2+}]$, to the ferrocenium cation, $[Cp_2Fe^{3+}]^+$, leading to an increase in the hydrodynamic diameter of the micelles (Figure 5.25). As the hydrophilic ferrocenium cation becomes hydrated in aqueous solution, the poly(2-acryloyloxyethyl ferrocenium carboxylate) core swells, resulting in an increase in the hydrodynamic diameter. The nature of the oxidizing agent influences the degree of swelling. For instance, $KMnO_4$, H_2O_2, NaClO, or $FeCl_3$ increases the hydrodynamic diameter from ~135 nm to 250, 350, 250 or 235 nm, respectively.[129] The addition of 1.8% m/v excess ascorbic acid, a reducing agent, to the oxidized micellar solution decreases the hydrodynamic diameter without changing the morphology of the micelles (Figure 5.25). Indeed, the diameter of the H_2O_2-oxidized micelles decreases from 350 nm to 170 nm after reduction of ferrocenium to ferrocene, suggesting a smart response to changes in the redox potential of their environment. TEM showed the reversible self-assembling and disassembling behavior of the micelles (Figure 5.25).

To demonstrate the stimulus-responsiveness of these ferrocene-based polymer micelles, the hydrophobic anticancer drug 10-hydroxycamptothecin (HCPT) was encapsulated in the core and triggered to release its cargo *via* exposure to an external oxidative environment. The oxidizing agent, H_2O_2, triggers the immediate and sustained release of the drug over an extended time (Figure 5.26).[129] In an H_2O_2-free medium maintained at pH 7.4 and 37 °C, HCPT slowly diffuses from the compactly-packed micellar core held together by aromatic interactions of the ferrocene. Disruption of the aromatic interactions *via* oxidation of ferrocene to the ferrocenium cation weakens the cohesive energy of the micelles, creating more channels and diffusion paths for the fast release of HCPT. Indeed, increasing the H_2O_2 concentration or pH of the external environment enhances the drug release (Figure 5.26).[129] As cancer cells are more oxidative and acidic than normal cells, the redox-responsiveness of these ferrocene-based polymer micelles may enhance the discriminatory release of HCPT in cancer cells, reducing systemic toxicity.

An external stimulus or concentration-dependent unimer/micelles equilibrium compromises micellar stability and leads to inadvertent release profiles. A challenge, therefore, is the design of stable nanocontainers that maintain their structural integrity irrespective of the effect of concentration or external stimulus. Redox-responsive patchy nanocapsules self-assembled from poly(vinyl ferrocene)-*block*-poly(methyl methacrylate) (Figure 5.27)[130] exemplify nanocontainers that maintain their structural integrity in the presence of an external stimulus. The microphase separation between the

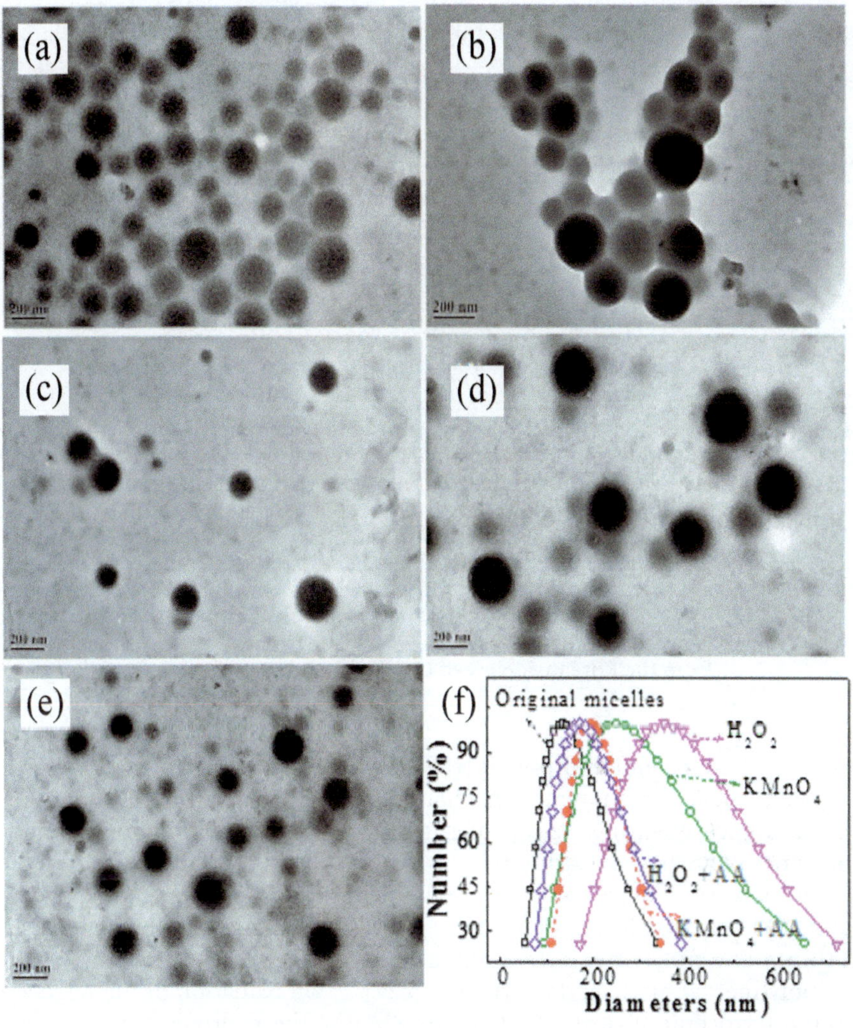

Figure 5.25 TEM micrographs of poly(2-acryloyloxyethyl ferrocenecarboxylate)-*block*-poly(2-(dimethylamino) ethyl methacrylate) micelles. (a) Original micelles; (b) micelles after oxidation with $KMnO_4$; (c) the $KMnO_4$-oxidized micelles after reduction by ascorbic acid (AA); (d) micelles after oxidation with H_2O_2; (e) H_2O_2-oxidized micelles after reduction with AA; (f) particle size distribution of the micelles imaged above. Reproduced from *Journal of Industrial and Engineering Chemistry*, **48**, Y. Wang, Y.-L. Luo, F. Xu *et al.*, Redox-responsive PAEFc-*b*-PDMAEMA amphiphilic block copolymer self-assembly micelles: Physicochemical properties and anticancer drug controlled release, 66–78.[129] Copyright (2017), with permission from the Korean Society of Industrial and Engineering Chemistry.

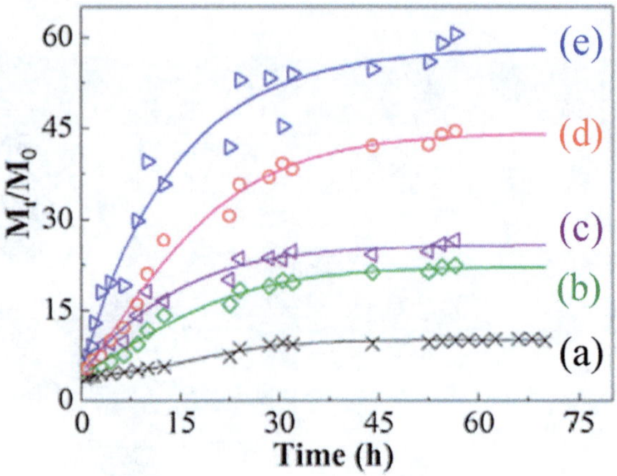

Figure 5.26 Oxidation-triggered release of the anticancer drug, 10-hydroxycamptothecin, from poly(2-acryloyloxyethyl ferrocenecarboxylate)-*block*-poly(2-(dimethylamino) ethyl methacrylate) micelles. (a) pH 7.4; (b) pH 7.4 and 0.15% H_2O_2; (c) pH 5.8 and 0.15% H_2O_2; (d) pH 7.4 and 0.60% H_2O_2; and (e) pH 5.8 and 0.60% H_2O_2. Reproduced from *Journal of Industrial and Engineering Chemistry*, **48**, Y. Wang, Y.-L. Luo, F. Xu *et al.*, Redox-responsive PAEFc-*b*-PDMAEMA amphiphilic block copolymer self-assembly micelles: Physicochemical properties and anticancer drug controlled release, 66–78.[129] Copyright (2017), with permission from the Korean Society of Industrial and Engineering Chemistry.

poly(vinyl ferrocene) and poly(methyl methacrylate) creates patches in the capsular shell. These patches are channels for efficient diffusion of encapsulated cargoes, ensuring the capsules maintain their structural integrity in the presence of an external stimulus.

The core–shell morphology of the nanocapsules was characterized by TEM and SEM. Apparently, TEM indicates a mean hydrodynamic diameter of 236 ± 110 nm and poly(vinyl ferrocene) patches of sizes 25 ± 3 nm surrounded by poly(methyl methacrylate) (Figure 5.27).[130] On oxidation of the ferrocene to the ferrocenium cation by H_2O_2 or $KMnO_4$, the hydrodynamic diameter increases from 236 to 310 or 282 nm, respectively.[130] The authors attribute the increase to swelling of the polar domains in aqueous solution after oxidation of hydrophobic poly(vinyl ferrocene) to polar, hydrophilic poly(vinyl ferrocenium).[130] The morphology of the patches depends on the oxidizing agent used. For instance, nanocapsules treated with H_2O_2 have indentations whereas those treated with $KMnO_4$ have outgrowths on the surfaces (Figure 5.28).[130] The sizes and distances between the indentation or outgrowths agreed reasonably well with the sizes and distances between the poly(vinyl ferrocene) domains before oxidation. The formation of the hydrophilic domains on the capsule shell influences the release kinetics of compartmentalized cargoes. The nanocapsule significantly increases its release of encapsulated pyrene with an increase in oxidation time compared with un-oxidized nanocapsules that exhibit a constant release profile (Figure 5.29).[130] Again,

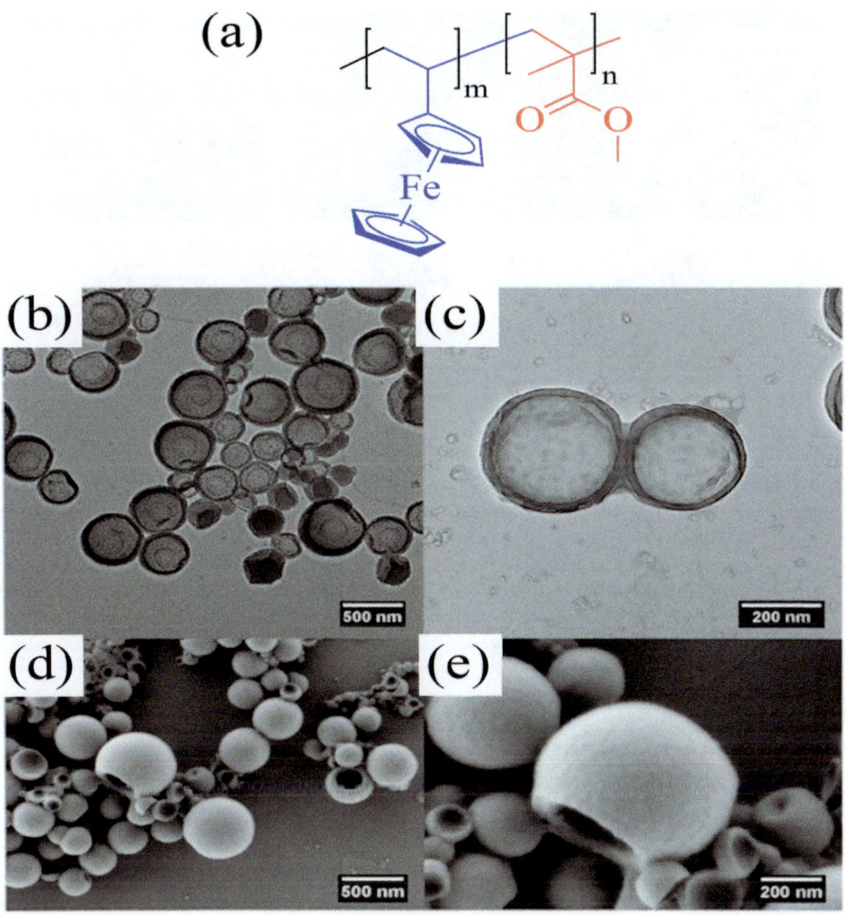

Figure 5.27 (a) Schematic of poly(vinyl ferrocene)-*block*-poly(methyl methacrylate). (b) and (c) TEM and (d) and (e) SEM of poly(vinyl ferrocene)-*block*-poly(methyl methacrylate) nanocapsules. The TEM images reveal the poly(vinyl ferrocene)—dark domains—in the poly(methyl methacrylate) matrix. Reprinted with permission from ref. 130. Copyright (2012) American Chemical Society.

the oxidizing agent influences the release kinetics, with H_2O_2 triggering more release than $KMnO_4$ (Figure 5.29).[130] The difference in the ability of the oxidizing agents to diffuse through the nanocapsules and oxidize the poly(vinyl ferrocene) domain accounts for the difference in the release kinetics. The authors presume that H_2O_2 diffuses swiftly into the shell, oxidizing more ferrocene moieties to ferrocenium cations, enabling enhanced diffusion of pyrene, whereas the ionic MnO^{4-} diffuses slowly, localizing its oxidative property to the surface of the shell, creating fewer channels for diffusion.[130]

Microphase separation provides a strategy to design porous as well as multi-compartment nanocontainers, which are special platforms to deliver multiple incompatible cargos.[131,132] In a copolymer constructed from a

Figure 5.28 SEM and SFM images show evidence of changes in the surface morphology of nanocapsules after oxidation. (a) SEM of nanocapsules before oxidation (b) indents or (c) outgrowths after oxidation with H_2O_2 or $KMnO_4$, respectively. (d) SFM of nanocapsules before oxidation (e) indents or (f) outgrowths after oxidation with H_2O_2 or $KMnO_4$, respectively. Reprinted with permission from ref. 130. Copyright (2012) American Chemical Society.

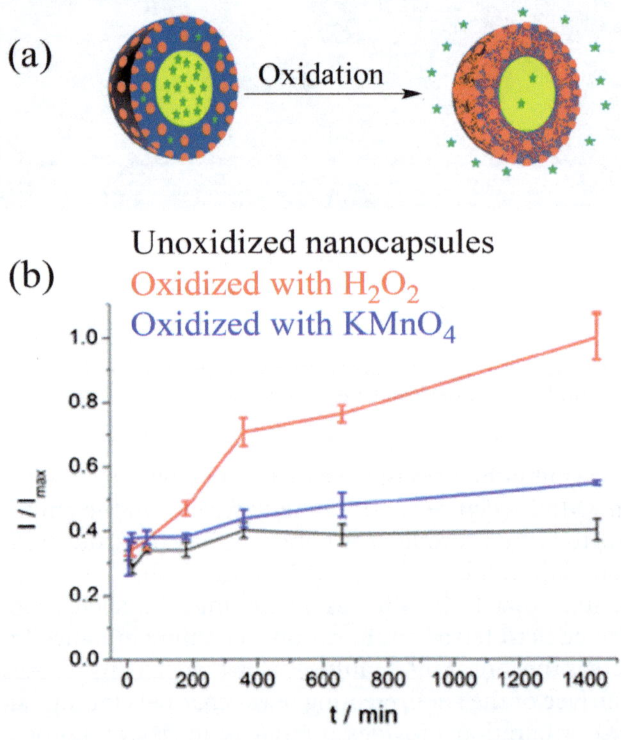

Figure 5.29 Oxidation-triggered cargo release by poly(vinyl ferrocene)-*block*-poly(methyl methacrylate) nanocapsules (a) and (b). Reprinted with permission from ref. 130. Copyright (2012) American Chemical Society.

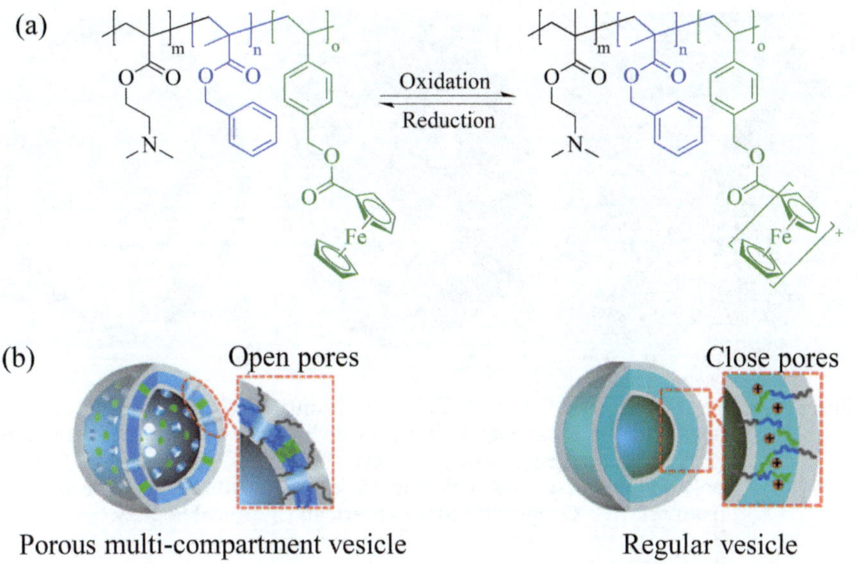

Scheme 5.5 (a) Schematic of the triblock copolymer, poly(2-(dimethylamino)ethyl methacrylate)-*block*-poly(benzyl methacrylate)-*block*-poly(4-vinylbenzyl ferrocenecarboxylate) and (b) redox-triggered morphological changes in the copolymer vesicles. Reprinted with permission from ref. 131. Copyright (2016) American Chemical Society.

metal-containing block and a wholly organic block, the unique chemistry of the metal drives its block to phase separate at the microscale from the organic block, yielding porous and multicompartment nanocontainers. As an example, a ferrocene-containing triblock copolymer, poly(2-(dimethylamino)ethyl methacrylate)-*block*-poly(benzyl methacrylate)-*block*-poly(4-vinylbenzyl ferrocenecarboxylate) obtained *via* seeded RAFT polymerization gives porous multicompartment nanoscale vesicles. The poly(4-vinylbenzyl ferrocenecarboxylate) and the poly(benzyl methacrylate) blocks phase separate at the microscale resulting in the formation of the porous multicompartment vesicles (Scheme 5.5).[131] The degree of polymerization of the poly(4-vinylbenzyl ferrocenecarboxylate) block tunes the microstructure of the vesicular membrane due to the interfacial tension between the incompatible poly(4-vinylbenzyl ferrocenecarboxylate) and poly(benzyl methacrylate) blocks. By control of the number of 4-vinylbenzyl ferrocenecarboxylate monomer units, the structure of the membrane is tuned (Figure 5.30).[131] The multicompartment vesicles are smart, reversibly responding to the redox potential of their external environment. Oxidation with $FeCl_3$ transforms the porous, phase-segregated multicompartment vesicles into conventional vesicles with a uniform membrane in which the poly(4-vinylbenzyl ferrocenecarboxylate) block is compatible with the poly(benzyl methacrylate) block (Scheme 5.5 and Figure 5.31).[131] A change in redox potential triggers a transition from hydrophobic ferrocene to the hydrophilic ferrocenium cation, forcing the poly(4-vinylbenzyl ferrocenecarboxylate) and poly(benzyl

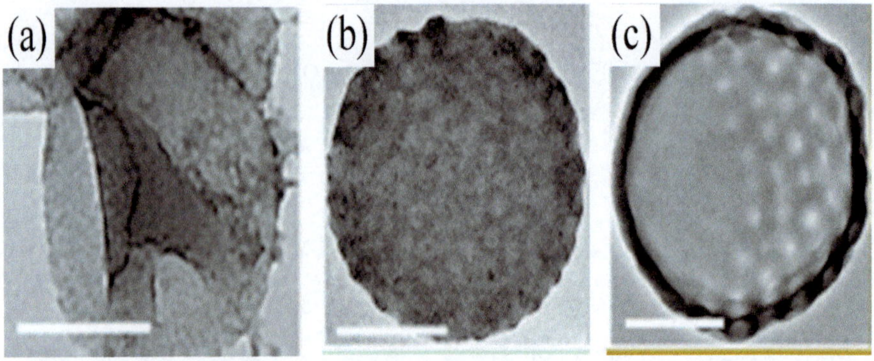

Figure 5.30 TEM images of poly(2-(dimethylamino)ethyl methacrylate)-*block*-poly(benzyl methacrylate)-*block*-poly(4-vinylbenzyl ferrocenecarboxylate) with different degrees of polymerization (a) 0, (b) 59, (c) 124 of the poly(4-vinylbenzyl ferrocenecarboxylate). Reprinted with permission from ref. 131. Copyright (2016) American Chemical Society.

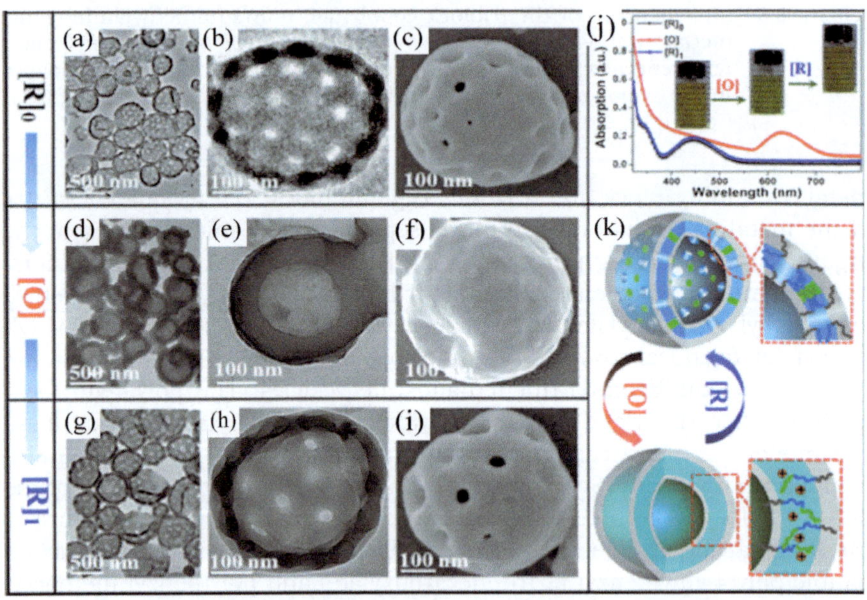

Figure 5.31 Redox-response of a poly(2-(dimethylamino)ethyl methacrylate)-*block*-poly(benzyl methacrylate)-*block*-poly(4-vinylbenzyl ferrocenecarboxylate). (a)–(c) vesicles before oxidation, (d)–(f) vesicles after oxidation, (g)–(i) oxidized vesicles after reduction, (j) UV-vis spectra of oxidized (red line) and reduced (blue line) vesicles, (k) representation of the redox-response of the vesicles. Reprinted with permission from ref. 131. Copyright (2016) American Chemical Society.

methacrylate) blocks to rearrange and form a layered membrane (Scheme 5.5 and Figure 5.31). Reduction with $SnCl_2$ reverts the morphology to the native porous, phase-segregated multicompartment vesicles. SEM, TEM, and UV-vis spectroscopy show the redox-responsive behavior of these vesicles (Figure 5.31). Evidently, the chemistry of metals enables the design of multicompartment nanocontainers with responsive behavior. This development expands the potential of drug delivery technologies towards combination therapy because multiple incompatible drugs can be encapsulated within discrete compartments of a nanocontainer.

In biological systems, enzymes catalyze many redox processes, modulating the redox potential of their environments. Therefore, designing enzyme-controlled redox-responsive drug delivery systems is an attractive research goal. Thomi *et al.* pursued this goal, constructing a ferrocene-based polymer nanocontainer from a poly(dimethylferrocenylsilane)$_{0.87}$-*block*-poly(methylvinyl-ferrocenylsilane)$_{0.13}$ block copolymer that releases encapsulated 2-propylpyridine when triggered by glucose oxidase (Figure 5.32).[133] This enzyme catalyzes the oxidation of β-D-glucose to D-glucono-δ-lactone, simultaneously generating H_2O_2.[134] This *in-situ*-generated oxidizing environment transforms ferrocene to the ferrocenium cation and alters the morphology of the nanocontainers, leading to cargo release. Cyclic voltammetry confirms that the nanocontainers are indeed oxidizable as the oxidation half-wave potential disappeared after subjecting the containers to electrolysis (Figure 5.32).

Redox-triggered transformation of hydrophobic ferrocene to the hydrophilic ferrocenium cation enables the design of cyclodextrin/ferrocene redox responsive hydrogels for drug delivery applications. Cyclodextrins are macrocyclic oligomers of glucose, possessing a hydrophilic exterior and hydrophobic cavity (Scheme 5.6).[135] Consequently, they concurrently encapsulate hydrophobic molecules within their cavity and dissolve in aqueous media, making them useful in drug delivery. The hydrophobic cavity of cyclodextrins encapsulates hydrophobic ferrocene, forming inclusion complexes that dissociate upon oxidation because of the low affinity of the cavity for the hydrophilic ferrocenium cation (Scheme 5.6).[136–142] Other redox-active metallocenes, such as cobaltocene, demonstrate this redox-controlled host–guest interaction, forming inclusion complexes with cyclodextrin while their cationic forms, such as cobaltocenium, lack affinity for the cyclodextrin cavity.[141,143]

Harada *et al.* applied the concept of an inclusion complex to design a supramolecular hydrogel with a redox-triggered reversible sol–gel phase transition.[144] They obtained the supramolecular hydrogel by mixing a β-cyclodextrin-modified poly(acrylic acid) host polymer with a ferrocene-modified poly(acrylic acid) guest polymer in a 1 : 1 ratio (Scheme 5.7).[144] Attempts to develop supramolecular hydrogels from α-cyclodextrin-modified poly(acrylic acid) and ferrocene-modified poly(acrylic acid) failed (Figure 5.33), and the authors attributed this failure to the importance of multipoint crosslinks, which in addition to "host–guest complementarity", are critical to hydrogelation.[144] Previously, Harada and Takahashi showed that

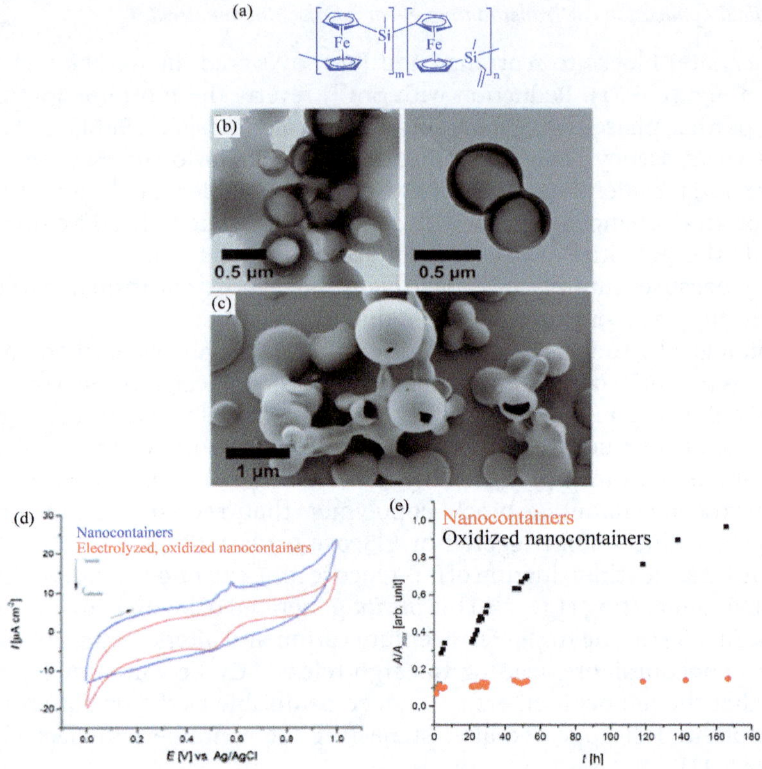

Figure 5.32 (a) Schematic of poly(dimethylferrocenylsilane)-*block*-poly(methylvinyl-ferrocenyl-silane), (b) TEM and (c) SEM of the block copolymer, (d) cyclic voltammograms of copolymeric nanocontainers and (e) redox-triggered release from nanocontainers. Reprinted with permission from ref. 133. Copyright (2016) American Chemical Society.

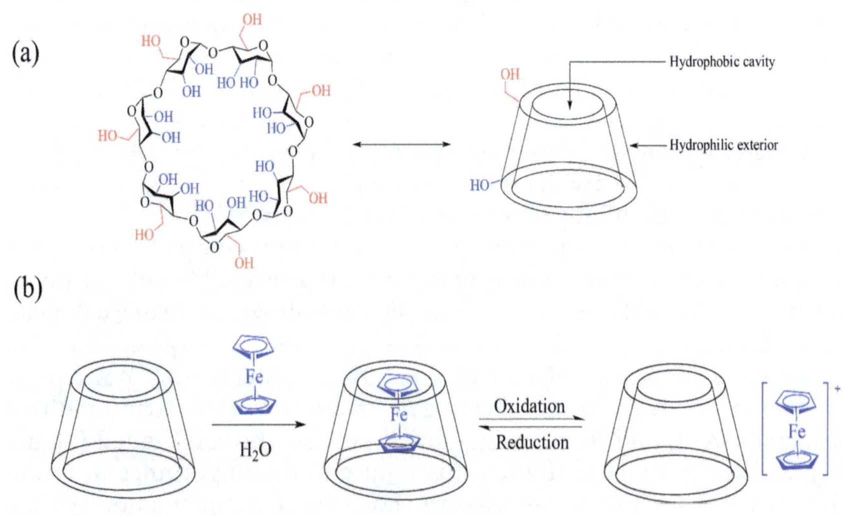

Scheme 5.6 Schematic of (a) a cyclodextrin, (b) encapsulation of ferrocene within the hydrophobic core of the cyclodextrin.

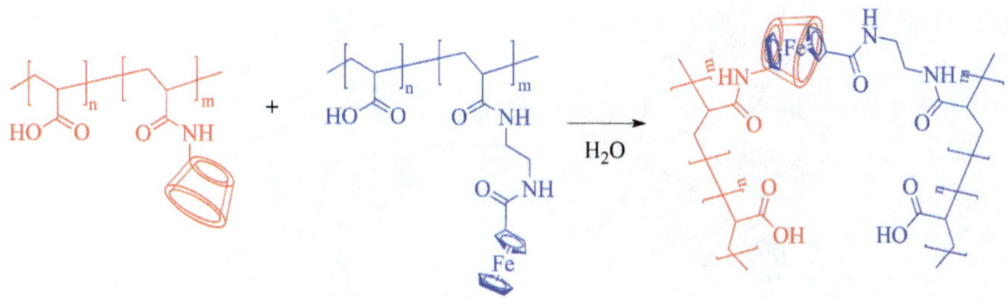

Scheme 5.7 Schematic of the hydrogelation of β-cyclodextrin-modified poly(acrylic acid) with ferrocene-modified poly(acrylic acid).

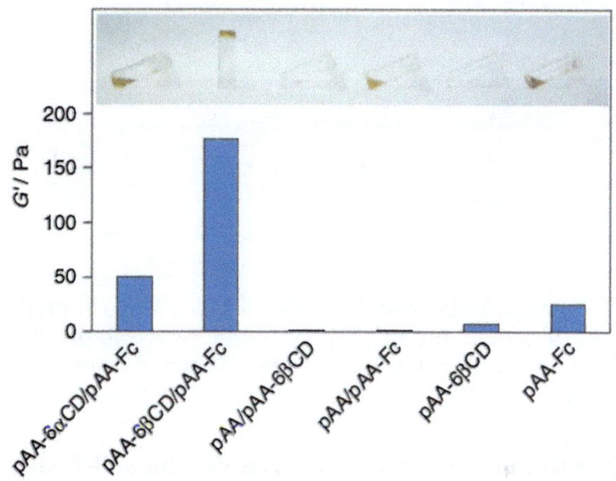

Figure 5.33 Storage elastic moduli show evidence of hydrogelation of α-cyclodextrin-modified poly(acrylic acid)/ferrocene-modified poly(acrylic acid) (pAA-6αCD/pAA-Fc), β-cyclodextrin-modified poly(acrylic acid)/ferrocene-modified poly(acrylic acid) (pAA-6βCD/pAA-Fc), β-cyclodextrin-modified poly(acrylic acid)/poly(acrylic acid) (pAA/pAA-6βCD), ferrocene-modified poly(acrylic acid)/poly(acrylic acid) (pAA/pAA-Fc), β-cyclodextrin-modified poly(acrylic acid) (pAA-6βCD), ferrocene-modified poly(acrylic acid) (pAA-Fc). Reprinted by permission from Macmillan Publishers Ltd: [*Nature Communications*] (ref. 144), copyright (2011).

inclusion complexes of ferrocene with β-cyclodextrin or α-cyclodextrin differ.[139] While one β-cyclodextrin binds one ferrocene to form a 1 : 1 complex, two α-cyclodextrins bind one ferrocene to form a 2 : 1 complex due to the smaller size of the α-cyclodextrin cavity.[139] Perhaps this difference explains the inability of the α-cyclodextrin-modified poly(acrylic acid) to form a gel. Characteristic of ferrocene-based materials, the hydrogel is redox responsive, transforming to "sol" upon addition of an oxidant, NaClO, and reverting to "gel" in the presence of glutathione, a reductant (Figure 5.34).[144] Also, the hydrogel responds to electrochemical redox processes, exhibiting sol–gel phase transitions upon changes in electrochemical redox potential

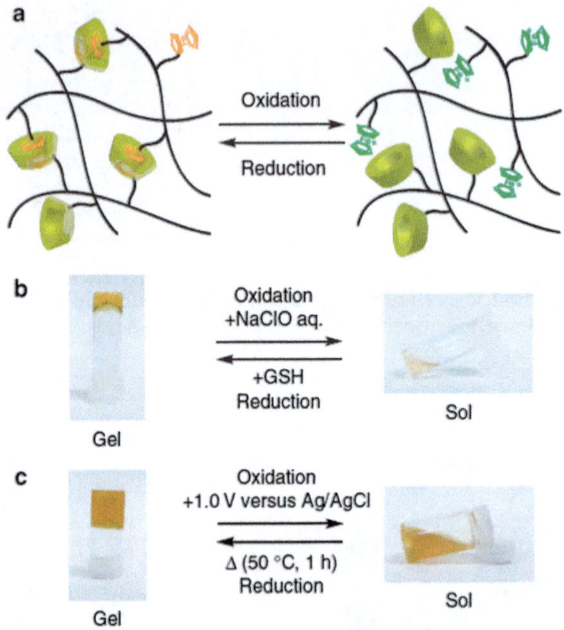

Figure 5.34 (a–c) Redox-responsive behavior of β-cyclodextrin-modified poly(acrylic acid)/ferrocene-modified poly(acrylic acid). Reprinted by permission from Macmillan Publishers Ltd: [*Nature Communications*] (ref. 144), copyright (2011).

(Figure 5.34). Although the authors focus on the self-healing property of this hydrogel, it is obvious that this technology is potentially useful in drug delivery applications.

Another strategy that has been explored in the design of ferrocene-based hydrogels is the encapsulation of the hydrophilic electron-deficient ferrocenium cation in a hydrophilic host, such as the cavity of calixarene, or in an electron-rich host, such as the cavity of pillararene. Such an inclusion-complex-based hydrogel disassembles on reduction to the hydrophobic ferrocene, inclusion complexes that phase transition from gel to sol upon oxidation in an aqueous medium.[144] Rheology experiments on the pillar[6]arene-ferrocenium cation copolymer gel provide evidence of transient networks of supramolecular assembly with a storage elastic modulus that is higher than the loss modulus at a high oscillatory frequency and *vice versa* at low frequency (Scheme 5.8).

The supramolecular gel is responsive to the presence of a competitive host or guest because excess competitive host (per-butylated pillar[6]arene), or competitive guest (cobaltocenium cation) disassembled the pillar[6]arene-ferrocenium cation copolymer gel, yielding a low viscosity solution (Figure 5.35).[145] In contrast, the gel is unresponsive to a non-competing host (per-butylated pillar[5]arene), which has a smaller cavity than pillar[6]arene to encapsulate the ferrocenium cation (Figure 5.35). Furthermore,

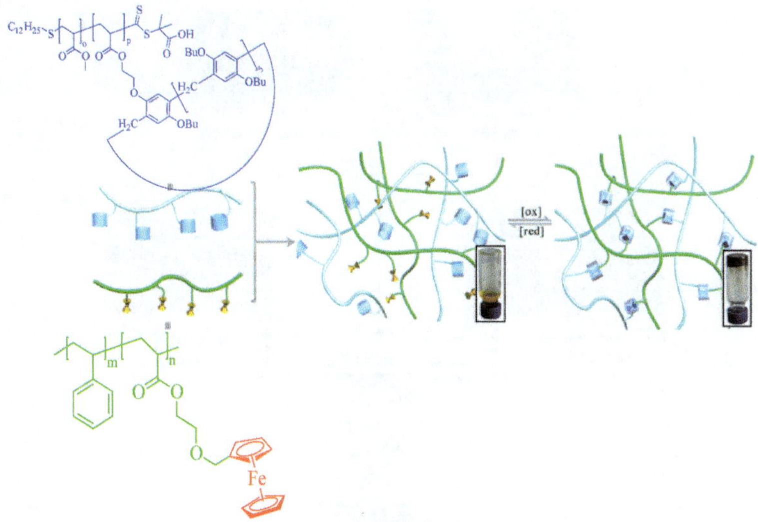

Scheme 5.8 Schematic of the gelation of ferrocene-functionalized polystyrene-*co*-poly(acrylate) copolymer pillar[6]arene-functionalized poly(acrylate) copolymer. Reprinted with permission from ref. 145. Copyright (2015) American Chemical Society.

a redox-responsive behavior was evidenced by the phase transition of the gel to free-flowing solution upon adding an excess amount of hydrazine, which reduces the ferrocenium cation to ferrocene, disassembling the pillar[6]arene-ferrocenium networks.[145] The responsive behavior is reversible because the supramolecular network re-assembled on oxidation of ferrocene to ferrocenium using $Ag\{B[C_6H_3(CF_3)_2]_4\}$ (Figure 5.35).

While ferrocenium cation-pillar[6]arene gelation in organic solvents[146,147] is fascinating, it is attractive to explore the development of ferrocene-pillar[6]arene-based hydrogels in aqueous solutions.[148–157] Indeed, hydrogels, a class of 3D hydrophilic crosslinked polymers, are increasingly being investigated in drug delivery. A fundamental property of hydrogels is their outstanding ability to swell and shrink in aqueous solution, a characteristic that is essential to their structure and biomedical application.[158] One research area focuses on controlling the swelling–shrinking behavior, a goal, if realized, should afford smart hydrogels for drug delivery applications.

Ni *et al.* designed ferrocene-pillar[6]arene-based hydrogels with outstanding swelling–shrinking behavior that smartly responds to temperature, pH, redox, and competitive guests *via* the formation and dissociation of ferrocene-pillar[6]arene inclusion complexes.[148] Their approach involves the synthesis of a series of ferrocene-functionalized cross-linked random copolymers with varying (0–20%) molar ratio of ferrocene (Scheme 5.9).[148] In pure water, the swelling ratio of disc-shaped samples of these copolymers decreases with an increase in the molar ratio of ferrocene.[148] In an aqueous solution containing hydrophilic carboxylate-functionalized pillar[6]arene

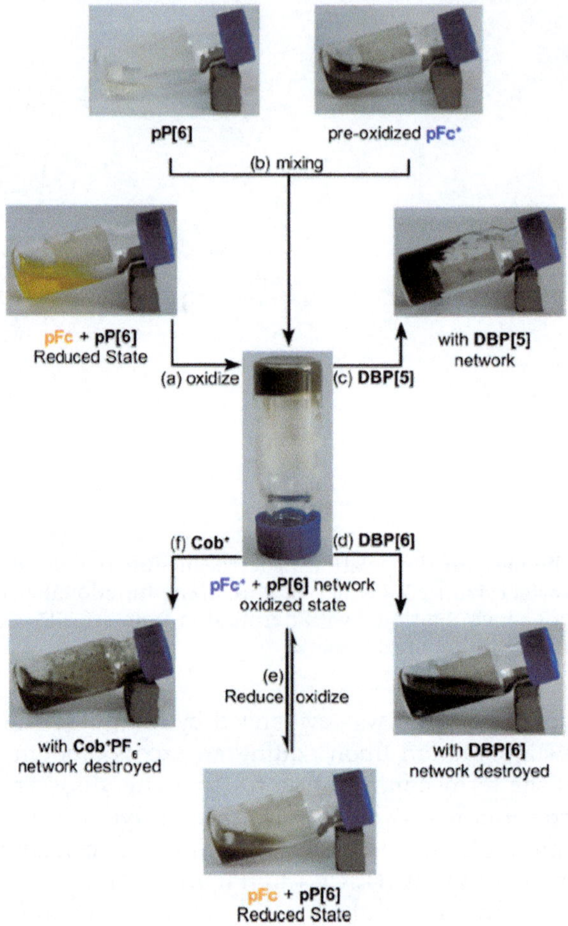

Figure 5.35 (a–f) Redox-triggered gelation of ferrocene- or ferrocenium-functionalized polystyrene-*co*-poly(acrylate) copolymer (pFc or pFc^+, respectively) and pillar[6]arene-functionalized poly(acrylate) copolymer (pP[6]). The gel responds to a competitive host, per-butylated pillar[6]arene (DBP[6]) or per-butylated pillar[6]arene (DBP[5]), or guest, cobaltocenium cation (Cob^+PF6^-). Reprinted with permission from ref. 145. Copyright (2015) American Chemical Society.

with a hydrophobic cavity, the ferrocene-functionalized copolymers drastically swell (Figure 5.36).[148] The encapsulation of ferrocene within the hydrophobic cavity of pillar[6]arene and the resulting enhanced hydrophilicity of the copolymer, as well as the strong electrostatic repulsion between the negatively charged carboxylated polymer chains, drive the samples to swell.[148] The molar ratio of ferrocene in the copolymer tunes the swelling ratio due to an interplay of the ferrocene hydrophobicity and electrostatic repulsion from the carboxylate groups (Figure 5.36). Increasing the molar ratio from 0 to 10% leads to a dramatic increase in the swelling ratio, but a molar ratio

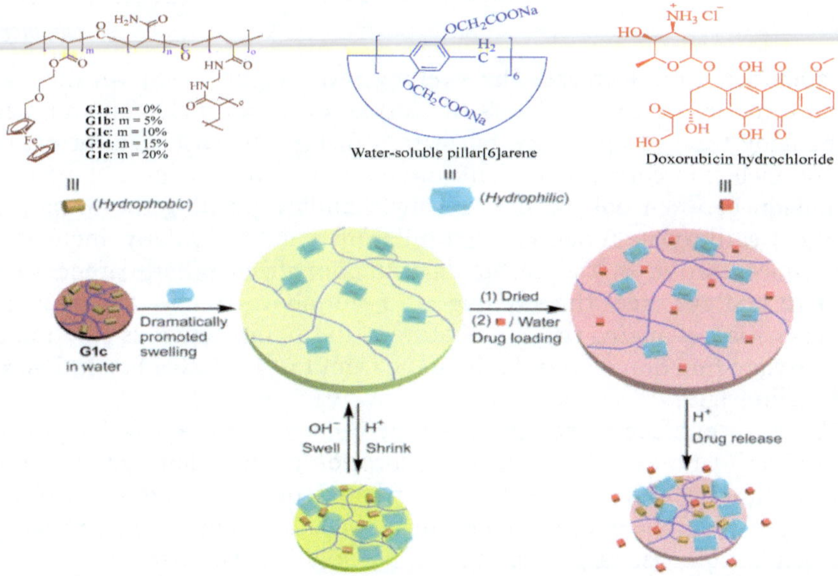

Scheme 5.9 Schematic of the pH-controlled swelling–shrinking and drug release of ferrocene polymer-pillar[6]arene inclusion complexes. Reprinted with permission from ref. 148. Copyright (2016) American Chemical Society.

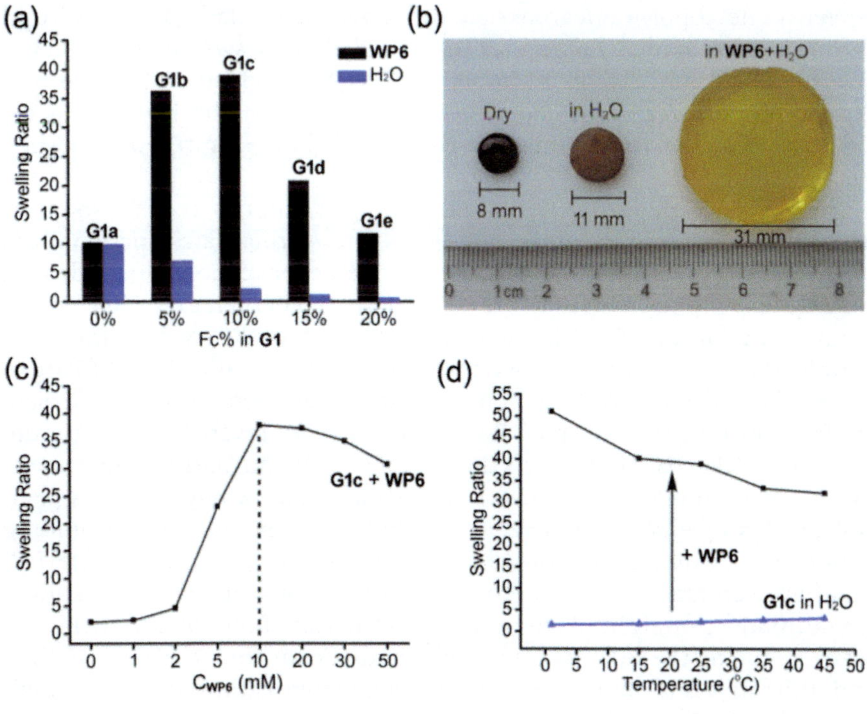

Figure 5.36 (a) The molar ratio of ferrocene in the polymer tunes the swelling ratio, (b) dry polymer hydrogels swell in pillar[6]arene solution, (c) the concentration of pillar[6]arene tunes the swelling ratio (d) temperature tunes the swelling ratio. Reprinted with permission from ref. 148. Copyright (2016) American Chemical Society.

greater than 10% decreases the swelling ratio (Figure 5.36). An increase from 0 to 10% ferrocene favors the formation of an inclusion complex, enhancing the hydrophilicity, but with a further increase, the formation of the inclusion complex was unfavorable due to increasing electrostatic repulsion between polymeric ferrocenyl handles grasping the negatively charged carboxylate-functionalized pillar[6]arene.[148] Similarly, increasing the concentration of the carboxylate-functionalized pillar[6]arene from 0 to 10 mM enhances the swelling ratio, but above 10 mM, the hydrogel shrinks. Also, changing the temperature of the 10 mM aqueous solution of carboxylate-functionalized pillar[6]arene from 1 to 45 °C shrinks the hydrogel (Figure 5.36).[148]

The ferrocene copolymer absorbs water and the carboxylate-functionalized pillar[6]arene swells, with the absorption of the pillar[6]arene being the rate limiting step.[148] Thus, in a concentrated solution of carboxylate-functionalized pillar[6]arene, the copolymer swells at a higher rate, reaching the equilibrium size faster than in a dilute solution. For instance, in 5, 10, or 50 mM, the copolymer swells to the equilibrium size in ~180, 120, or 36 hours, respectively.[148] If the pH of the swelling carboxylate-functionalized pillar[6]arene medium is altered, the ferrocene copolymer/carboxylate-functionalized pillar[6]arene hydrogel or ferrocene copolymer hydrogel smartly responds, either swelling or shrinking.[148] For instance, the 35 mm-diameter ferrocene copolymer/carboxylate-functionalized pillar[6]arene hydrogel shrinks to 21 mm in 12 hours after HCl is added to the swelling medium but reverts to a 32 mm-disc in 12 hours after NaOH is added.[148] The hydrogel derived from the ferrocene copolymer responds irreversibly because the gel swells from 12 mm to 21 mm when HCl is added but is unresponsive when NaOH is added.[148]

The presence of ferrocene in the hydrogels implies redox-responsive behavior. Therefore, the ferrocene copolymer/carboxylate-functionalized pillar[6]arene hydrogel, as well as the ferrocene copolymer hydrogel, smartly respond to changes in the redox potential of their external environment. The addition of $AgNO_3$, an oxidizing agent, to the swelling medium, shrinks the ferrocene copolymer/carboxylate-functionalized pillar[6]arene hydrogel from 35 mm to 20 mm, but swells the ferrocene copolymer hydrogel from 13 mm to 23 mm.[148] Both responses are reversible because the addition of $NH_2NH_2{\cdot}H_2O$ reverts the hydrogel to 28 mm and 16 mm, respectively. The seemingly contradictory behavior of the two hydrogels is plausible, as Wang *et al.* previously proved that ferrocene has a lower binding affinity for carboxylate-functionalized pillar[6]arene ($K_a = (1.27 \pm 0.42) \times 10^5$) than ferrocenium ($K_a = (8.68 \pm 0.72) \times 10^7$).[149] Consequently, the cationic ferrocenium strongly binds the anionic carboxylate-functionalized pillar[6]arene, reducing the electrostatic repulsion that contributed to the swelling, and ultimately shrinking the ferrocene copolymer/carboxylate-functionalized pillar[6]arene hydrogel. In the ferrocene-copolymer hydrogel, oxidation to the hydrophilic ferrocenium cation increases swelling. Also, the presence of a competitive guest that binds the carboxylated pillar[6]arene stronger than ferrocene, such as paraquats, tunes the swelling–shrinking behavior of the ferrocene copolymer/carboxylate-functionalized pillar[6] arene hydrogel. Adding *N,N′*-dimethyl-4,4′-bipyridinium iodide or *N,N′*-dimethyl-4,4′-bipyridinium bromide shrinks the hydrogels from 36 mm to 21 mm or 25 mm, respectively. Interestingly, the ferrocene copolymer hydrogel is unresponsive to these competitive guests.

Apart from the fundamental interest in the swelling–shrinking behavior of these hydrogels, Ni and coworkers investigated the *in vitro* pH-responsive behavior.[148] They successfully encapsulated the anticancer drug doxorubicin hydrochloride by immersing the dried ferrocene copolymer/carboxylated pillar[6]arene hydrogel in an aqueous solution of the drug and triggered the release *via* control of the pH of the release medium.[148] Lowering the pH protonates the carboxylate groups and shrinks the hydrogels, releasing the drug. Evidently, adjusting the pH to 7 releases 5% of the drug within 36 hours; decreasing the pH to 4 releases 28% within 24 hours, and further reducing the pH to 2 releases 80% in 0.5 hours (Figure 5.37).[148] This pH-responsive ferrocene copolymer/carboxylate-functionalized pillar[6]arene hydrogel is, therefore, an attractive drug release technology for targeting cancer cells, which could trigger the release of encapsulated drugs, since the microenvironment of these cells is acidic.

The concept of ferrocene–macrocyclic guest inclusion complexes enables the construction of polymeric nanoscale vesicles as well. For instance, a cyclodextrin-functionalized polystyrene and ferrocene-functionalized poly(ethylene glycol) form into a supramolecular polymer *via* host–guest interactions (Scheme 5.10).[159] In an aqueous solution, this polymer self-assembled into voltage-responsive vesicles (Scheme 5.10).[159] TEM reveals a dark ~19 nm peripheral wall enclosing a light central core,[159] a typical vesicular

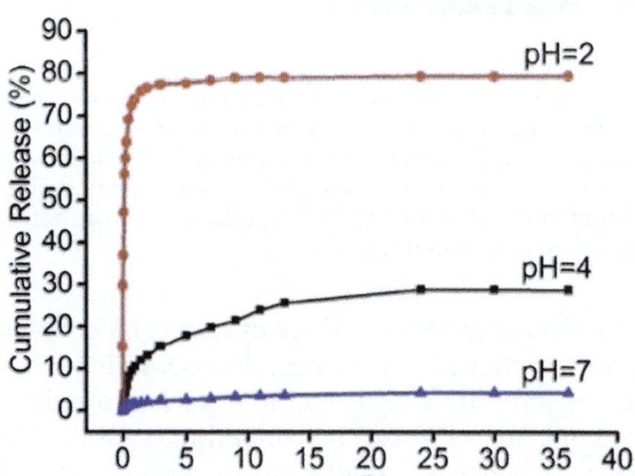

Figure 5.37 pH-Controlled release of doxorubicin hydrochloride from a ferrocene polymer. Reprinted with permission from ref. 148. Copyright (2016) American Chemical Society.

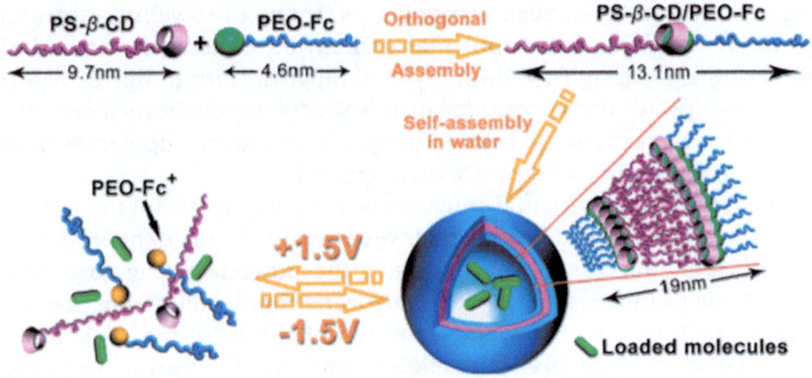

Scheme 5.10 Schematic of the self-assembly of vesicles from a supramolecular polymer from cyclodextrin-functionalized polystyrene (PS-β-CD) and ferrocene-functionalized poly(ethylene glycol) (PEO-Fc). PEO-Fc^+ is ferrocenium-functionalized poly(ethylene glycol). Reprinted with permission from ref. 159. Copyright (2010) American Chemical Society.

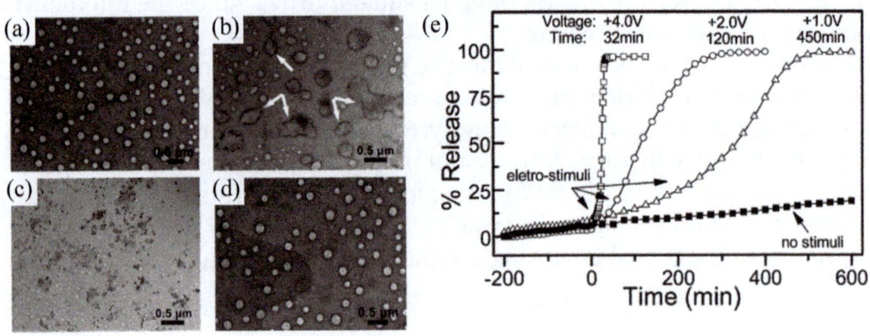

Figure 5.38 TEM micrographs reveal the voltage-controlled assembly and disassembly of vesicles as well as voltage-controlled release of Rhodamine B from vesicles. (a) Original vesicles, (b) vesicles after applying +1.5 V for two hours, (c) vesicles after applying +1.5 V for five hours, (d) vesicles after −1.5 V for five hours, (e) voltage-triggered controlled release of Rhodamine B. Reprinted with permission from ref. 159. Copyright (2010) American Chemical Society.

morphology. The reversible redox activity of ferrocene enables the vesicles to smartly respond to changes in voltage, disassembling when exposed to an oxidative +1.5 V stimulus and reassembling when a reductive −1.5 V is applied (Figure 5.38). These ferrocene-containing supramolecular vesicles function as smart drug delivery platforms, encapsulating and releasing a model drug, Rhodamine B, upon changes in voltage (Figure 5.38).

Ferrocene host–guest chemistry is versatile, enabling the formation of inclusion complexes with various macrocyclic molecules. Numerous examples demonstrate this, including supramolecular diblock copolymer vesicles constructed *via* host–guest interactions of a pillar[6]arene-functionalized poly(*N*-isopropylacrylamide) polymer with a ferrocene-functionalized methoxypoly(ethylene glycol) polymer (Scheme 5.11).[151] The diblock copolymer is hydrophilic at 25 °C since the two polymer blocks are hydrophilic but transforms into an amphiphilic diblock copolymer at 37 °C because the poly(*N*-isopropylacrylamide) block is hydrophobic above its cloud point, 32 °C.[151] TEM and dynamic light scattering show evidence of the formation of supramolecular vesicles at 37 °C (Figure 5.39). The TEM images reveal vesicles with a wall thickness of 4.6 nm and an average diameter of 98 nm, which is close to the average diameter of 104 nm obtained with DLS (Figure 5.39). Typical of poly(*N*-isopropylacrylamide) polymers, the supramolecular polymer vesicles are thermal-responsive. TEM and DLS reveal that the vesicles

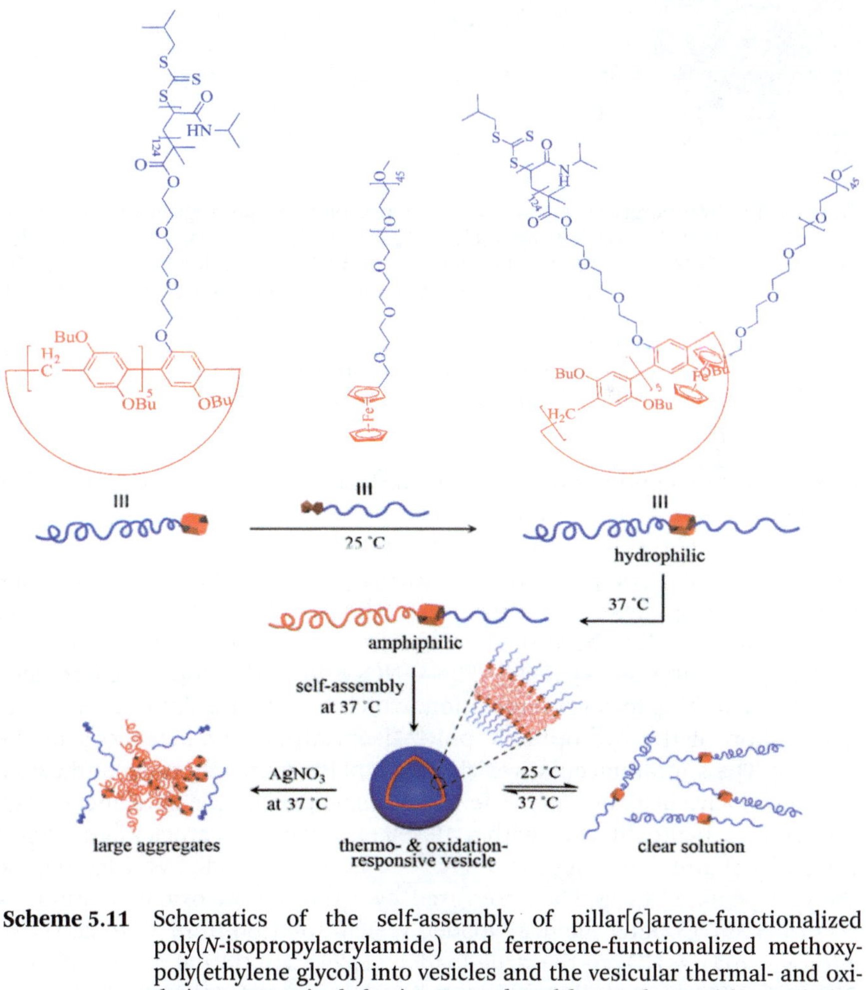

Scheme 5.11 Schematics of the self-assembly of pillar[6]arene-functionalized poly(*N*-isopropylacrylamide) and ferrocene-functionalized methoxypoly(ethylene glycol) into vesicles and the vesicular thermal- and oxidation-responsive behavior. Reproduced from ref. 151 with permission from The Royal Society of Chemistry.

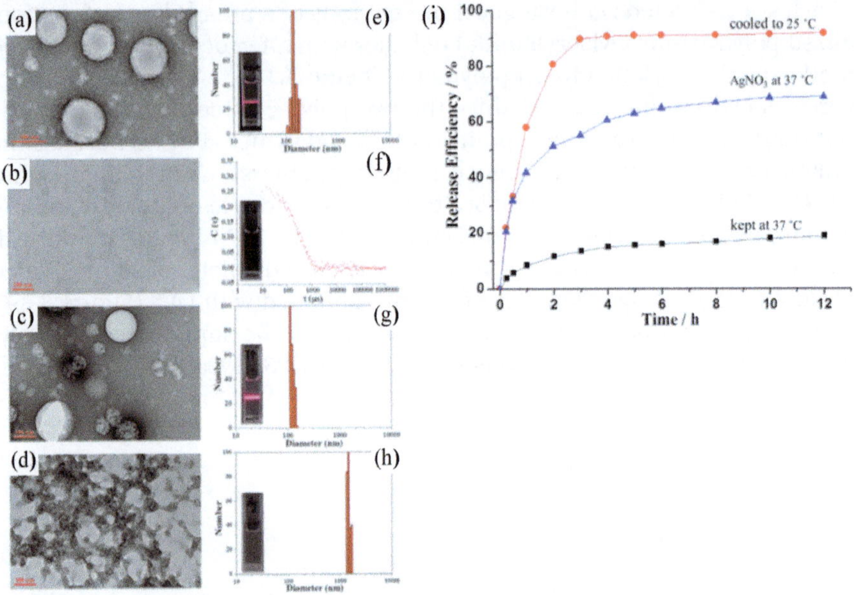

Figure 5.39 Temperature- and oxidation-controlled morphological changes (a–h) and doxorubicin hydrochloride release (i) in supramolecular vesicles. TEM micrographs of vesicles (a) at 37 °C, (b) cooled from 37 to 25 °C, (c) heated from 25 to 37 °C, (d) upon addition of $AgNO_3$ at 37 °C. Vesicle size distribution (e) at 37 °C, (f) cooled from 37 to 25 °C, (g) heated from 25 to 37 °C, (h) upon addition of $AgNO_3$ at 37 °C. (i) Temperature- and oxidation-controlled release of doxorubicin hydrochloride. Reproduced from ref. 151 with permission from The Royal Society of Chemistry.

disassemble on cooling from 37 °C to 25 °C (Figure 5.39)[151] since the hydrophobic poly(*N*-isopropylacrylamide) becomes hydrophilic at 25 °C. On heating the solution, the disassembled vesicles aggregated, reforming the vesicles as evidenced by TEM and DLS experiments (Figure 5.39). Furthermore, the vesicles are responsive to an oxidative stimulus, because the addition of $AgNO_3$ disassembled the vesicles, transforming the hollow walled spheres into irregular large aggregates.[151] The $AgNO_3$ oxidizes hydrophobic ferrocene to the hydrophilic ferrocenium cation, which becomes hydrated, leading to aggregation of the hydrophobic poly(*N*-isopropylacrylamide) block of the vesicles. The supramolecular vesicles are a platform for thermal- and oxidation-responsive drug delivery. The vesicles encapsulate the anticancer drug, doxorubicin hydrochloride, with a 48.7% encapsulation efficiency as determined by fluorescence spectroscopy.[151] *In vitro*, the loaded vesicles release the encapsulated drug when triggered by a thermal or oxidative stimulus (Figure 5.39). Cooling the drug-loaded vesicle solution from 37 to 25 °C, or adding 4 mM $AgNO_3$ to the drug-loaded vesicle solution at 37 °C, releases 91.8 or 68.9%, respectively, within two hours compared with only 18.9% from an unstimulated vesicle (Figure 5.39).[151]

5.4.2 Stimulus-responsive Coordination Drug Delivery Systems

Nature exploits the dynamics of metal coordination bonds to control the mechanical properties of biomaterials.[160–163] Inspired by this, materials scientists have explored the coordination chemistry of transition metals, such as copper, iron, and cobalt, to construct stimuli-responsive hydrogels. A fundamental understanding of the principle of Hard and Soft Acid and Base theory assists in the design of these materials as it allows scientists to generalize and predict the selective binding of ligands to transition metals of different oxidation states. Exemplarily, the hard Lewis acid Co^{3+} prefers to form strong coordination bonds with hard Lewis bases, such as ligands with O- and N-donor groups. Therefore, reduction to Co^{2+}, an intermediate Lewis acid, is expected to weaken the coordination bonds. A coordination polymer hydrogel assembled *via* the Co^{3+} coordination-bond-induced cross-linking of histidine-functionalized poly(ethylene glycol) transforms from an elastic solid to a viscoelastic liquid upon oxidation of the Co^{3+} to Co^{2+}.[164] Recently, $Co^{2+/3+}$-, $Fe^{2+/3+}$-, or $Cu^{+/2+}$-triggered sol–gel phase transitions in a coordination polymer hydrogel have emerged as a strategy[164–174] to allow for the design of stimulus-responsive drug delivery technologies. For instance, Fe^{3+} induces the cross-linking of four-armed poly(caprolactone)-*block*-poly(acrylic acid) star polymers to form a smart, light-responsive nanogel.[174] The Fe^{3+} coordinates to the carboxyl groups of the acrylic acid.[174] Exposure of the nanogel to laser light in the presence of lactic acid reduces Fe^{3+} to Fe^{2+}, disassembling the cross-links and weakening the strength of the gel as evidenced by a decrease in the storage modulus from 6488 to 113 Pa.[174] The laser-light-triggered de-crosslinking of the gel enables the controlled release of encapsulated doxorubicin hydrochloride. The doxorubicin-loaded nanogel significantly reduces tumor volume better than free doxorubicin (Figure 5.40) and is less likely to cause toxic side effects in healthy cells. The requirement of lactic acid as a photo-reductant to trigger release localizes the drug delivery function of these nanogels to lactic-acid-rich microenvironments, such as cancer cells, and ensures target specificity but limits the scope of application.

A stimulus-responsive coordination polymer hydrogel constructed *via* Fe^{3+}-induced cross-linking of poly(guluronan-*co*-mannuronan), (alginate), and poly(galacturonan), (pectate), is more versatile, requiring only simple visible light to trigger release.[168] This Fe^{3+}-alginate and Fe^{3+}-pectate hydrogel encapsulates drug model compounds, such as Congo Red, folic acid, and chloramphenicol, and releases these cargoes upon visible light irradiation.[168] The Fe^{3+}-carboxylate complex in the hydrogel is excited by visible light, then decays *via* a non-radiative pathway, transferring an electron from the carboxylate to the Fe^{3+}, which reduces to Fe^{2+}.[168] Because Fe^{2+}, an intermediate Lewis acid, weakly interacts with carboxylate ligand, a hard Lewis base, the cross-linking weakens, disassembling the polymer hydrogel, and eventually releasing the encapsulated cargo.

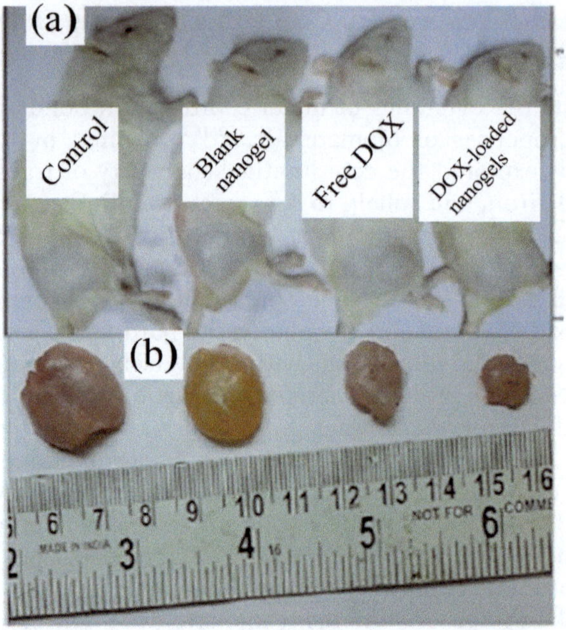

Figure 5.40 Light-triggered release of doxorubicin hydrochloride (DOX) in the presence of a lactic-acid-rich microenvironment of cancer cells. (a) Rat model (b) size of tumor. Reprinted with permission from ref. 174. Copyright (2016) American Chemical Society.

5.5 Biocompatibility of Transition-metal-containing Macromolecules

Whereas the unique chemistry of transition metals enables their use in the design of biomedical technologies, it also triggers deleterious biochemical processes that are harmful to organisms. It is, therefore, imperative that these technologies be safe and selective, localizing their functions at the desired body site to increase the therapeutic window. Indeed, researchers are interested in the safety profile of their biomaterials. For instance, Abd-El-Aziz *et al.* examined the *in vitro* cytotoxicity of their iron-based antimicrobial macromolecules and found them compatible with human epidermal cell lines at their minimum inhibitory concentrations against MRSA and VRE.[106–109] Specifically, human epidermal keratinocyte cells and human foreskin BJ fibroblast cells remain viable after exposure to these iron-containing antimicrobial dendrimers for 24 hours as evidenced by AlamarBlue cell viability assay results.[106,107] Also, confocal laser fluorescence microscopy reveals intact cells after 24 hours' exposure to these iron-containing antimicrobial dendrimers (Figure 5.41).[106] Furthermore, haemolysis of mammalian, specifically sheep, red blood cells is insignificant after a two-hour treatment with these dendrimers even at concentrations higher than their minimum

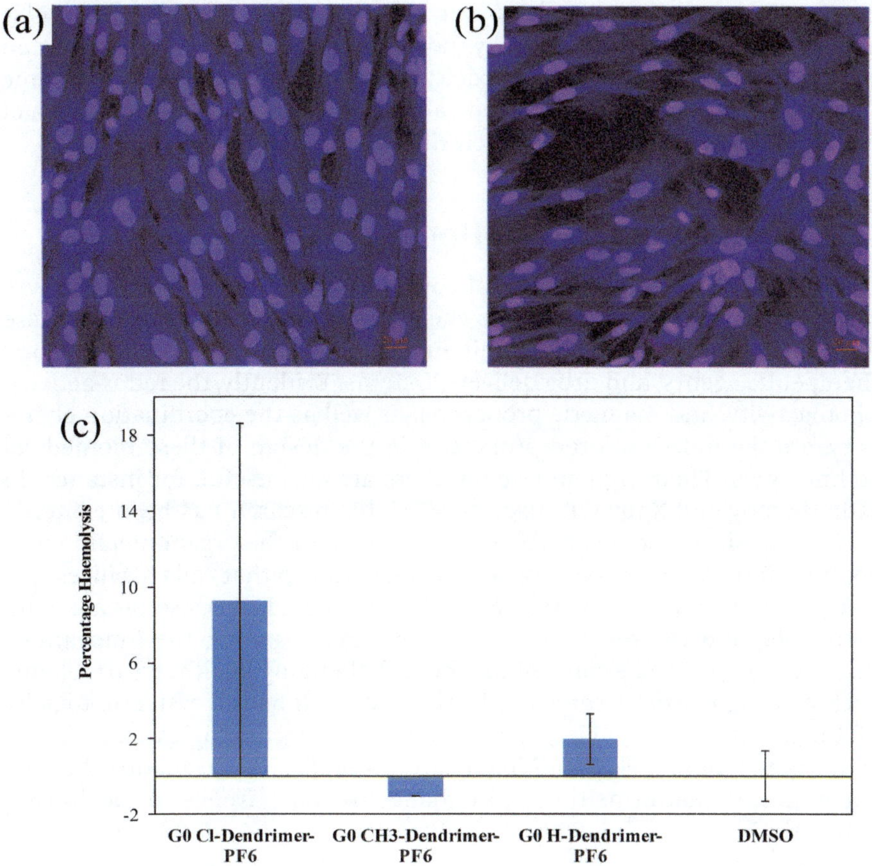

Figure 5.41 (a) and (b) Confocal laser fluorescence microscopy of human BJ fibroblast cells (a) before incubation and (b) after incubation with antimicrobial iron-containing dendrimers at concentrations higher than the minimum inhibitory concentration against MRSA. No remarkable cytotoxicity is evident after incubation. (c) Cytotoxicity of antimicrobial iron-containing dendrimers to sheep red blood cells. Percentage haemolysis of sheep red blood cells treated with 128 μg mL^{-1} of the dendrimers. Reprinted with permission from ref. 106. Copyright (2015) American Chemical Society.

inhibitory concentrations (Figure 5.41).[106] Tang and coworkers proved the biocompatibility of their series of cationic cobaltocenium-containing polymers using poly(2-(methacryloyloxy)ethyl) cobaltocenium-carboxylate.[112] These polymers selectively inhibit the growth of different strains of MRSA with little hemolytic activity against mammalian red blood cells. Also, the prodrug strategy explored in the design of various transition-metal-based macromolecular anticancer therapeutics ensures cytotoxicity of the active drug is localized in cancerous tissues, reducing systemic toxicity.[56,58] Typical of macromolecular therapeutics, transition-metal-containing biosensors, contrasting agents, therapeutics, and drug delivery vectors preferentially accumulate in cancerous tissues by the enhanced permeation and retention effects, thus reducing systemic toxicity.[76] It is, therefore, logical to assume that these macromolecular biomedical technologies are preferred to small molecule drugs that indiscriminately diffuse into healthy cells.

5.6 Conclusion and Outlook

It is evident that transition-metal-containing macromolecules occupy a niche in the biomedical field. The chemistry of the metal centers in these macromolecules is pivotal to their function as contrast agents, sensors, therapeutic agents, and drug delivery systems. Evidently, the redox-activity, photoactivity, and magnetic properties, as well as the coordination chemistry, are the most explored properties in the design of these biomedical technologies. Their high atomic numbers are also useful, for instance, in radiotherapy and X-ray CT. Also, most of this research has been centered on few transition metals, particularly platinum or the organometallic iron complex ferrocene. This narrow focus creates a gap that calls for investigation of other transition metals. Nonetheless, the advances so far are commendable, and the results are promising. For instance, the innovations discussed above tackle some of the most challenging healthcare problems, such as drug-resistant cancer, infections, as well as the systemic toxicity of therapeutics. We envisage that biomedical technologies constructed from transition-metal-containing macromolecules may be among the next generation of platforms that will enhance the effectiveness of healthcare delivery.

References

1. J. A. Lemire, J. J. Harrison and R. J. Turner, *Nat. Rev. Microbiol.*, 2013, **11**, 371.
2. P. Ghalioungui, *Magic and Medical Science in Ancient Egypt*, Hodder and Stoughton, London, 1963.
3. L. Kelland, *Expert Opin. Invest. Drugs*, 2007, **16**, 1009.
4. L. Kelland, *Nat. Rev. Cancer*, 2007, **7**, 573.
5. R. Gracia and D. Mecerreyes, *Polym. Chem.*, 2013, **4**, 2206.
6. A. P. Turner, *Chem. Soc. Rev.*, 2013, **42**, 3184.
7. J. Kim, K. Park, H. Y. Nam, S. Lee, K. Kim and I. C. Kwon, *Prog. Polym. Sci.*, 2007, **32**, 1031.
8. World Health Organization, *Global Report on Diabetes*, World Health Organization, Geneva, 2016.
9. M. Saleem, H. Yu, L. Wang, Zain-ul-Abdin, H. Khalid, M. Akram, N. M. Abbasi and J. Huang, *Anal. Chim. Acta*, 2015, **876**, 9.
10. W. Schuhmann, T. J. Ohara, H. L. Schmidt and A. Heller, *J. Am. Chem. Soc.*, 1991, **113**, 1394.

11. M. Şenel and M. F. Abasıyanık, *Electroanalysis*, 2010, **22**, 1765.
12. M. Şenel, *Synth. Met.*, 2011, **161**, 1861.
13. P. D. Hale, L. I. Boguslavsky, T. Inagaki, H. I. Karan, H. S. Lee, T. A. Skotheim and Y. Okamoto, *Anal. Chem.*, 1991, **63**, 677.
14. O. Reynes, J. Moutet, J. Pecaut, G. Royal and E. Saint-Aman, *New J. Chem.*, 2002, **26**, 9.
15. O. Reynes, G. Royal, E. Chaînet, J. Moutet and E. Saint-Aman, *Electroanalysis*, 2003, **15**, 65.
16. O. Reynes, T. Gulon, J. Moutet, G. Royal and E. Saint-Aman, *J. Organomet. Chem.*, 2002, **656**, 116.
17. D. Astruc, M. Daniel and J. Ruiz, *Chem. Commun.*, 2004, 2637.
18. C. Ornelas, J. Ruiz Aranzaes, E. Cloutet, S. Alves and D. Astruc, *Angew. Chem.*, 2007, **119**, 890.
19. J. R. Aranzaes, C. Belin and D. Astruc, *Angew. Chem., Int. Ed.*, 2006, **45**, 132.
20. C. Valério, J. Fillaut, J. Ruiz, J. Guittard, J. Blais and D. Astruc, *J. Am. Chem. Soc.*, 1997, **119**, 2588.
21. G. Chatelain, A. Meyer, F. Morvan, J. Vasseur and C. Chaix, *New J. Chem.*, 2011, **35**, 893.
22. L. V. Wang and H. Wu, *Biomedical Optics: Principles and Imaging*, John Wiley & Sons, Hoboken, NJ, 2007.
23. S. Gigan, *Nat. Photonics*, 2017, **11**, 14.
24. D. Liu, R. C. Huxford and W. Lin, *Angew. Chem., Int. Ed.*, 2011, **50**, 3696.
25. V. Balzani, G. Bergamini, S. Campagna and F. Puntoriero, Photochemistry and photophysics of coordination compounds: overview and general concept, in *Photochemistry and Photophysics of Coordination Compounds I*, ed. V. Balzani and S. Campagna, Springer-Verlag, Berlin Heidelberg, 2007, p. 1.
26. F. L. Thorp-Greenwood, R. G. Balasingham and M. P. Coogan, *J. Organomet. Chem.*, 2012, **714**, 12.
27. K. E. deKraff, W. S. Boyle, L. M. Burk, O. Z. Zhou and W. Lin, *J. Mater. Chem.*, 2012, **22**, 18139.
28. D. Chen, D. Yang, C. A. Dougherty, W. Lu, H. Wu, X. He, T. Cai, M. E. Van Dort, B. D. Ross and H. Hong, *ACS Nano*, 2017, **11**, 4315.
29. H. Gong, Z. Dong, Y. Liu, S. Yin, L. Cheng, W. Xi, J. Xiang, K. Liu, Y. Li and Z. Liu, *Adv. Funct. Mater.*, 2014, **24**, 6492.
30. Y. Wang, W. Liu and X. Yin, *Chem. Sci.*, 2017, **8**, 3891.
31. W. Liu, Y. Wang, Y. Li, S. Cai, X. Yin, X. He and Y. Zhang, *Small*, 2017, **13**, 1603459.
32. X. Qian, S. Shen, T. Liu, L. Cheng and Z. Liu, *Nanoscale*, 2015, **7**, 6380.
33. J. Li, F. Jiang, B. Yang, X. Song, Y. Liu, H. Yang, D. Cao, W. Shi and G. Chen, *Sci. Rep.*, 2013, **3**, 1998.
34. J. R. Rice, J. L. Gerberich, D. P. Nowotnik and S. B. Howell, *Clin. Cancer Res.*, 2006, **12**, 2248.
35. P. Sood, K. B. Thurmond, J. E. Jacob, L. K. Waller, G. O. Silva, D. R. Stewart and D. P. Nowotnik, *Bioconjugate Chem.*, 2006, **17**, 1270.

36. J. M. Rademaker-Lakhai, C. Terret, S. B. Howell, C. M. Baud, R. F. De Boer, D. Pluim, J. H. Beijnen, J. H. Schellens and J. P. Droz, *Clin. Cancer Res.*, 2004, **10**, 3386.
37. D. P. Nowotnik and E. Cvitkovic, *Adv. Drug Delivery Rev.*, 2009, **61**, 1214.
38. Z. Song, H. Chang, N. Han, Z. Liu, Y. Liu, H. Wang, J. Shao, Z. Wang, H. Gao and J. Yin, *RSC Adv.*, 2017, 7, 19794.
39. J. Hao, Q. Lou, Q. Wei, S. Mei, L. Li, G. Wu, Q. S. Mi, C. Mei and Z. Dong, *J. Biol. Chem.*, 2017, **292**, 4571.
40. J. Tang and S. Zhuang, *Kidney Int.*, 2016, **89**, 267.
41. T. John, N. Lomeli and D. A. Bota, *Behav. Brain Res.*, 2017, **319**, 200.
42. S. G. Chaney, S. L. Campbell, E. Bassett and Y. Wu, *Crit. Rev. Oncol. Hematol.*, 2005, **53**, 3.
43. C. Luo, J. Sun, B. Sun and Z. He, *Trends Pharmacol. Sci.*, 2014, **35**, 556.
44. R. G. Kenny, S. W. Chuah, A. Crawford and C. J. Marmion, *Eur. J. Inorg. Chem.*, 2017, **2017**, 1596.
45. J. B. Zawilska, J. Wojcieszak and A. B. Olejniczak, *Pharmacol. Rep.*, 2013, **65**, 1.
46. D. P. Nowotnik, *Curr. Bioact. Compd.*, 2011, **7**, 21.
47. B. Pelaz, C. Alexiou, R. A. Alvarez-Puebla, F. Alves, A. M. Andrews, S. Ashraf, L. P. Balogh, L. Ballerini, A. Bestetti, C. Brendel, S. Bosi, M. Carril, W. C. W. Chan, C. Chen, X. Chen, X. Chen, Z. Cheng, D. Cui, J. Du, C. Dullin, A. Escudero, N. Feliu, M. Gao, M. George, Y. Gogotsi, A. Grünweller, Z. Gu, N. J. Halas, N. Hampp, R. K. Hartmann, M. C. Hersam, P. Hunziker, J. Jian, X. Jiang, P. Jungebluth, P. Kadhiresan, K. Kataoka, A. Khademhosseini, J. Kopecěk, N. A. Kotov, H. F. Krug, D. S. Lee, C. Lehr, K. W. Leong, X. Liang, M. L. Lim, L. M. Liz-Marzań, X. Ma, P. Macchiarini, H. Meng, H. Möhwald, P. Mulvaney, A. E. Nel, S. Nie, P. Nordlander, T. Okano, J. Oliveira, T. H. Park, R. M. Penner, M. Prato, V. Puntes, V. M. Rotello, A. Samarakoon, R. E. Schaak, Y. Shen, S. Sjöqvist, A. G. Skirtach, M. G. Soliman, M. M. Stevens, H. Sung, B. Z. Tang, R. Tietze, B. N. Udugama, J. S. VanEpps, T. Weil, P. S. Weiss, I. Willner, Y. Wu, L. Yang, Z. Yue, Q. Zhang, Q. Zhang, X. Zhang, Y. Zhao, X. Zhou and W. J. Parak, *ACS Nano*, 2017, **11**, 2313.
48. D. Peer, J. M. Karp, S. Hong, O. C. Farokhzad, R. Margalit and R. Langer, *Nat. Nanotechnol.*, 2007, **2**, 751.
49. V. Wagner, A. Dullaart, A. Bock and A. Zweck, *Nat. Biotechnol.*, 2006, **24**, 1211.
50. F. Caruso, T. Hyeon and V. M. Rotello, *Chem. Soc. Rev.*, 2012, **41**, 2537.
51. C. Agatemor and M. P. Shaver, *Biomacromolecules*, 2013, **14**, 699.
52. H. Maeda, G. Bharate and J. Daruwalla, *Eur. J. Pharm. Biopharm.*, 2009, **71**, 409.
53. M. Oh and C. A. Mirkin, *Nature*, 2005, **438**, 651.
54. W. Lin, W. J. Rieter and K. M. Taylor, *Angew. Chem., Int. Ed.*, 2009, **48**, 650.
55. J. Della Rocca, D. Liu and W. Lin, *Acc. Chem. Res.*, 2011, **44**, 957.
56. C. He, D. Liu and W. Lin, *Biomaterials*, 2015, **36**, 124.

57. C. He, X. Duan, N. Guo, C. Chan, C. Poon, R. R. Weichselbaum and W. Lin, *Nat. Commun.*, 2016, **7**, 12499.
58. C. He, C. Poon, C. Chan, S. D. Yamada and W. Lin, *J. Am. Chem. Soc.*, 2016, **138**, 6010.
59. C. He, D. Liu and W. Lin, *ACS Nano*, 2015, **9**, 991.
60. D. Liu, C. Poon, K. Lu, C. He and W. Lin, *Nat. Commun.*, 2014, **5**, 4182.
61. C. Poon, C. He, D. Liu, K. Lu and W. Lin, *J. Controlled Release*, 2015, **201**, 90.
62. C. Poon, X. Duan, C. Chan, W. Han and W. Lin, *Mol. Pharmaceutics*, 2016, **13**, 3665.
63. W. J. Rieter, K. M. Pott, K. M. Taylor and W. Lin, *J. Am. Chem. Soc.*, 2008, **130**, 11584.
64. K. J. Gagnon, H. P. Perry and A. Clearfield, *Chem. Rev.*, 2012, **112**, 1034.
65. F. Greco and M. J. Vicent, *Adv. Drug Delivery Rev.*, 2009, **61**, 1203.
66. P. Wong, F. Demircioglu, E. Ghazaly, W. Alrawashdeh, M. R. Stratford, C. L. Scudamore, B. Cereser, T. Crnogorac-Jurcevic, S. McDonald, G. Elia, T. Hagemann, H. M. Kocher and K. M. Hodivala-Dilke, *Cancer Cell*, 2015, **27**, 123.
67. A. C. Faber, A. F. Farago, C. Costa, A. Dastur, M. Gomez-Caraballo, R. Robbins, B. L. Wagner, W. M. Rideout III, C. T. Jakubik, J. Ham, E. J. Edelman, H. Ebi, A. T. Yeo, A. N. Hata, Y. Song, N. U. Patel, R. J. March, A. T. Tam, R. J. Milano, J. L. Boisvert, M. A. Hicks, S. Elmiligy, S. E. Malstrom, M. N. Rivera, H. Harada, B. E. Windle, S. Ramaswamy, C. H. Benes, T. Jacks and J. A. Engelman, *Proc. Natl. Acad. Sci. U. S. A.*, 2015, **112**, E1288.
68. R. Kanasty, J. R. Dorkin, A. Vegas and D. Anderson, *Nat. Mater.*, 2013, **12**, 967.
69. S. B. Brown, E. A. Brown and I. Walker, *Lancet Oncol.*, 2004, **5**, 497.
70. J. Xuan, Z. Wang, Y. Chen, D. Liang, L. Cheng, X. Yang, Z. Liu, R. Ma, T. Sasaki and F. Geng, *Angew. Chem.*, 2016, **128**, 14789.
71. S. S. Lucky, K. C. Soo and Y. Zhang, *Chem. Rev.*, 2015, **115**, 1990.
72. L. B. Josefsen and R. W. Boyle, *Met.-Based Drugs*, 2008, 276109.
73. N. J. Patel, Y. Chen, P. Joshi, P. Pera, H. Baumann, J. R. Missert, K. Ohkubo, S. Fukuzumi, R. R. Nani, M. J. Schnermann, P. Chen, J. Zhu, K. M. Kadish and R. K. Pandey, *Bioconjugate Chem.*, 2016, **27**, 667.
74. H. Ali and J. E. Van Lier, *Chem. Rev.*, 1999, **99**, 2379.
75. S. H. Selman, J. A. Hampton, A. R. Morgan, R. W. Keck, A. Balkany and D. Skalkos, *Photochem. Photobiol.*, 1993, **57**, 681.
76. K. Lu, C. He and W. Lin, *J. Am. Chem. Soc.*, 2014, **136**, 16712.
77. K. Lu, C. He and W. Lin, *J. Am. Chem. Soc.*, 2015, **137**, 7600.
78. K. Lu, C. He, N. Guo, C. Chan, K. Ni, R. R. Weichselbaum and W. Lin, *J. Am. Chem. Soc.*, 2016, **138**, 12502.
79. R. Ideta, F. Tasaka, W. Jang, N. Nishiyama, G. Zhang, A. Harada, Y. Yanagi, Y. Tamaki, T. Aida and K. Kataoka, *Nano Lett.*, 2005, **5**, 2426.
80. K. Sugisaki, T. Usui, N. Nishiyama, W. Jang, Y. Yanagi, S. Yamagami, S. Amano and K. Kataoka, *Invest. Ophthalmol. Visual Sci.*, 2008, **49**, 894.

81. P. M. Pereira, S. Silva, J. A. Cavaleiro, C. A. Ribeiro, J. P. Tomé and R. Fernandes, *PLoS One*, 2014, **9**, e95529.
82. R. Chen, J. Zhang, J. Chelora, Y. Xiong, S. V. Kershaw, K. F. Li, P. Lo, K. W. Cheah, A. L. Rogach, J. A. Zapien and C. Lee, *ACS Appl. Mater. Interfaces*, 2017, **9**, 5699.
83. T. Liu, C. Wang, W. Cui, H. Gong, C. Liang, X. Shi, Z. Li, B. Sun and Z. Liu, *Nanoscale*, 2014, **6**, 11219.
84. S. Wang, K. Li, Y. Chen, H. Chen, M. Ma, J. Feng, Q. Zhao and J. Shi, *Biomaterials*, 2015, **39**, 206.
85. T. Liu, C. Wang, X. Gu, H. Gong, L. Cheng, X. Shi, L. Feng, B. Sun and Z. Liu, *Adv. Mater.*, 2014, **26**, 3433.
86. S. S. Chou, B. Kaehr, J. Kim, B. M. Foley, M. De, P. E. Hopkins, J. Huang, C. J. Brinker and V. P. Dravid, *Angew. Chem., Int. Ed.*, 2013, **125**, 4254.
87. G. Delaney, S. Jacob, C. Featherstone and M. Barton, *Cancer*, 2005, **104**, 1129.
88. S. Her, D. A. Jaffray and C. Allen, *Adv. Drug Delivery Rev.*, 2017, **109**, 84.
89. D. R. Wahl and T. S. Lawrence, *Adv. Drug Delivery Rev.*, 2017, **109**, 74.
90. J. Liu, Y. Yang, W. Zhu, X. Yi, Z. Dong, X. Xu, M. Chen, K. Yang, G. Lu, L. Jiang and Z. Liu, *Biomaterials*, 2016, **97**, 1.
91. World Health Organization, *World Health Statistics 2017: Monitoring Health for the SDGs Sustainable Development Goal*, World Health Organization, Geneva, 2017.
92. World Health Organization, *Antimicrobial Resistance: Global Report on Surveillance*, World Health Organization, Geneva, 2014.
93. Jim O'Neill (Chair), Review on Antimicrobial Resistance, *Tackling Drug Resistant Infections Globally: Final Report and Recommendations*, 2016.
94. L. J. Shallcross, S. J. Howard, T. Fowler and S. C. Davies, *Philos. Trans. R. Soc., B*, 2015, **370**, 20140082.
95. P. J. Sadler, H. Li and H. Sun, *Coord. Chem. Rev.*, 1999, **185–186**, 689.
96. N. P. Barry and P. J. Sadler, *ACS Nano*, 2013, **7**, 5654.
97. Z. Guo and P. J. Sadler, *Angew. Chem., Int. Ed.*, 1999, **38**, 1512.
98. G. Gasser and N. Metzler-Nolte, *Curr. Opin. Chem. Biol.*, 2012, **16**, 84.
99. A. S. Abd-El-Aziz, C. Agatemor and N. Etkin, *Biomaterials*, 2017, **118**, 27.
100. P. J. Sadler and Z. Guo, *Pure Appl. Chem.*, 2009, **70**, 863.
101. N. P. Barry and P. J. Sadler, *Chem. Commun.*, 2013, **49**, 5106.
102. Y. Yan, J. Zhang, L. Ren and C. Tang, *Chem. Soc. Rev.*, 2016, **45**, 5232.
103. J. S. McCarthy, T. Rückle, E. Djeriou, C. Cantalloube, D. Ter-Minassian, M. Baker, P. O'Rourke, P. Griffin, L. Marquart, R. H. van Huijsduijnen and J. J. Möhrle, *Malar. J.*, 2016, **15**, 469.
104. C. Pierrot, S. Lafitte, D. Dive, L. Fraisse, J. Brocard and J. Khalife, *Int. J. Parasitol.*, 2005, **35**, 1601.
105. F. Dubar, C. Slomianny, J. Khalife, D. Dive, H. Kalamou, Y. Guérardel, P. Grellier and C. Biot, *Angew. Chem., Int. Ed.*, 2013, **52**, 7690.
106. A. S. Abd-El-Aziz, C. Agatemor, N. Etkin, D. P. Overy, M. Lanteigne, K. McQuillan and R. G. Kerr, *Biomacromolecules*, 2015, **16**, 3694.

107. A. S. Abd-El-Aziz, C. Agatemor, N. Etkin, R. Bissessur, D. Overy, M. Lanteigne, K. McQuillan and R. G. Kerr, *Macromol. Biosci.*, 2017, **17**, 1700020.
108. A. S. Abd-El-Aziz, A. A. Abdelghani, S. K. El-Sadany, D. P. Overy and R. G. Kerr, *Eur. Polym. J.*, 2016, **82**, 307.
109. A. S. Abd-El-Aziz, A. A. Abdelghani, J. K. Pearson, M. K. Awad, D. P. Overy and R. G. Kerr, *Macromol. Chem. Phys.*, 2016, **217**, 987.
110. A. S. Abd-El-Aziz, C. Agatemor, N. Etkin, D. P. Overy and R. G. Kerr, *RSC Adv.*, 2015, **5**, 86421.
111. J. Zhang, J. Yan, P. Pageni, Y. Yan, A. Wirth, Y. P. Chen, Y. Qiao, Q. Wang, A. W. Decho and C. Tang, *Sci. Rep.*, 2015, **5**, 11914.
112. J. Zhang, Y. P. Chen, K. P. Miller, M. S. Ganewatta, M. Bam, Y. Yan, M. Nagarkatti, A. W. Decho and C. Tang, *J. Am. Chem. Soc.*, 2014, **136**, 4873.
113. F. Noor, A. Wüstholz, R. Kinscherf and N. Metzler-Nolte, *Angew. Chem., Int. Ed.*, 2005, **44**, 2429.
114. J. T. Chantson, M. V. V. Falzacappa, S. Crovella and N. Metzler-Nolte, *J. Organomet. Chem.*, 2005, **690**, 4564.
115. H. B. Albada, P. Prochnow, S. Bobersky, J. E. Bandow and N. Metzler-Nolte, *Chem. Sci.*, 2014, **5**, 4453.
116. B. C. Hoffknecht, P. Prochnow, J. E. Bandow and N. Metzler-Nolte, *J. Inorg. Biochem.*, 2016, **160**, 246.
117. H. B. Albada, A. Chiriac, M. Wenzel, M. Penkova, J. E. Bandow, H. Sahl and N. Metzler-Nolte, *Beilstein J. Org. Chem.*, 2012, **8**, 1753.
118. J. T. Chantson, M. Vittoria Verga Falzacappa, S. Crovella and N. Metzler-Nolte, *ChemMedChem*, 2006, **1**, 1268.
119. M. W. Tibbitt, J. E. Dahlman and R. Langer, *J. Am. Chem. Soc.*, 2016, **138**, 704.
120. L. Jiang, P. Vader and R. Schiffelers, *Gene Ther.*, 2017, **24**, 157.
121. Y. Zhang, J. Yu, H. N. Bomba, Y. Zhu and Z. Gu, *Chem. Rev.*, 2016, **116**, 12536.
122. D. Gu, A. J. O'Connor, G. G. H Qiao and K. Ladewig, *Expert Opin. Drug Delivery*, 2017, **14**, 879.
123. H. Sies, *Redox Biol.*, 2015, **4**, 180.
124. D. Becker, M. Dickman, V. Gladyshev, S. Ragsdale and R. Banerjee, *Redox Biochemistry*, John Wiley & Sons, Hoboken, NJ, 2007.
125. K. Ono, T. Akaike, T. Sawa, Y. Kumagai, D. A. Wink, D. J. Tantillo, A. J. Hobbs, P. Nagy, M. Xian, J. Lin and J. M. Fukuto, *Free Radical Biol. Med.*, 2014, 77, 82.
126. M. Schwarzländer, T. P. Dick, A. J. Meyer and B. Morgan, *Antioxid. Redox Signaling*, 2016, **24**, 680.
127. T. C. Jorgenson, W. Zhong and T. D. Oberley, *Cancer Res.*, 2013, **73**, 6118.
128. C. Greening, F. H. Ahmed, A. E. Mohamed, B. M. Lee, G. Pandey, A. C. Warden, C. Scott, J. G. Oakeshott, M. C. Taylor and C. J. Jackson, *Microbiol. Mol. Biol. Rev.*, 2016, **80**, 451.

129. Y. Wang, Y. Luo, F. Xu, Y. Chen and W. Tang, *J. Ind. Eng. Chem.*, 2017, **48**, 66.
130. R. H. Staff, M. Gallei, M. Mazurowski, M. Rehahn, R. Berger, K. Landfester and D. Crespy, *ACS Nano*, 2012, **6**, 9042.
131. P. Shi, Y. Qu, C. Liu, H. Khan, P. Sun and W. Zhang, *ACS Macro Lett.*, 2016, **5**, 88.
132. A. O. Moughton, M. A. Hillmyer and T. P. Lodge, *Macromolecules*, 2012, **45**, 2.
133. L. Thomi, P. Schaefer, K. Landfester and F. R. Wurm, *Macromolecules*, 2016, **49**, 105.
134. S. B. Bankar, M. V. Bule, R. S. Singhal and L. Ananthanarayan, *Biotechnol. Adv.*, 2009, **27**, 489.
135. M. E. Davis and M. E. Brewster, *Nat. Rev. Drug Discovery*, 2004, **3**, 1023.
136. R. Castro, I. Cuadrado, B. Alonso, C. M. Casado, M. Morán and A. E. Kaifer, *J. Am. Chem. Soc.*, 1997, **119**, 5760.
137. A. Harada, Y. Takashima and M. Nakahata, *Acc. Chem. Res.*, 2014, **47**, 2128.
138. A. Harada, Y. Hu, S. Yamamoto and S. Takahashi, *J. Chem. Soc., Dalton Trans.*, 1988, 729.
139. A. Harada and S. Takahashi, *J. Chem. Soc., Chem. Commun.*, 1984, 645–646.
140. A. Harada, *Acc. Chem. Res.*, 2001, **34**, 456.
141. B. González, I. Cuadrado, B. Alonso, C. M. Casado, M. Morán and A. E. Kaifer, *Organometallics*, 2002, **21**, 3544.
142. M. Nakahata, Y. Takashima and A. Harada, *Angew. Chem., Int. Ed.*, 2014, **53**, 3617.
143. B. González, C. M. Casado, B. Alonso, I. Cuadrado, M. Morán, Y. Wang and A. E. Kaifer, *Chem. Commun.*, 1998, 2569.
144. M. Nakahata, Y. Takashima, H. Yamaguchi and A. Harada, *Nat. Commun.*, 2011, **2**, 511.
145. W. Xia, M. Ni, C. Yao, X. Wang, D. Chen, C. Lin, X. Hu and L. Wang, *Macromolecules*, 2015, **48**, 4403.
146. W. Xia, X. Hu, Y. Chen, C. Lin and L. Wang, *Chem. Commun.*, 2013, **49**, 5085.
147. X. Wu, Y. Yu, L. Gao, X. Hu and L. Wang, *Org. Chem. Front.*, 2016, **3**, 966.
148. M. Ni, N. Zhang, W. Xia, X. Wu, C. Yao, X. Liu, X. Hu, C. Lin and L. Wang, *J. Am. Chem. Soc.*, 2016, **138**, 6643.
149. Q. Duan, Y. Cao, Y. Li, X. Hu, T. Xiao, C. Lin, Y. Pan and L. Wang, *J. Am. Chem. Soc.*, 2013, **135**, 10542.
150. L. Peng, A. Feng, M. Huo and J. Yuan, *Chem. Commun.*, 2014, **50**, 13005.
151. S. Wang, C. Yao, M. Ni, Z. Xu, M. Cheng, X. Hu, Y. Shen, C. Lin, L. Wang and D. Jia, *Polym. Chem.*, 2017, **8**, 682.
152. M. J. Hardie, *Supramol. Chem.*, 2002, **14**, 7.
153. L. Zhang, A. Macías, T. Lu, J. I. Gordon, G. W. Gokel and A. E. Kaifer, *J. Chem. Soc., Chem. Commun.*, 1993, 1017.

154. L. Zhang, A. Macias, R. Isnin, T. Lu, G. W. Gokel and A. E. Kaifer, *J. Inclusion Phenom. Macrocyclic Chem.*, 1994, **19**, 361.
155. R. M. Nielson and J. T. Hupp, *Inorg. Chem.*, 1996, **35**, 1402.
156. M. V. Rekharsky, T. Mori, C. Yang, Y. H. Ko, N. Selvapalam, H. Kim, D. Sobransingh, A. E. Kaifer, S. Liu, L. Isaacs, W. Chen, S. Moghaddam, M. K. Gilson, K. Kim and Y. Inoue, *Proc. Natl. Acad. Sci. U. S. A.*, 2007, **104**, 20737.
157. T. Komura, T. Yamaguchi, K. Kura and J. Tanabe, *J. Electroanal. Chem.*, 2002, **523**, 126.
158. W. Hennink and C. Van Nostrum, *Adv. Drug Delivery Rev.*, 2012, **64**, 223–236.
159. Q. Yan, J. Yuan, Z. Cai, Y. Xin, Y. Kang and Y. Yin, *J. Am. Chem. Soc.*, 2010, **132**, 9268.
160. C. N. Schmitt, Y. Politi, A. Reinecke and M. J. Harrington, *Biomacromolecules*, 2015, **16**, 2852.
161. M. Enke, S. Bode, J. Vitz, F. H. Schacher, M. J. Harrington, M. D. Hager and U. S. Schubert, *Polymer*, 2015, **69**, 274.
162. M. J. Sever, J. T. Weisser, J. Monahan, S. Srinivasan and J. J. Wilker, *Angew. Chem.*, 2004, **116**, 454.
163. E. Degtyar, M. J. Harrington, Y. Politi and P. Fratzl, *Angew. Chem., Int. Ed.*, 2014, **53**, 12026.
164. S. V. Wegner, F. C. Schenk, S. Witzel, F. Bialas and J. P. Spatz, *Macromolecules*, 2016, **49**, 4229.
165. F. Peng, G. Li, X. Liu, S. Wu and Z. Tong, *J. Am. Chem. Soc.*, 2008, **130**, 16166.
166. R. D. Harris, J. T. Auletta, S. A. M. Motlagh, M. J. Lawless, N. M. Perri, S. Saxena, L. M. Weiland, D. H. Waldeck, W. W. Clark and T. Y. Meyer, *ACS Macro Lett.*, 2013, **2**, 1095.
167. J. T. Auletta, G. J. LeDonne, K. C. Gronborg, C. D. Ladd, H. Liu, W. W. Clark and T. Y. Meyer, *Macromolecules*, 2015, **48**, 1736.
168. G. E. Giammanco, C. T. Sosnofsky and A. D. Ostrowski, *ACS Appl. Mater. Interfaces*, 2015, **7**, 3068.
169. G. E. Giammanco and A. D. Ostrowski, *Chem. Mater.*, 2015, **27**, 4922.
170. R. P. Narayanan, G. Melman, N. J. Letourneau, N. L. Mendelson and A. Melman, *Biomacromolecules*, 2012, **13**, 2465.
171. Z. Jin, G. Güven, V. Bocharova, J. Halámek, I. Tokarev, S. Minko, A. Melman, D. Mandler and E. Katz, *ACS Appl. Mater. Interfaces*, 2012, **4**, 466.
172. B. J. Kim, H. Cheong, B. H. Hwang and H. J. Cha, *Angew. Chem., Int. Ed.*, 2015, **54**, 7318.
173. H. Lee, J. H. Lee, S. Kang, J. Y. Lee, G. John and J. H. Jung, *Chem. Commun.*, 2011, **47**, 2937.
174. S. Panja, G. Dey, R. Bharti, P. Mandal, M. Mandal and S. Chattopadhyay, *Chem. Mater.*, 2016, **28**, 8598.

CHAPTER 6

Transition-metal-containing Macromolecules as Materials for Carbon Dioxide Capture, Catalysis, and Data Storage

6.1 Introduction

Transition metals are indispensable to the function and structure of many materials. This realization, as well as the serendipitous discovery of an exceptionally stable transition-metal-containing organometallic molecule, ferrocene, fosters the idea of incorporating transition metals into macromolecules with the goal of expanding knowledge, addressing a myriad of problems or improving the quality of life. Indeed, the chemistry of transition metals enables their applications in diverse fields including environmental cleanup, information storage, and catalysis.[1,2] For instance, many transition metals have rich redox chemistry,[3] a ubiquitous reaction in industry, biology, and geology. Also, the coordination chemistry of transition metals drives the chelation of diverse organic ligands with implications for environmental cleanup. Many engineering fields, especially information technology, rely on magnetic materials; some transition metals are the spin carrier in these materials.[4] The incorporation of transition metals into macromolecules, therefore, provides a single platform where the attractive properties of organic polymers complement the unique chemistries of transition metals to realize functional materials with wide-ranging applications. This chapter discusses frontline developments in the potential of

Polymer Chemistry Series No. 27
Macromolecules Incorporating Transition Metals: Tackling Global Challenges
By Alaa S. Abd-El-Aziz, Christian Agatemor and Wai-Yeung Wong

Published by the Royal Society of Chemistry, www.rsc.org

transition-metal-containing macromolecules as technologies to sequester carbon dioxide (CO_2) from the environment, store information, and sustainably promote chemical reactions.

6.2 As Carbon Dioxide Capture Platforms

Almost all scientists agree on the reality and adverse impact of global warming (Figure 6.1).[5–11] Empirical evidence convincingly implicates increasing levels of greenhouse gases, especially carbon dioxide (CO_2), in the current global warming phenomenon (Figure 6.1).[5–13] Scientists are confident that global warming threatens the biosphere and could disrupt the food chain and biodiversity, with negative impacts on ecosystems, quality of life, and the global economy.[12] To achieve sustainable development, policymakers and scientists have proposed the reduction of CO_2 emission, expressly advocating the adoption of energy-efficient technologies, energy conservation, and the use of low-carbon-footprint energy technologies (Figure 6.1).[14] Whereas these measures may lower future emissions, they do not mitigate the impact of previous emissions. Therefore, technologies that sequester CO_2 from the environment, or capture it from industrial emissions, are an integral component of the strategy to combat global warming.

CO_2 sequestering and capture technologies include solid-state adsorbents, such as ion exchange membranes and solid amines, or aqueous alkaline scrubber systems. Amine-based systems are an efficient and low-cost chemisorption technology but they face the challenge of an inefficient and energy consuming regeneration step due to the high temperatures (100–150 °C) required to desorb the CO_2 from the absorbent.[5,15] Researchers, therefore, seek low-cost technologies that combine efficient CO_2 capture with facile desorption. Porous coordination polymers, also known as metal–organic frameworks (MOFs), are a promising platform for CO_2 capture, being intrinsically sorptive due to their porosity and high surface area.[15–26] By design, weak basic amines, which sequester CO_2 *via* physisorption, can be grafted onto MOFs to create platforms that could efficiently capture and facilely release CO_2. Many researchers have demonstrated the enhanced CO_2 capture property of amine-functionalized solid-state materials.[15–19] Shimizu *et al.*, for instance, constructed an amine-functionalized porous zinc-aminotriazolato-oxalate MOF (Figure 6.2) that efficiently sequesters CO_2 at low pressures.[15,18] The spatial arrangement of atoms in this MOF creates a 3D network that has amine-lined cubic-shaped pores of approximately 4.0 × 4.0 × 4.0 Å size and 31% free volume (Figure 6.2).[18] After activation, this MOF sequesters CO_2 rapidly at low pressure, with the sorption profile fitting a reversible Type 1 isotherm at 273 K. The ability to sequester CO_2 at low pressure is desirable, since the gas is typically present at low pressure in some industrial emissions. Shimizu and coworkers noted the selectivity of this MOF towards CO_2 in the presence of other gases, such as N_2, Ar, and H_2.[18] The small pore size of the MOF precludes the adsorption of gases, but the favorable interaction of CO_2 with the host lattice results

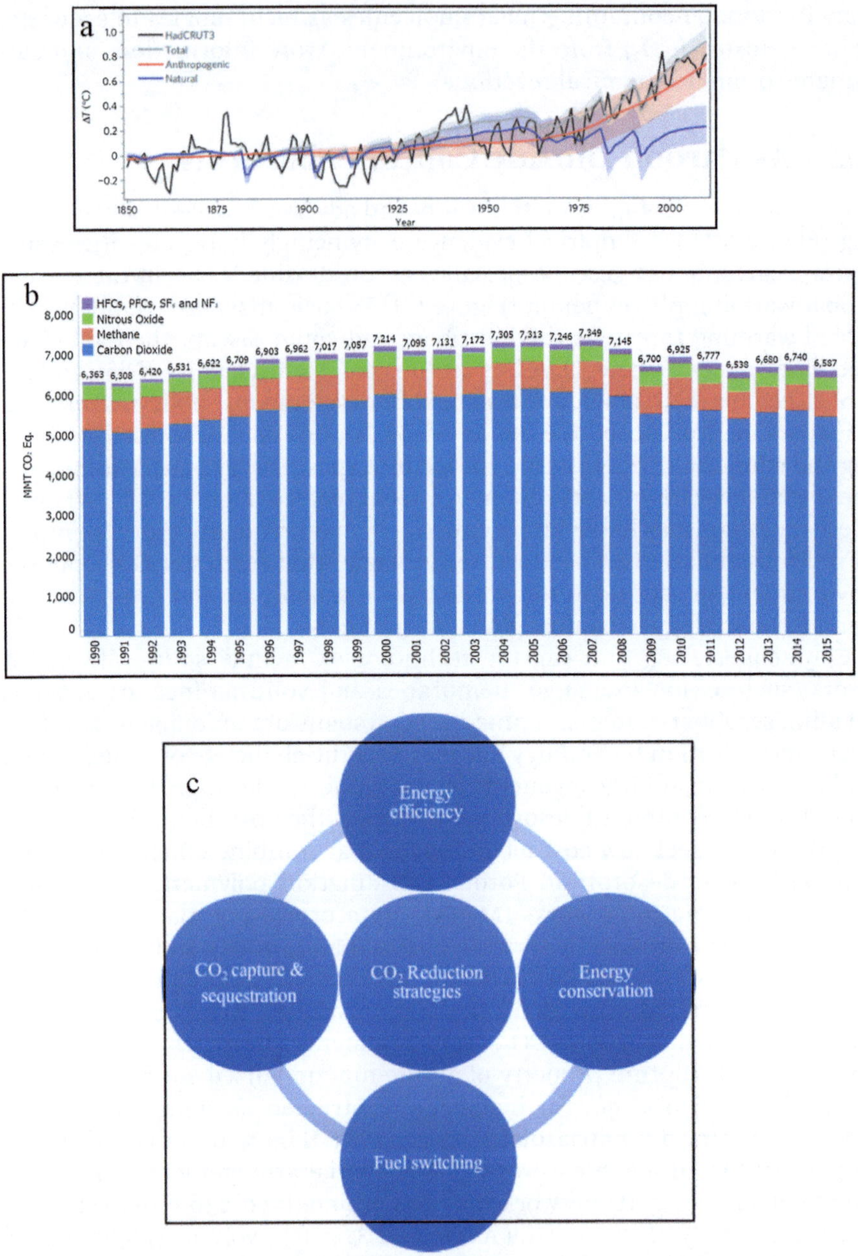

Figure 6.1 (a) Global warming evidenced by a rising global temperature; (b) CO_2 is the most emitted greenhouse gas in the U.S.; (c) strategies to reduce the CO_2 footprint. Figure sources: (a) reprinted by permission from Macmillan Publishers Ltd: [*Nature Geoscience*] (ref. 9), copyright (2012). (b) Courtesy of United States Environmental Protection Agency. Inventory of U.S. greenhouse gas emissions and sinks: 1990–2015 (www.epa.gov/sites/production/files/2017-02/documents/2017_complete_report.pdf).

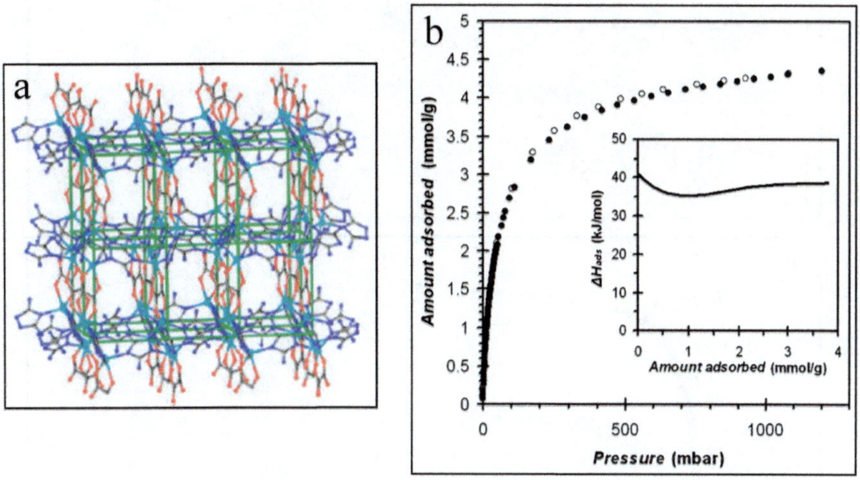

Figure 6.2 (a) Three-dimensional structure of amine-functionalized zinc-aminotriazolato-oxalate MOF (cyan: Zn, blue: N, red: O, gray: C). (b) CO_2 adsorption–desorption isotherm of the MOF at 273 K (closed circle adsorption, open circle desorption, inert enthalpy of adsorption ΔH_{ads}). Reproduced from ref. 18 with permission from The Royal Society of Chemistry.

in its preferential adsorption. The MOF proves to be robust, exhibiting up to eight adsorption–desorption cycles without disruption of the structural integrity. The enthalpy of adsorption (ΔH_{ads}) of CO_2 is high at zero and maximum loading, presumably because of favorable CO_2–amine interactions and the small pores of the MOF.[18] The ΔH_{ads} initially decreases, then increases with gas loading, an indication of uptake dynamics that involve a cooperativity effect. X-Ray diffraction experiments involving a CO_2-loaded single crystal of the MOF reveal host–guest interactions of CO_2 with free amine groups and oxalate groups, as well as guest–guest interactions with neighboring CO_2 molecules.[15] The cooperative effect of dispersion and electrostatic interactions between CO_2 molecules probably contributes to the increase in ΔH_{ads} with loading. Theoretical calculations corroborate empirical results, which suggest that cooperative guest–guest interactions contribute to increasing CO_2 uptake in MOFs.[15]

In MOFs, bridging ligands control flexibility, dimensionality, and pore sizes, which are properties that influence selectivity, as well as adsorption and desorption mechanisms. Suh *et al.* for instance, attributed pore opening dynamics to flexibility in a 3D pillared MOF designed by connecting 2D networks constructed from square-planer Ni^{2+} macrocycles and a 1,1-biphenyl-3,3,5,5-tetracarboxylate ligand *via* ethyl- or butyl-bridged Ni^{2+} bismacrocycles (Figure 6.3).[19] The dimensionality, 2D *vs.* 3D, of this MOF depends on steric effects in the pillars and the pore size of the 2D networks. The constructed MOF features 3D multilayer pillared networks with interlayer distances of 8.72 and 6.80 Å for the ethyl- and butyl-bridged networks,

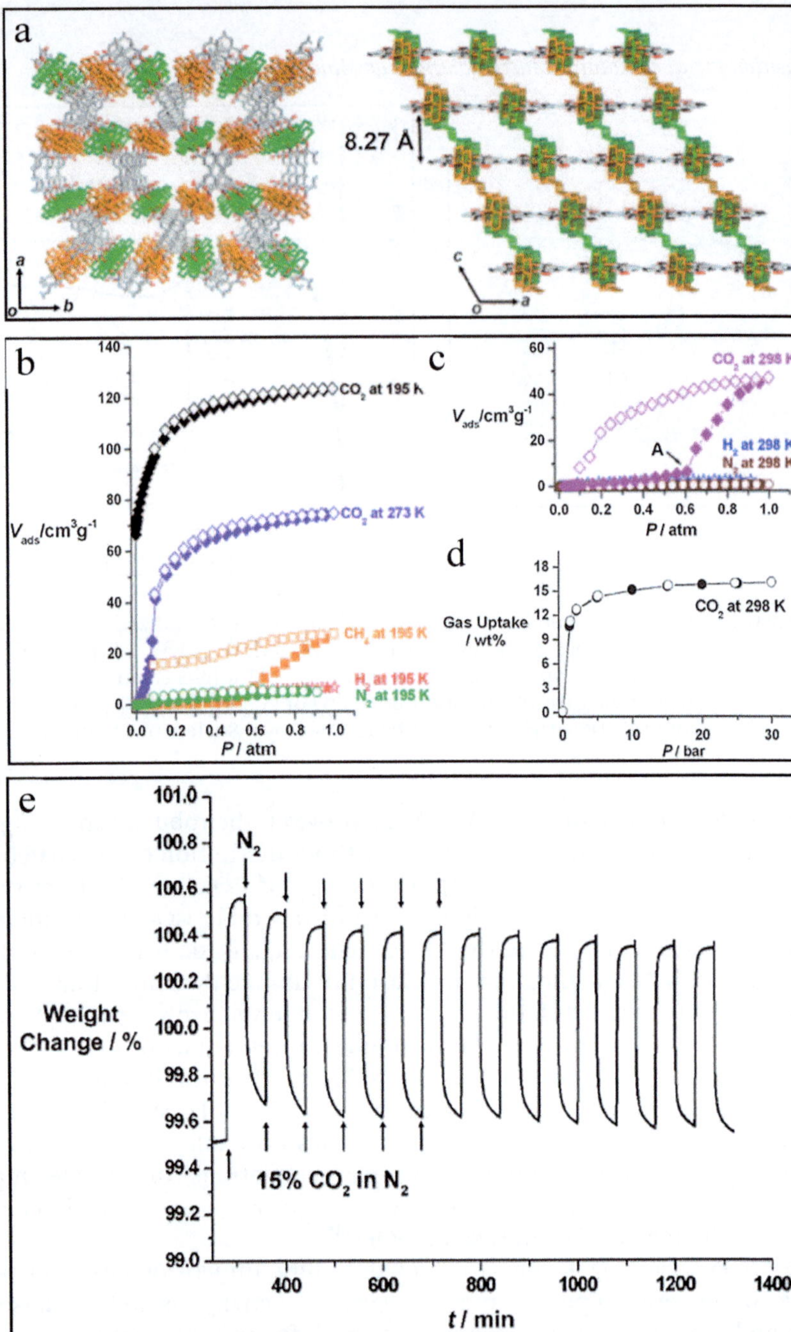

Figure 6.3 (a) and (b) X-ray crystal structure showing the effect of the bridging ligand on the pore size of the Ni^{2+} MOF. (a) Ethyl-bridged MOF (brown and green: ethyl-bridged Ni^{2+} bismacrocycles, light gray: 1,1-biphenyl-3,3,5,5-tetracarboxylate ligand). (b)–(d) CO_2, CH_4, H_2, and N_2 adsorption–desorption isotherms (closed shapes: adsorption, open shapes: desorption). (e) TGA measured gas cycling data showing the reversibility and selectivity of CO_2 in a CO_2–N_2 mixture gas stream. Reproduced from H.-S. Choi and M. P. Suh, Highly Selective CO_2 Capture in Flexible 3D Coordination Polymer Networks, *Angewandte Chemie, International Edition*,[19] John Wiley and Sons © 2009 WILEY-VCH Verlag GmbH & Co. KGaA, Weinheim.

respectively. The dimensionality and sizes of pores, as well as the free volume, depend on the length of the bridging ligand. The ethyl-bridged MOF, for instance, has 55.8% free volume and 4.85×12.12 Å^2 and 1.21×4.37 Å^2 3D pores along the [010] and [100] axes, respectively. In contrast, the butyl-bridged MOF features 35% free volume and 1.16×6.29 Å^2 1D pores. The small size of the pore impedes capture of N_2, H_2, and CH_4 gases; however, the quadrupole moment of CO_2 drives its interactions with MOFs, opening the pores and leading to the adsorption of CO_2 (Figure 6.3).[19] Upon reducing the pressure, the adsorbed CO_2 facilely desorbs from the MOFs. Presumably, the flexibility of the MOFs controls the adsorption and the hysteretic desorption,[19] properties that facilitate CO_2 sequestering, storing, and sensing. In these MOFs, gas uptake dynamics is selective towards CO_2 in the presence of N_2, H_2, and CH_4 as indicated, for instance, by the high selectivity ratio of 98 : 1 for CO_2–N_2 at 298 K and 1.0 atm.[19] Gas cycling experiments with an alternating flow of flue gas mimic (15% (v/v) CO_2 in N_2) and pure N_2 confirmed the selectivity and reversibility of CO_2 uptake by these MOFs, as evidenced by TGA measured weight gain and loss that correspond to adsorption and desorption of CO_2, respectively (Figure 6.3).

A real CO_2 gas stream may contain water molecules, which can interact with coordinatively unsaturated metal sites in Lewis-acid-derived MOFs. The result is irreversible degradation, loss of sorptive property or competitive binding of water molecules to reduce the sorptive capacity for CO_2 molecules. The structural diversity of available ligands offers the opportunity to tailor the moisture sensitivity of MOFs. A small ligand, for example, affords ultra-micropores, which enhance cooperative CO_2 guest–guest interactions, increasing uptake and ensuring stability in a moist environment.[15,23,27–30] Also, a tiny pore size endows MOFs with a molecular sieving capability,[23] allowing the selective capture of CO_2 *via* its favorable quadrupole moment. To address the instability of MOFs, especially in a moist environment, Vaidhyanathan *et al.* constructed a robust Ni-(4-pyridylcarboxylate)$_2$-derived MOF with ultra-micropores of 3.5 and 4.8 Å that selectively and efficiently captured CO_2 from a humidified gas mixture.[23] At high pressure (10 bar), this ultra-microporous MOF showed exceptionally high CO_2 uptake due to positive, cooperative guest–guest interactions. Also, the sorptive property and structural integrity were maintained despite exposure to extreme conditions of moisture or pressure, indicating exceptional hydrolytic and hydrostatic robustness. The weak interactions of CO_2 gas with the framework facilitated facile desorption of the gas from the adsorbent.[23]

Replacing unsaturated metal centers in MOFs with coordinatively saturated metal centers is a strategy to reduce competitive binding of water molecules. In this case, uptake occurs *via* other mechanisms including interactions between the host MOF lattice and the CO_2 guest molecules.[26] Well-implemented crystal engineering can tune the pore size of MOFs to control the CO_2 adsorption capacity or functionalize the interior of the pore for selective uptake of CO_2 in the presence of water. For instance, Yaghi *et al.* lined the interior of MOFs with CH_3, NH_2, CH_2NH_2, CH_2NHMe, CH_2NHBoc, and

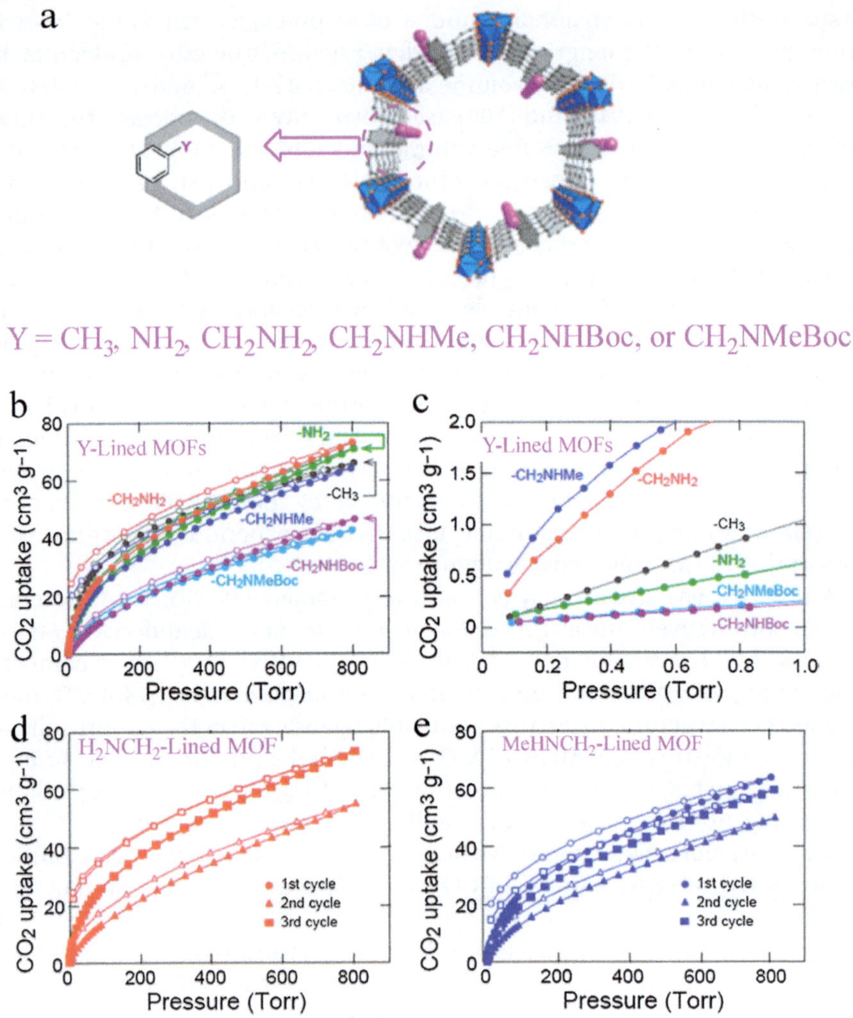

Figure 6.4 (a) Functionalized MOF (purple: functional group). (b) and (c) Effect of functional group on the CO_2 adsorption–desorption isotherm at 298 K. (d) and (e) Recyclability of amine-functionalized MOF. Reprinted with permission from ref. 24. Copyright (2014) American Chemical Society.

$CH_2NMeBoc$ (Boc is a protecting group, *tert*-butyloxycarbonyl) to capture CO_2 in a moist environment (Figure 6.4).[22,24] These MOFs proved their ability to adsorb CO_2, although the protection reduces the uptake capacity due to steric effects.[24] Cross polarization magic angle spinning ^{13}C NMR experiments showed evidence of the formation of carbamate species in the NH_2CH_2-, and $NHMeCH_2$-lined MOFs at room temperature and 1 atm, an indication that chemisorption contributes to the mechanism of uptake.[24] This finding may explain the superior uptake of CO_2 by the NH_2CH_2-, $NHMeCH_2$-lined MOF at low pressure compared with the other functionalized MOFs. Interestingly, the NH_2CH_2-lined MOF selectively sequesters CO_2 from 65% humidified N_2 gas streams.[24]

Chemisorption limits the efficient regeneration of adsorbents but offers an opportunity to convert captured CO_2 to chemicals of industrial relevance. Indeed, researchers are exploring this opportunity by grafting catalysts on porous polymer frameworks to promote the conversion of adsorbed CO_2 to useful chemicals. This is exemplified by the work of Zhao *et al.* on Tb/Co or Tb/Zn heterometallic MOFs ($\{[TbZn(BPDC)_2(\mu_2\text{-}H_2O)Cl(H_2O)_3]\cdot 5H_2O\cdot 0.5DMA\}_n$ and $\{[TbCo(BPDC)_2(\mu_2\text{-}H_2O)Cl(H_2O)_3]\cdot 8H_2O\cdot 0.5DMA\}_n$) that capture CO_2 and efficiently convert it to cyclic carbonates (Figure 6.5).[25] Both MOFs are isostructural, with a monoclinic crystal structure, and 6 Å channels that constitute 46% of the total volume. The MOF exhibits a nominal maximum CO_2 adsorption capacity of 25 $cm^3\ g^{-1}$ and a hysteretic desorption profile. The chemisorption mechanism of uptake prevents release of all captured CO_2 molecules. The Lewis acidity of the metal centers enables the conversion of the chemisorbed CO_2 to cyclic carbonates *via* Lewis acid activation of epoxides in the presence of a co-catalyst, tetrabutylammonium bromide, in over 90% yield at 70 °C (Figure 6.5).[25]

Most heterogeneous catalyzed synthesis of cyclic carbonates needs a high temperature, a requirement that questions the sustainability of the

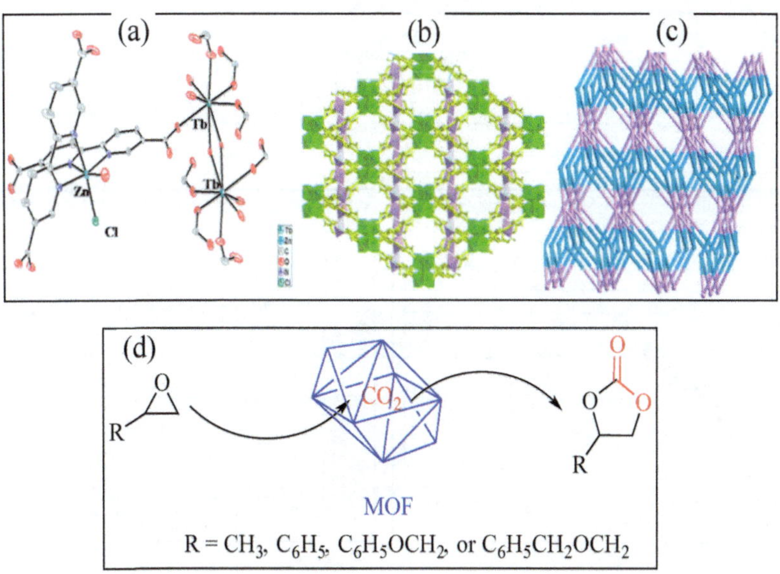

Figure 6.5 (a) Single crystal X-ray structure of a MOF showing the coordination environment of Tb^{3+}, Zn^{2+} and 5,5′-dicarboxylate-2,2′-dipyridine ($BPDC^{2-}$) ligand. (b) 3D lattice and (c) *flu* topology of the MOF. (d) Schematic of MOF-catalyzed conversion of sequestered CO_2 to cyclic carbonate. Reprinted with permission from ref. 25. Copyright (2017) American Chemical Society.

chemistry.[31] Undoubtedly, a sustainable process should achieve this synthesis at room temperature and atmospheric pressure. Co or Al salen-complex-based homogeneous catalysts enable the low-temperature synthesis of cyclic carbonates from CO_2 and cyclic epoxides.[31–39] The grafting of these homogeneous catalysts on porous macromolecular frameworks results in heterogeneous catalysts that sequester CO_2, and prompt the efficient coupling of CO_2 with cyclic epoxides to form cyclic carbonates under mild conditions. Xie *et al.* proved these concepts by grafting Co or Al salen complexes on porous conjugated polymers to construct heterogeneous catalysts for the room temperature and atmospheric pressure coupling of CO_2 with epoxides to form cyclic carbonates (Figure 6.6).[31] Electron microscopic techniques reveal the uniform and porous morphology of these salen polymers with ~0.5 nm pore size. The CO_2 uptake capacity of these polymers is high and reversible. As CO_2 pressure approaches atmospheric pressure,

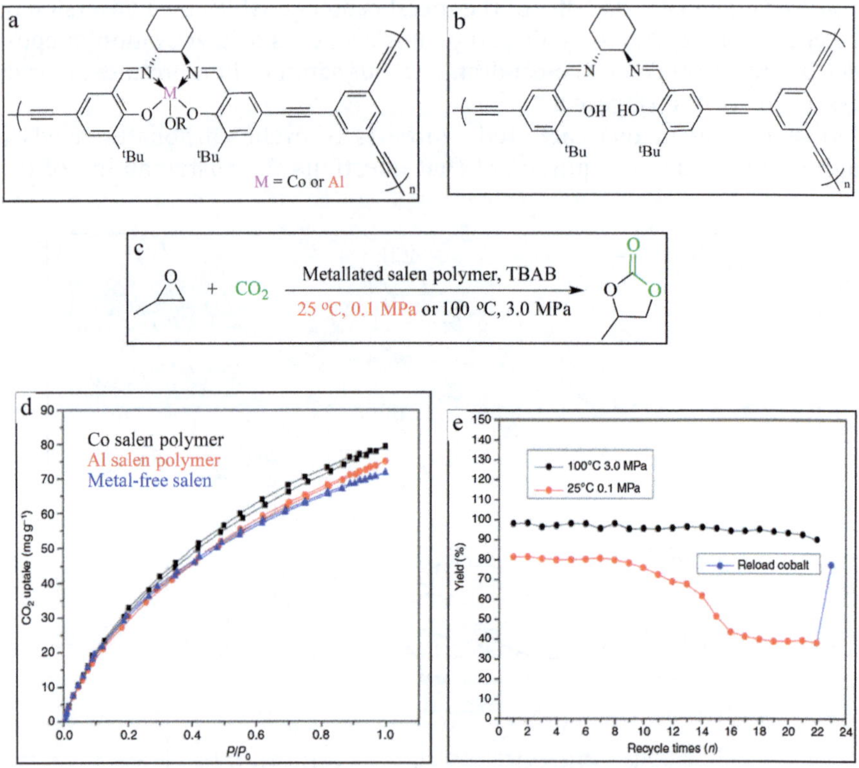

Figure 6.6 Schematics of (a) a metalated salen polymer and (b) metal-free salen polymer. (c) Schematic of salen-catalyzed conversion of captured CO_2. (d) CO_2 adsorption and desorption isotherms for the metalated (black squares: Co, red circles: Al) and metal-free polymer (blue triangles). (e) Catalytic activity and recyclability of the Co salen polymer. Reprinted by permission from Macmillan Publishers Ltd: *Nature Communications* (ref. 31), copyright (2013).

its adsorption reaches 79.3 mg g^{-1} and 76.5 mg g^{-1} into the Co and Al salen polymers, respectively, which are slightly higher than 71.0 mg g^{-1} for the metal-free salen polymer (Figure 6.6). Under mild reaction conditions (25 °C), the metalated salen polymers catalyze the synthesis of propylene carbonate from propylene oxide and the adsorbed CO_2 with good yield (Figure 6.6). The performance of these polymeric catalysts is superior to the molecular Co salen homogeneous catalyst under identical conditions.[31] Specifically, the Co salen polymer catalyzed the coupling of CO_2 and propylene oxide to afford propylene carbonate in 81.5%, which is higher than the 77.1% obtained with the molecular Co salen homogeneous catalyst. This superior performance is probably due to entrapment of CO_2 within the pores of the polymer, a phenomenon that increases the CO_2 concentration at the catalytic center. Remarkably, these salen polymers outperform the standard industrial catalysts for the coupling of CO_2 and propylene oxide. At 100 °C and high CO_2 pressure, the Co salen polymer is recyclable up to 22 times without loss of activity, but drastic loss of activity is evidenced at room temperature and atmospheric pressure (Figure 6.6).[31]

6.3 As Catalysts and Reaction Mediators

Catalysts are an indispensable component of energy-efficient chemical and biochemical systems,[40–42] and therefore, are crucial to a sustainable future. The pursuit of active and robust catalysts is as old as chemistry and the quest continues with unrelenting vigor. An emerging frontier explores macromolecular catalysts with the objective of designing a single platform that features the inherent recyclability and phase separation of heterogeneous catalysts, and the reactivity and selectivity of homogeneous catalysts. Indeed, a macromolecular framework provides a platform to control solubility, enabling a transition from homogeneous to heterogeneous catalysis. For instance, by grafting three, four, or nine palladium complexes on a nona-branched dendrimer (Figure 6.7), Astruc *et al.* modulated the solubility of the resulting metal-containing dendrimers, switching from homogeneous trimetallic or tetrametallic dendrimeric catalysts to a heterogeneous nona-metallic dendrimeric catalyst.[43] These palladium-based dendrimeric catalysts (Figure 6.7) efficiently catalyze Suzuki–Miyaura, Heck, and Sonogashira carbon–carbon cross-coupling reactions (Scheme 6.1). Under similar reaction conditions, the heterogeneous nona-metallic dendrimeric catalyst affords products in conversions that rival those of the homogeneous dendrimeric catalysts and is recyclable up to ten times without a significant drop in activity.[43]

Carbon–carbon cross-coupling reactions occupy an important niche in the synthesis of many useful chemicals, including pharmaceuticals and agrochemicals. It is attractive to employ heterogeneous catalysts to promote these reactions because their phase separation and recyclability ensure purity and sustainability. Various macromolecular architectures are used to access heterogeneous catalysts for carbon–carbon cross-coupling reactions.[44,45]

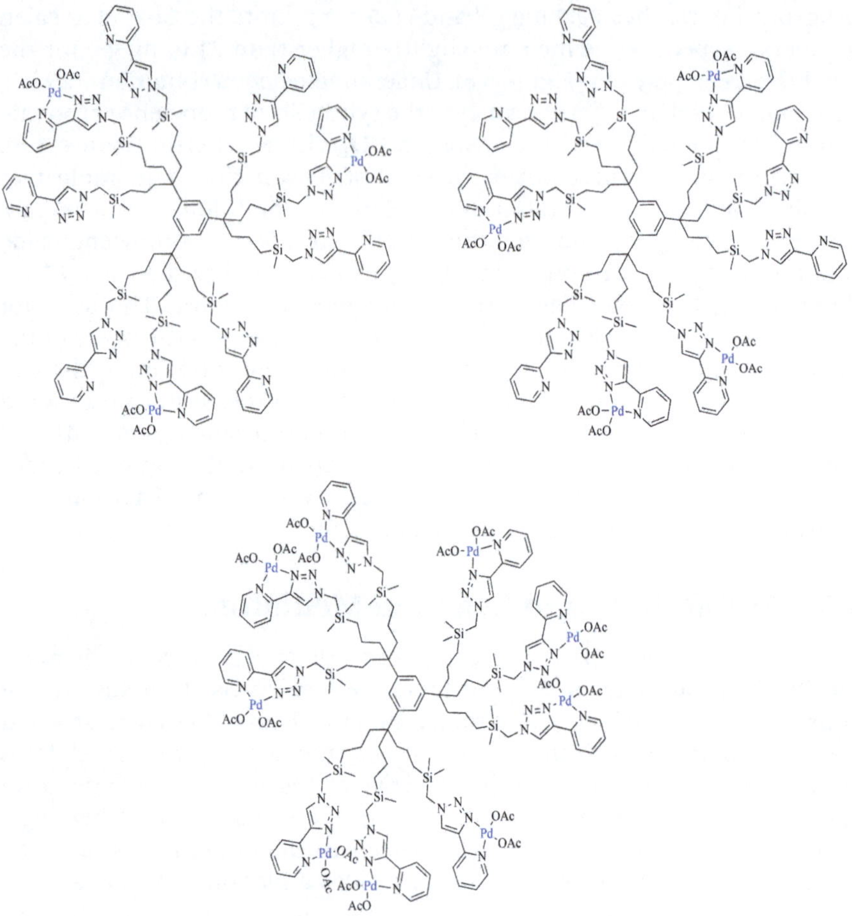

Figure 6.7 Schematics of palladium-containing dendrimeric catalysts.

A linear polymer or MOF, for instance, can be designed to provide an insoluble framework with a functional handle to graft transition-metal-based catalysts, yielding a heterogeneous catalytic system. A triazolyl ligand on linear polysiloxane polymer chelates Pd^{2+}, giving a heterogeneous catalyst for the Suzuki–Miyuara coupling of 2-bromotoluene and phenylboronic acid in a propanol–water mixture (Figure 6.8).[44] The catalyst proves to be highly efficient, yielding 100% conversion of the 2-bromotoluene to 2-methylbiphenyl, and is recyclable up to nine times after separation from the reaction product (Figure 6.8). The recovered catalyst features excellent stability and activity, affording products in over 95% conversion, even after storage in air for a week.[44] The amount of recovered catalyst decreases after each run, and the percent conversion drops after the ninth cycle, presumably due to leaching of the palladium (Figure 6.8). The authors presumed that the leached Pd^{2+} reduces to Pd^0, a conclusion supported by TEM and SEM micrographs of

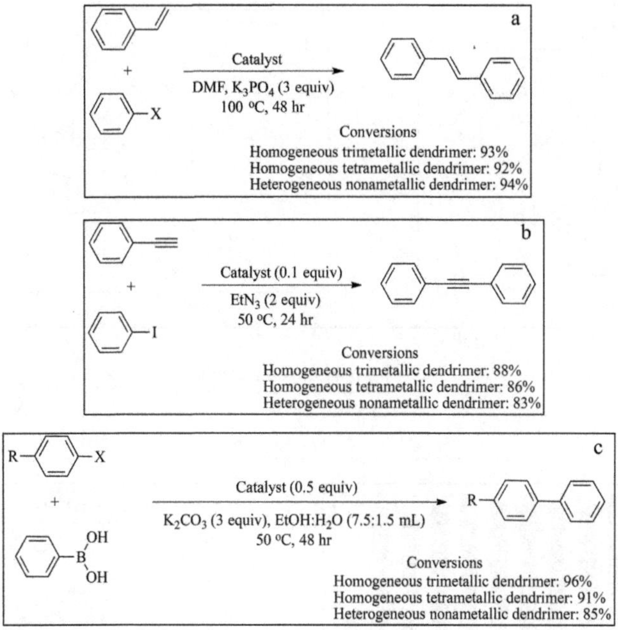

Scheme 6.1 Palladium-containing dendrimeric catalysts promote carbon–carbon cross-coupling (a) Heck, (b) Sonogashira, and (c) Suzuki–Miyaura reactions.

the recovered polymer catalyst that reveal the presence of nanoparticles.[44] However, in a Pd-based MOF catalyst obtained by anchoring Pd^0 nanoparticles on a cobalt(II) salicylate-derived MOF, the catalyst did not leach from the framework.[45] This MOF-based catalyst enables efficient carbon–carbon cross-coupling *via* Suzuki–Miyaura and Sonogashira reactions in water and is recyclable up to six times.[45]

Various concepts have been developed by researchers to design macromolecular heterogeneous catalysts, with the aim of promoting chemical reactions efficiently. For instance, the channels in porous coordination polymers were employed to sequester and concentrate CO_2, enabling a catalyst grafted on the polymer to better promote the chemical transformation of CO_2 to a value-added product. Duan *et al.* demonstrated this concept with a porous coordination polymer framework containing catalytic silver(I) complex grafts that trigger triple bonds to react with sequestered CO_2 to form cyclic carbonate (Figure 6.9).[46] Also, coordinatively unsaturated metal centers in an evacuated 3D Zn^{2+} MOF heterogeneously catalyze a carbon–carbon-bond-forming Henry reaction (Figure 6.10).[47] This recyclable 3D catalyst facilitates the coupling of several substituted benzaldehydes with nitroalkane, with yields that depend on the nature of the substrate. A single crystal to single crystal transmetalation reaction of this 3D Zn^{2+} MOF catalyst with Cu^{2+} affords its Cu^{2+} analog, which demonstrated comparable catalytic activity.[47]

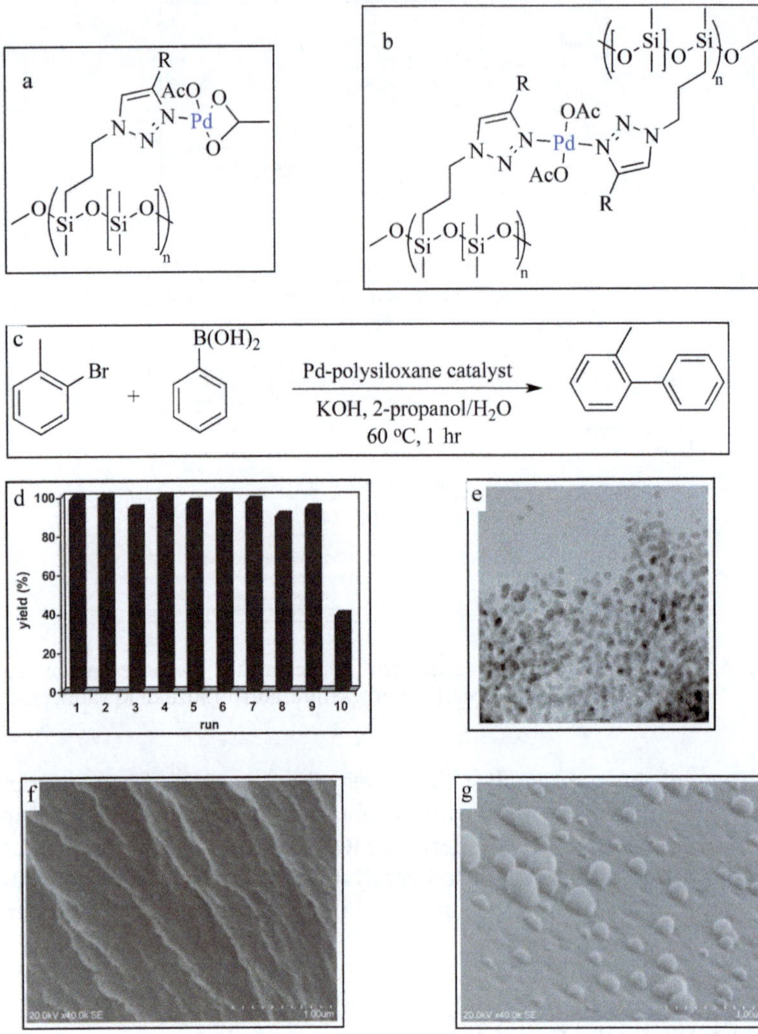

Figure 6.8 (a) and (b) Schematics of Pd-containing polysiloxane. (c) Schematics of the Suzuki–Miyaura reaction catalyzed by the Pd-containing polysiloxane. (d) Yield of the reaction at various runs. (e) TEM micrograph showing Pd^0 nanoparticle in Pd-containing polysiloxane after reaction. SEM micrographs of Pd-containing polysiloxane (f) before and (g) after reaction. Reproduced from *Applied Catalysis A: General*, **470**, E. Mieczyńska, T. Borkowski, M. Cypryk *et al.*, Palladium supported on triazolyl-functionalized polysiloxane as recyclable catalyst for Suzuki–Miyaura cross-coupling, 24–30.[44] Copyright (2014), with permission from Elsevier.

Also, the dendrimer framework is a suitable platform to graft transition-metal-based catalysts.[48–52] The group of Astruc amply demonstrated this, as exemplified by grafting ruthenium carbene or benzylidene complexes on a dendritic scaffold to catalyze the room temperature ring-opening metathesis polymerization (ROMP) of norbornene.[49,50] These dendrimeric

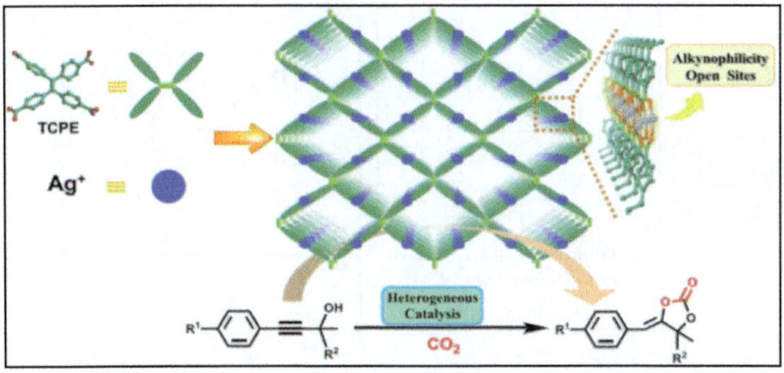

Figure 6.9 A silver(I)-based MOF catalyzes the transformation of sequestered CO_2 to cyclic carbonate. Reprinted with permission from ref. 46. Copyright (2017) American Chemical Society.

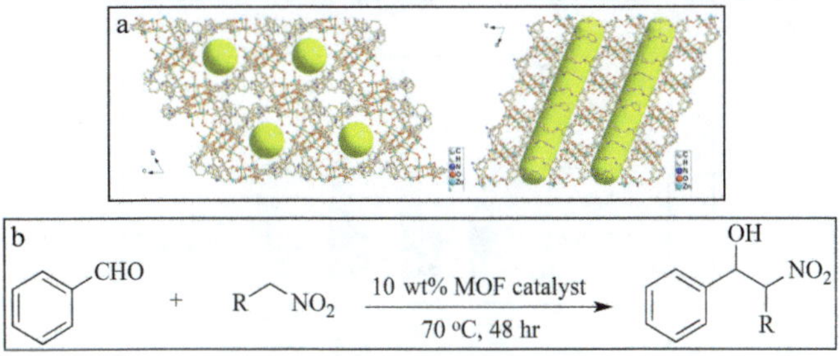

Figure 6.10 (a) 3D Representation of a Zn^{2+} MOF with coordinated H_2O in the channels. Evacuation of the H_2O molecules activates the MOF to catalyze a Henry reaction (b). Reproduced from ref. 47 with permission from The Royal Society of Chemistry.

catalysts more actively promote the ROMP of norbornene than the molecular ruthenium model catalyst, although the activity decreases with an increase in dendrimer generation. Apart from increasing reactivity, macromolecular catalysts also ensure enantioselectivity.[53–55] The importance of enantioselectivity cannot be overemphasized, especially in the synthesis of pharmaceuticals, where physiologic effects strongly depend on enantiomerism. In catalysis, enantioselectivity can arise if the steric effect in a catalyst framework biases its stereoselectivity. It is reasonable to assume that a bulky macromolecular scaffold can shield the microenvironment or alter the conformation and preferred orientation of the catalytic site to tune the stereoselectivity of a grafted catalyst. Indeed, many researchers demonstrated the feasibility to tune stereoselectivity using macromolecular catalysts as exemplified by Gade *et al.*, who employed palladium-containing dendrimers to tune the enantiomeric excess (ee) in the allylic amination of 1,3-diphenyl-1-acetoxypropene (Figure 6.11).[53] They achieved a 9% ee with a palladium-based molecular catalyst but up to 69% ee with a palladium-based dendrimeric catalyst with a positive dendritic effect (Figure 6.11).

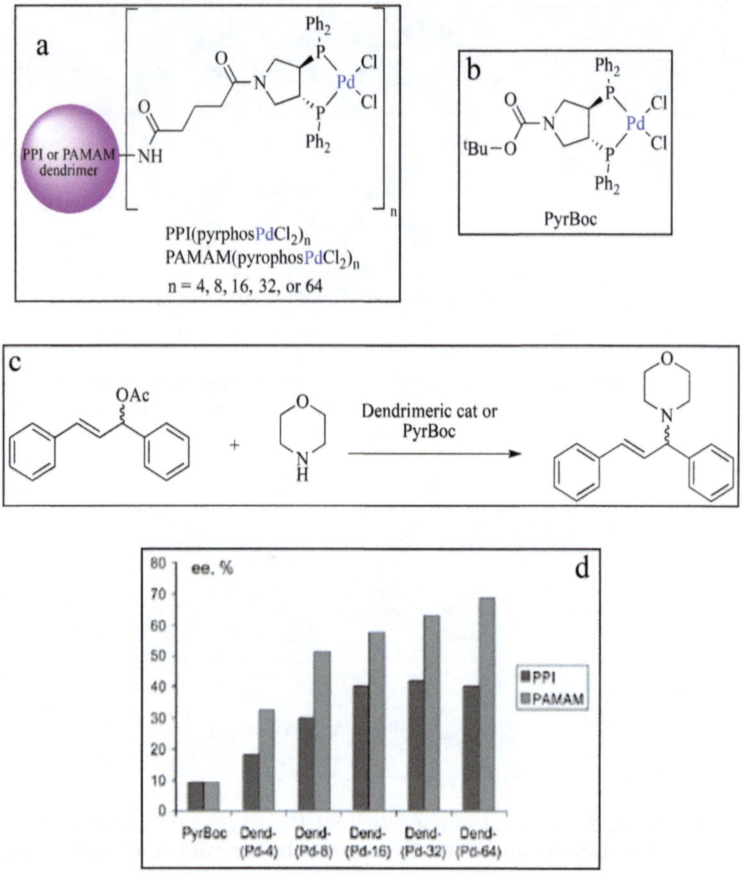

Figure 6.11 (a) Schematic of an enantioselective Pd-containing dendrimeric catalyst, (b) molecular Pd model catalyst, (c) schematic of allylic amination, (d) enantiomeric excess (ee) of the allylic amination reaction. Reproduced from ref. 53 with permission from The Royal Society of Chemistry.

Higher enantioselectivity is attainable, for instance, in transition-metal-cored dendrimer catalyzed hydrogenation reactions. A typical example is the over 93% ee observed with iridium-cored[54] or rhodium-cored[55] dendrimeric catalysts that promote the asymmetric hydrogenation of quinolines or α-acetamido cinnamic acid, respectively.

A transition metal moiety, such as $(CO)_3Cr$, $(CO)_3Mn^+$, $(C_6H_6)Fe^{2+}$, or $(C_5H_5)Fe^+$, complexed to an arene ligand activates the ligand towards nucleophilic aromatic substitution or an addition reaction,[56–59] resulting in facile synthetic routes to some chemicals of importance in the pharmaceutical and agrochemical industries. As an example, Sutherland *et al.* demonstrated this with the $(C_5H_5)Fe^+$-mediated room temperature addition of a cyanide ion to the arene ligand to obtain benzonitrile or phthalonitrile after oxidative demetallation with 2,3-dichloro-5,6-dicyano-1,4-benzoquinone

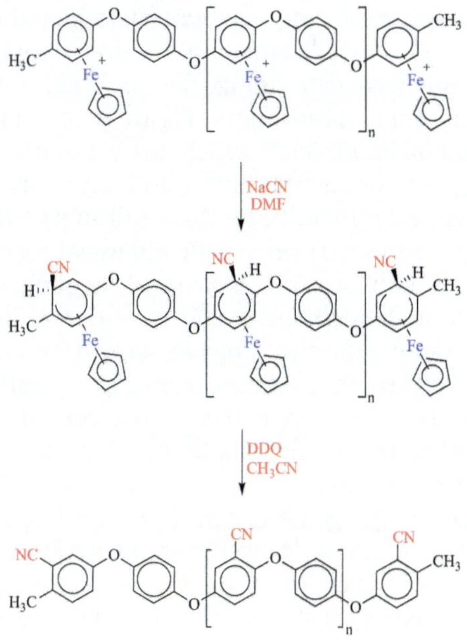

Scheme 6.2 Organoiron-mediated synthesis of nitrile-functionalized poly(aromatic ether).

(DDQ).[59] The nitrile group is integral to the activity of many pharmaceuticals, agrochemicals, and dyes,[60–64] so a facile synthetic route to these compounds is attractive, given that conventional and most emerging synthetic methods employ stringent conditions such as high or low temperatures.[60–63] Abd-El-Aziz *et al.* explored the $(C_5H_5)Fe^+$-mediated room temperature nucleophilic aromatic addition to construct a nitrile-functionalized poly(aromatic ether) after oxidative demetallation with 2,3-dichloro-5,6-dicyano-1,4-benzoquinone (Scheme 6.2).[56] The addition of the nitrile group to the polymer backbone is selective, being added only at the position *meta* to the ether bridge.[56] This room-temperature metal-mediated selective addition of a nitrile group to poly(aromatic ether) has implications for the synthesis of polymers with various functional groups including amines and carboxylic acids and therefore the construction of functional materials.

6.4 As Data Storage Materials

6.4.1 As Precursors of Magnetic Ceramics for Data Storage

Data storage is central to all facets of life as exemplified by the storage of biological data in DNA, and digital data in the modern computer. Indeed, social scientists acknowledge that society's growth and development depends, among other factors, on its technological capacity to communicate, compute and store data.[65,66] Materials scientists are pushing the frontiers of information technology, exploring new materials including DNA, semiconductors, optical, magnetic or redox materials as precursors for the fabrication of high-density and long-term data storage devices.[67] Recently, Abd-El-Aziz *et al.* constructed organometallic dendrimers (Figure 6.12) and explored them as precursors for magnetic ceramics,[68] which are materials for the fabrication of magnetic storage media such as disks and tapes. They obtained these homometallic Fe and heterometallic Fe/Co dendrimers from η^6-dichlorobenzene-η^5-cyclopentadienyliron(II) *via* simple chemical reactions that include nucleophilic aromatic substitution (S_NAr) and Steglich esterification reactions. Pyrolysis of these dendrimers at 900 °C under a nitrogen atmosphere yields soft ferromagnetic ceramics. Powder X-ray diffraction indicated that the ceramics are polycrystalline, with the degree of crystallinity ranging from 37 to 56%, and electron microscopic techniques confirmed the polycrystallinity, precisely revealing the nanocrystallites of Fe and Co species embedded in the amorphous carbonaceous matrix (Figure 6.13).[68] The size of the crystallites is tunable by changing the generation of the dendrimers. The powder X-ray diffraction of ceramics derived from the heterometallic dendrimers indicated the presence of {111} Co, {200} Co, {400} Co_3O_4, {311} Fe_3O_4, and {110} α-Fe species, which suggests that the Fe and Co species probably exist in separate crystalline domains. In contrast, in some magnetic ceramics derived from Fe/Co linear polymers, the magnetic species coexist in the same crystalline domain as an alloy,[69] presumably because of their proximity within the polymer framework. In the heterometallic Fe/Co dendrimers, the Co moieties are at the periphery and the Fe is in the interior branches (Figure 6.12), a framework that creates distinct regions of these moieties and could favor their crystallization in separate domains. Abd-El-Aziz and coworkers posited that the presence of Fe, as well as Co in the dendrimer, gives rise to ferromagnetism after pyrolysis and, indeed, magnetization experiments proved their postulation (Figure 6.14). The presence of a hysteresis loop in the magnetization curves of the ceramics implies ferromagnetism and the low coercivity values (H_c) suggest soft magnetism (Figure 6.14). The dendrimer generation tunes the magnetic properties of the homometallic Fe dendrimers, specifically increasing softness with an increase in dendrimer generation. Grafting cobalt on the dendrimers' periphery changes the magnetic behavior of the homometallic dendrimers.

Magnetic ceramics have a significant niche in the magnetic materials market because of their use as permanent magnets in several devices, including magnetic recording heads.[4] Compared with other magnetic materials, magnetic ceramics feature low electrical conductivity, in addition to their low production cost, factors that inspire scientists and engineers to explore their use in these applications. Several synthetic routes exist, with the pyrolytic transformation of transition-metal-containing macromolecules to magnetic ceramics emerging with several advantages. For instance, the use of these macromolecules results in clean magnetic ceramics, in contrast with the traditional powder technology that requires sintering additives which may contaminate the target product, limiting the scope of application. Pyrolysis of

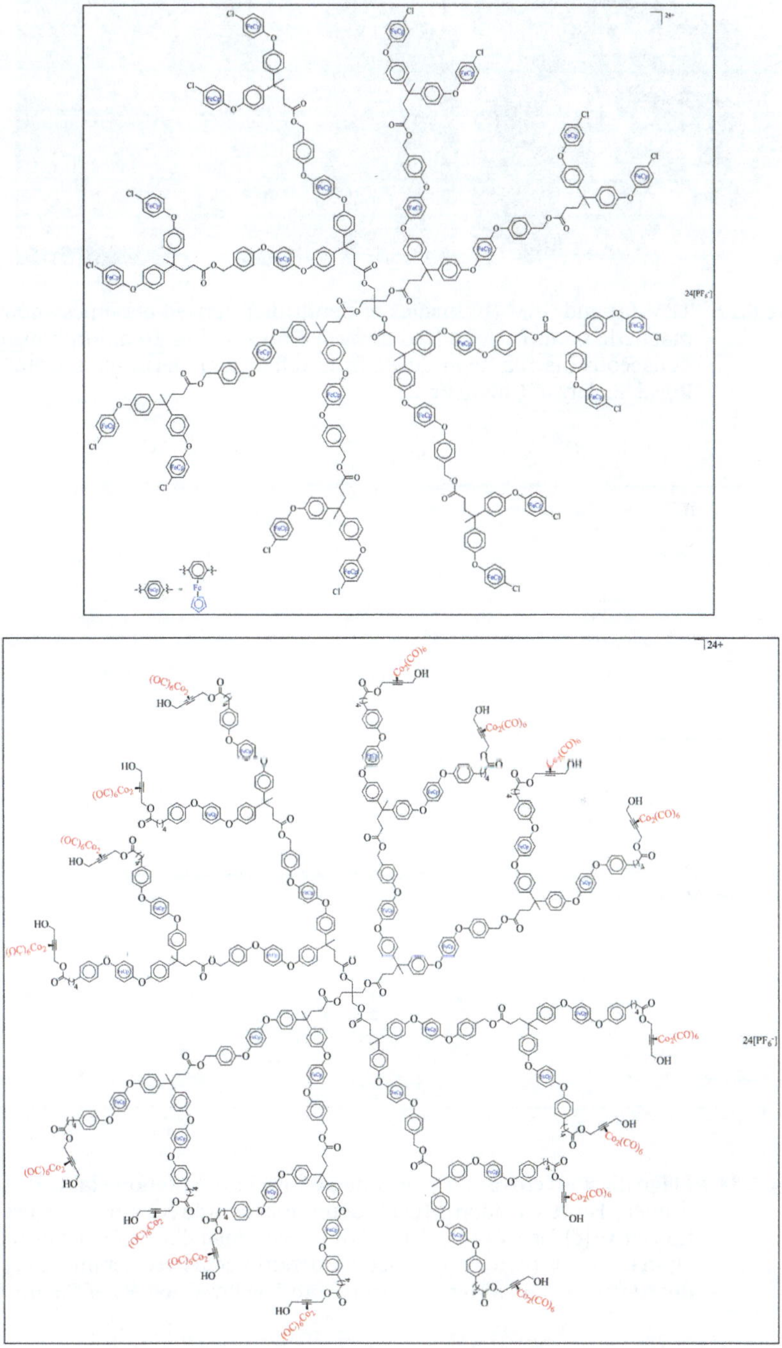

Figure 6.12 Schematics of homometallic Fe (top) and heterometallic Fe/Co (bottom) dendrimer precursors for magnetic ceramics.

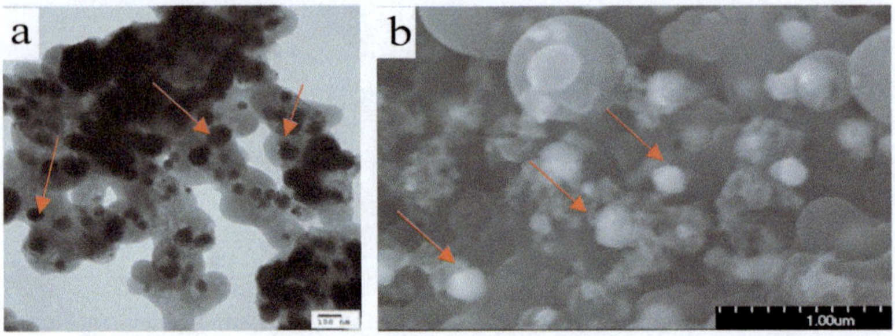

Figure 6.13 TEM (a) and SEM (b) images of dendrimer-derived ceramics showing magnetic nanoparticles (red arrows) embedded in an amorphous carbonaceous matrix. Reproduced from ref. 68 with permission from The Royal Society of Chemistry.

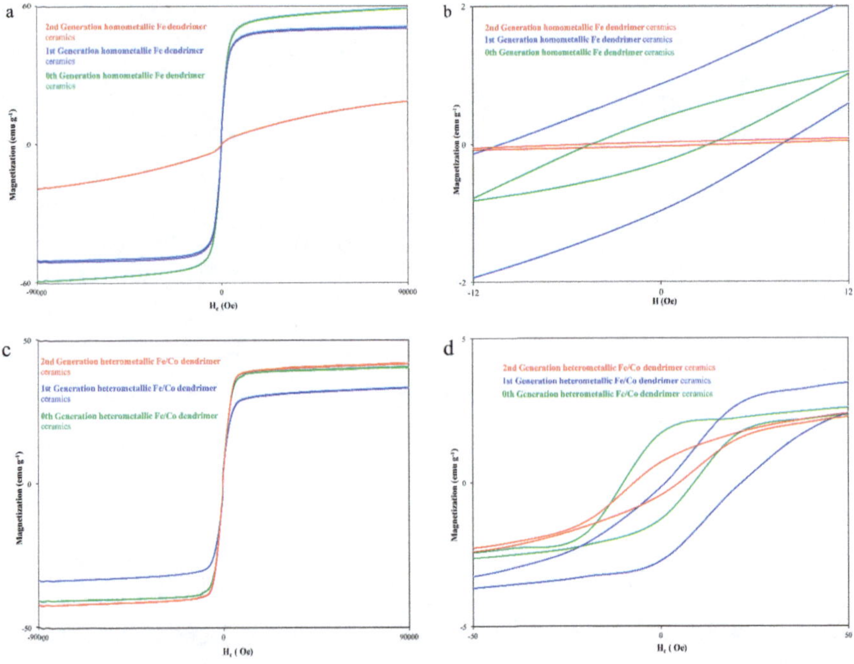

Figure 6.14 Magnetization curve (a) for ceramics derived from homometallic Fe dendrimers. Hysteresis loop (b) of the dendrimer derived ceramics. Magnetization curve (c) for ceramics derived from heterometallic Fe/Co dendrimers. Hysteresis loop (d) for heterometallic dendrimer derived ceramics. Reproduced from ref. 68 with permission from The Royal Society of Chemistry.

transition-metal-containing macromolecules ensures homogeneity by blending the magnetic species at the molecular level and allows control over the macromolecular structure and concentration of magnetic species per unit volume, to tune the magnetic property and ceramic yield. Also, researchers explore this synthetic route to integrate the chemical resistance of ceramics with the magnetic property of metals into a single platform. From an engineering perspective, it is possible to employ traditional polymer processing techniques to shape the precursor macromolecule to obtain magnetic ceramic fibers, films, and nanoparticles. For instance, Wong *et al.* obtained magnetic nanoparticles from homometallic Fe, Co, or Ni as well as heterometallic Fe/Pt, Fe/Ni, or Fe/Cu porphyrin-derived polymers (Figure 6.15).[70]

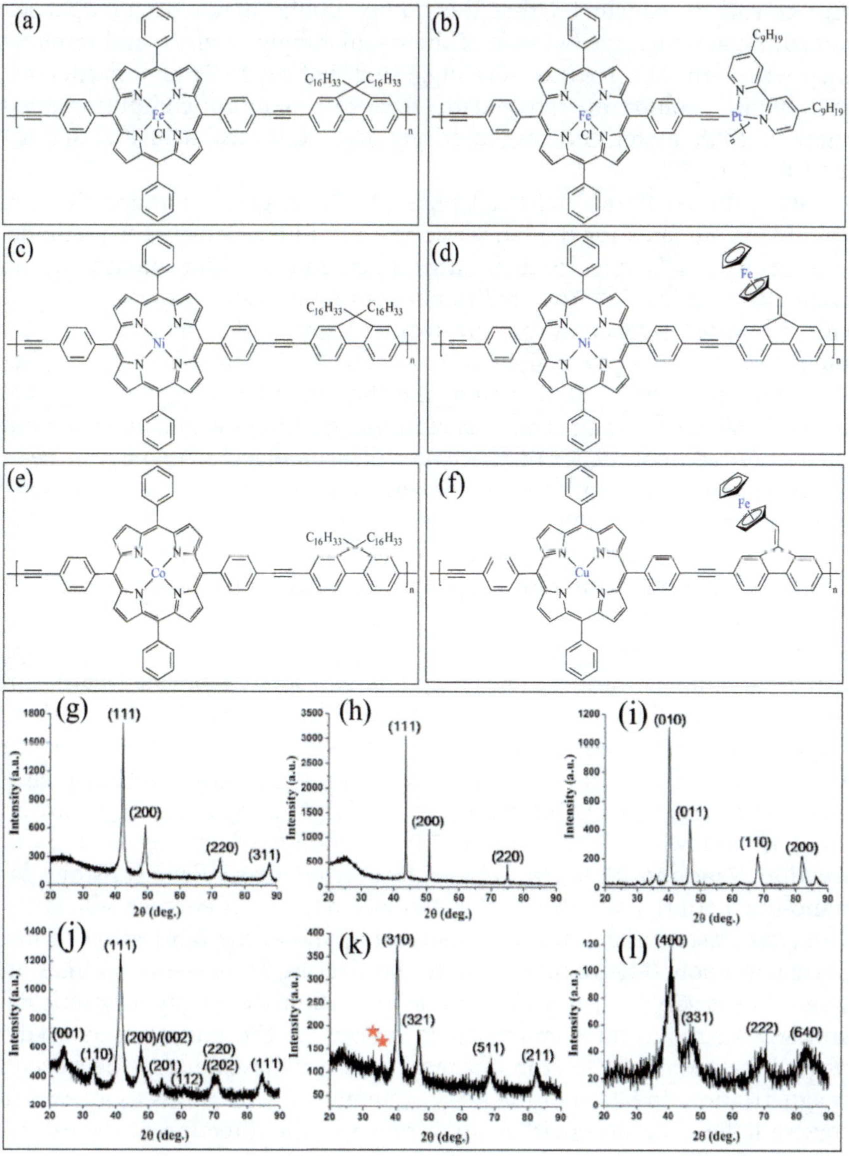

Figure 6.15 (a–f) Schematics of homometallic Fe, Co, and Ni and heterometallic Fe/Pt, Fe/Ni, and Fe/Cu porphyrin-derived polymers. Powder X-ray diffraction pattern of Fe (g), Co (h), and Ni (i), Fe/Pt (j), Fe/Ni (k), and Fe/Cu (l) porphyrin-derived polymers. Reproduced from ref. 70 with permission from The Royal Society of Chemistry.

In contrast with the heterometallic Fe/Co dendrimers constructed by Abd-El-Aziz *et al.*,[68] these heterometallic polymers yield alloyed nanoparticles with the degree of alloying as well as chemical stability and composition of the alloy being dependent on spin–orbit coupling and hybridization of 3d and 5d electrons.[70] Powder X-ray diffraction, for instance, revealed that the Fe/Pt polymer gives more chemically stable alloyed nanoparticles than the Fe/Ni or Fe/Cu polymers, as evidenced by the diffraction patterns (Figure 6.15). Wong and coworkers concluded that the homometallic magnetic nanoparticles are superparamagnetic because of their vanishingly small H_c and remanent magnetization (M_r). Their results suggested that the Fe/Pt nanoparticles are suitable for room temperature data storage or quantum computing due to their small diameter and comparatively high H_c values, 0.54 T at 300 K or 1.12 T at 5 K.

Face-centered tetragonal ($L1_0$) phase Fe/Pt magnetic nanoparticles are suitable candidates for the fabrication of ultra-high-density magnetic data storage materials due to their large magnetocrystalline anisotropy and chemical stability.[71–74] The traditional synthetic route to these nanoparticles is a two-step reaction that involves refluxing individual Fe and Pt compounds together to give a superparamagnetic face-centered cubic (A_1) phase Fe/Pt nanoparticle in the first step, and the target ferromagnetic $L1_0$ phase in the final step.[73] Wong *et al.* reasoned that grafting the Fe and Pt moieties on a single compound, for instance on a macromolecule, might give the target product in one step.[73] Indeed, polymers with grafted Fe and Pt moieties afford the $L1_0$ phase Fe/Pt in a one-step pyrolysis under an inert atmosphere (Figure 6.16).[71–73] Interestingly, these transition-metal-containing polymers can be molded into different shapes using polymer processing techniques such as electron-beam lithography, UV photolithography, or nanoimprint lithography, and then pyrolyzed to obtain magnetic shapes. As an example, using nanoimprint lithography, Wong and coworkers fabricated hard ferromagnetic nano-lines and nano-dots from a heterometallic Fe/Pt polymer (Figure 6.16).[72] Pyrolysis of the bulk polymer yields a magnetic ceramic with Fe/Pt nanoparticles embedded in the carbonaceous matrix. A transmission electron microscope estimated the size of the nanoparticle to be around 4.6 ± 1.2 nm, which is close to the 5.0 nm obtained from powder X-ray diffraction. Nanoimprint lithography of the polymer on a silicon substrate gave nano-line patterns, which transformed into hard magnetic nano-lines after pyrolysis. Also, the nanoimprint patterning allows the fabrication of magnetic nano-dots (Figure 6.16). Room temperature H_c of the nano-lines and nano-dots is very high (1.4 T), suitable for ultra-high-density magnetic data storage. Magnetic force microscopic images of the patterned nano-dots proved the possibility to store magnetic data on these patterns because the magnetization direction of the array systematically responds to an external magnetic field, pointing either up or down as the direction of the external field changes (Figure 6.16).[72]

Nanoimprint patterning of polymers into magnetic shapes is feasible with either heterometallic polymers, such as a Co/Pt polymer (Figure 6.17)[74] or a

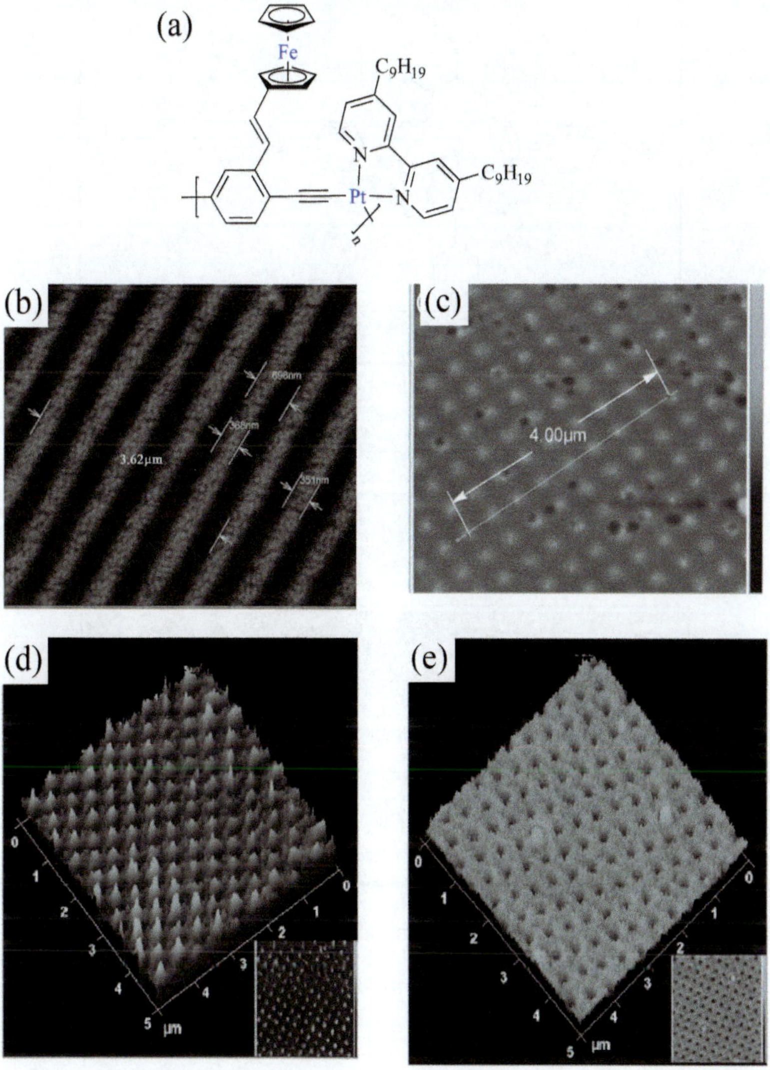

Figure 6.16 (a) Schematic of the Fe/Pt-containing polymer precursor for the $L1_0$ phase Fe/Pt nanoparticles. (b) SEM micrograph of the nano-lines pattern of the Fe/Pt nanoparticles after pyrolysis of the nanoimprinted line pattern of the polymer. (c) AFM micrograph of the nano-dots pattern of the Fe/Pt nanoparticles after pyrolysis of the nanoimprinted dots pattern of the polymer. (d) MFM micrograph of the nano-dot pattern of the Fe/Pt nanoparticles in an external magnetic field in the −*Z* direction. (e) MFM micrograph of the nano-dot pattern of the Fe/Pt nanoparticles in an external magnetic field in the +*Z* direction. Reproduced from Q. Dong, G. Li, C.-L. Ho *et al.*, A Polyferroplatinyne Precursor for the Rapid Fabrication of $L1_0$-FePt-type Bit Patterned Media by Nanoimprint Lithography, *Advanced Materials*,[72] John Wiley and Sons, © 2012 WILEY-VCH Verlag GmbH & Co. KGaA, Weinheim.

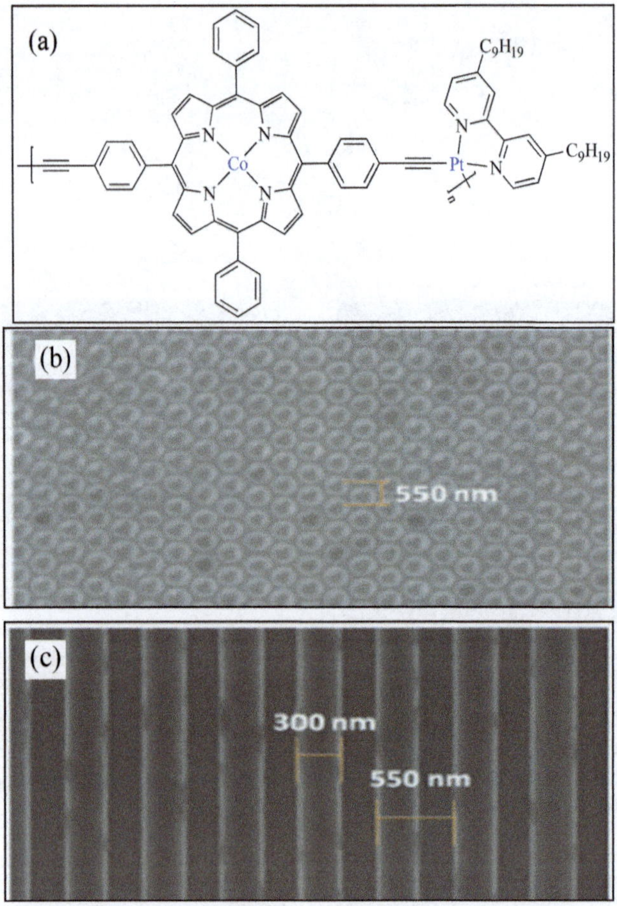

Figure 6.17 (a) Schematic of a heterometallic Co/Pt polymer. SEM of the nanoimprinted polymer nano-dots (b) and nano-lines (c). Reproduced from ref. 74 with permission from The Royal Society of Chemistry.

blend of homometallic Co and Pt polymers (Figure 6.18).[75] Pyrolysis of the heterometallic Co/Pt polymer yields magnetic ceramics with $L1_0$ phase Co/Pt alloyed nanoparticles of sizes 3.4 ± 0.58 nm, which feature ferromagnetic and paramagnetic behavior at room temperature.[74] On the other hand, blending homometallic Co and Pt polymers affords larger Co/Pt alloyed nanoparticles of sizes 6.54 ± 1.28 nm. The nanoparticles derived from the heterometallic Co/Pt polymer have good H_c values of 0.54 T at 300 K and 2.75 T at 5 K, suggesting their potential as candidates for fabricating room temperature magnetic data storage devices.[74] Both the heterometallic Fe/Co polymer and blends of homometallic Fe and Co polymers can be molded and pyrolyzed into magnetic shapes.[74,75]

Shape retention after pyrolysis of molded metallopolymers is essential to obtain the target shape, but this ultimately depends on the ceramic

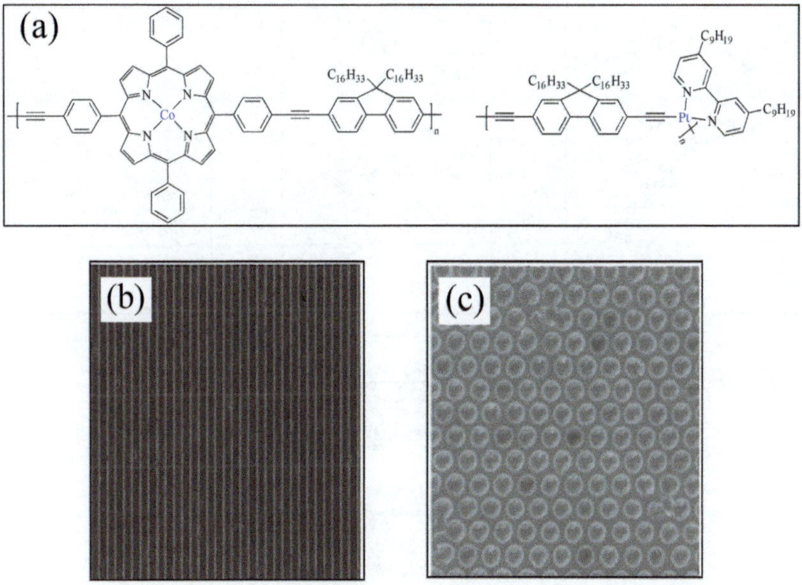

Figure 6.18 (a) Schematic of homometallic Co and Pt polymers for blend. SEM of the nanoimprinted polymer blend nano-lines (b) and nano-dots (c). Reproduced from *Journal of Organometallic Chemistry*, **819**, T. Gong, Z. Gao, W. Bian *et al.*, Co-containing and Pt-containing polymer blend to ferromagnetic CoPt NPs: Synthesis, characterization and patterning study by nanoimprint lithography, 237–241.[75] Copyright (2016) with permission from Elsevier.

yield. Manners *et al.* pioneered the use of cross-linked polymer networks to increase the ceramic yield and eventually retain the shape of molded polymers after pyrolysis.[76,77] They demonstrated that high-temperature polymerization of [1]silaferrocenophane yields insoluble, cross-linked polymer networks that give the shaped magnetic ceramics (Figure 6.19).[76,77] Indeed, high-temperature pyrolysis at about 1000 °C transforms the molded polymer networks into highly rigid magnetic ceramics in 90% yield with good shape retention (Figure 6.19). Low-temperature treatment of some transition-metal-containing polymers is also a route to generating magnetic nanoparticles that are embedded in a functional polymer matrix. For instance, Co block copolymers yield Co nanoparticles when heated at a temperature that is below the copolymer decomposition temperature, a strategy that gives a magnetic polymer[78,79] in contrast to the high-temperature approach that yields magnetic ceramics.[76] An example is an amphiphilic poly(4-(phenylethynyl)styrene)-*block*-poly(ethylene oxide)-*block*-poly(4-(phenylethynyl)styrene) that contains pendant cobalt moieties (Figure 6.20).[78] This polymer forms magnetic Co nanoparticles within self-assembled poly(4-(phenylethynyl)styrene) domains when heated at 120 °C for 24 hours under vacuum. Indeed, the polymeric films derived from the polymer become ferromagnetic after thermal treatment at 120 °C.

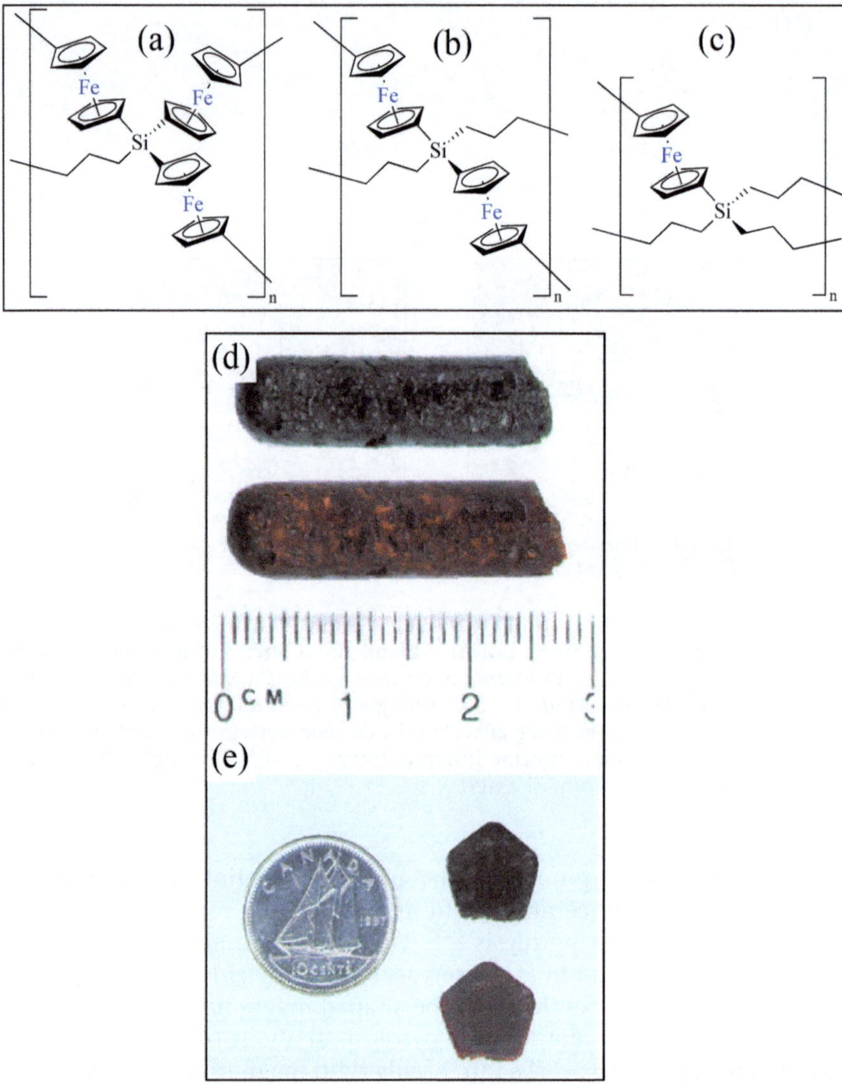

Figure 6.19 (a–c) Schematic of the three proposed structures of the cross-linked polyferrocenylsilane polymer. Photographs of a shaped-polymer rod (d, below) and its pyrolyzed shaped ceramic (d, top) as well as a pentagon-shaped polymer (e, below) and its pyrolyzed shaped ceramic (e, top). Reprinted with permission from ref. 77. Copyright (2002) American Chemical Society.

The degree of polymerization (DP) of the hydrophobic poly(4-(phenylethynyl)styrene) block tunes the morphology of the assembled polymer from an irregular poly(4-(phenylethynyl)styrene) domain with embedded Co nanoparticles to a cylindrical domain (Figure 6.20). Increasing the DP of the poly(4-(phenylethynyl)styrene) block decreases the saturation

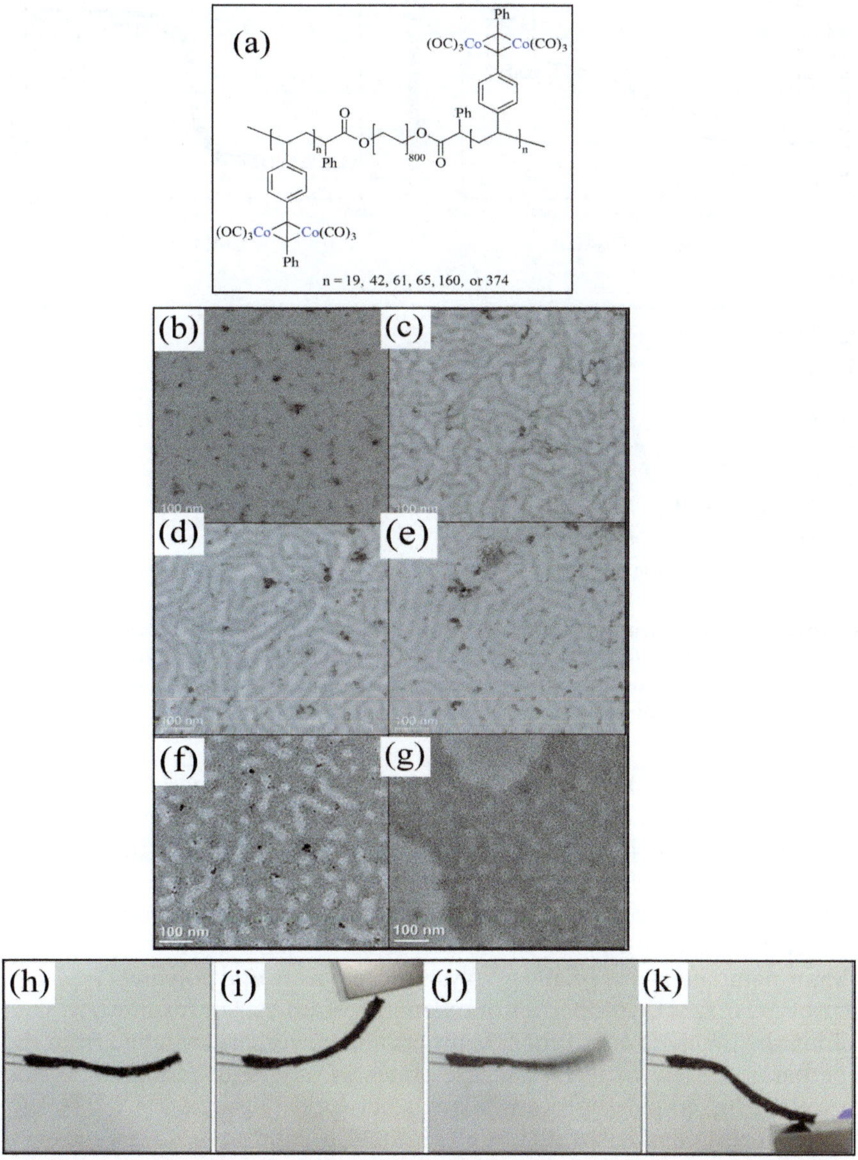

Figure 6.20 (a) Schematic of a Co-functionalized block copolymer for the fabrication of a magnetic hydrogel. TEM showing the morphology of a thermally annealed block copolymer magnetic film (b): n = 19, (c): n = 42; (d): n = 61; (e): n = 65; (f): n = 160; (g): n = 374. (h)–(k) Response of the hydrogel to a laboratory magnet. Reprinted with permission from ref. 78. Copyright (2016) American Chemical Society.

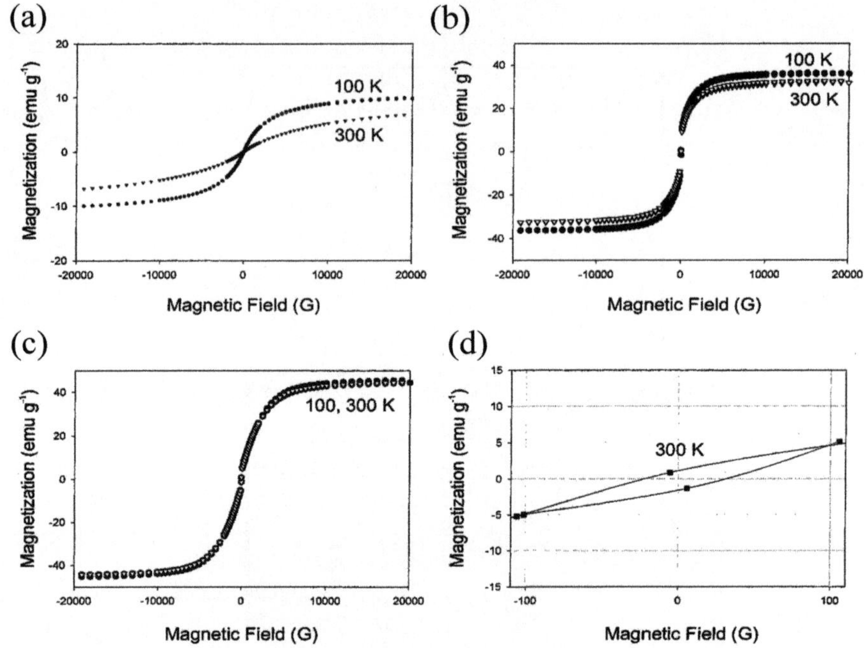

Figure 6.21 Magnetization curves of ceramics derived from cross-linked poly(ferrocenylsilane) pyrolyzed at 650 °C (a), 850 °C (b), 1000 °C (c) and (d). Reprinted with permission from ref. 77. Copyright (2002) American Chemical Society.

magnetization (M_s) and M_r because of an increase in the concentration of Co nanoparticles. When immersed in water, the films form hydrogels, with storage and loss moduli that increase with an increase in the DP of the hydrophobic poly(4-(phenylethynyl)styrene) block. The hydrogels are magnetic as evidenced by their magnetization to a laboratory neodymium magnet (Figure 6.20).[78]

Many parameters, especially pyrolysis temperature and polymer structure, control the magnetic properties of magnetic ceramics. For instance, by controlling the pyrolytic conditions, Manners and coworkers switched from the superparamagnetic to ferromagnetic states in magnetic ceramics derived from cross-linked poly(ferrocenylsilane)s networks (Figure 6.21).[76,77] Pyrolysis of a fabricated polymer disk at 650 °C, 850 °C, and 1000 °C in a nitrogen atmosphere gave small (31 ± 6 nm), bimodal particle size distribution (92 ± 11 nm and 286 ± 9 nm), and large (123 ± 19 nm and 637 ± 37 nm) α-Fe nanoparticles.[77] Indeed, powder X-ray diffraction showed that the intensity of the {110} α-Fe reflection increased with an increase in the pyrolysis temperature. The M_s of the disk was tuned by control of the pyrolysis temperature. Specifically, increasing the pyrolysis temperature from 650 to 850 °C increased M_s from 7 emu g^{-1} to 32 emu g^{-1} at 300 K. Also, pyrolysis at 650 and 850 °C produced a superparamagnetic ceramic disk, which did not exhibit hysteresis at 100 or 300 K (Figure 6.21). Interestingly, the disk obtained from pyrolysis at 1000 °C was ferromagnetic, exhibiting room-temperature hysteresis as well as a higher M_s of 44 emu g^{-1} (Figure 6.21). Recently, Wong and coworkers varied the pyrolysis temperature to control the H_c of magnetic ceramics derived from some Fe/Pt polymers.

The polymer structure is also an important parameter that controls the magnetic property. For instance, Abd-El-Aziz *et al.* tuned the H_c of magnetic ceramics derived from homometallic Fe dendrimers by changing the dendrimer generation,[68] while Tang *et al.* realized high M_s by using hyperbranched polymers in place of linear polymers.[80] Also, the ability to self-assemble into nanostructured morphologies, which confines the nanoparticles, is critical to the magnetic property.[81–83] As an example, in contrast with a Co homopolymer that yields superparamagnetic Co nanoparticles after heat treatment, a block copolymer with a Co block self-assembles into nanostructures that confine the Co nanoparticles, inducing ferromagnetic behavior (Figure 6.22).[82] This finding reinforces the importance of structure on the magnetic property because the homopolymer with higher Co concentration (26%) is superparamagnetic with a H_c of zero, whereas the block copolymer with lower Co concentration (14%) is ferromagnetic with a H_c of 200 Oe (Figure 6.22).

6.4.2 As Memristors

Resistors, capacitors, and inductors are the known elements of electrical circuits. In 1976, Leon Chua hypothesized the existence of a fourth element, which he named "memristor" from "memory resistor".[84–88] A memristor is a passive two-terminal nonvolatile memory device whose resistance depends on its history of electrical charges. Thus, until new electrical charges flow through it, a memristor remembers its previous resistance, so conceptually, it stores electrical data. Memristors are a type of two-terminal nonvolatile resistive random-access memory (RRAM) device.[86] This emerging technology, which could replace flash memory, stores data by stimulating structural changes in the active material to switch the resistance.

Many materials including metal-containing macromolecules[89–96] exhibit this resistive switching effect as exemplified by the storage of electrical information in a single crystal of a Rb MOF by Grzybowski and coworkers.[89] Although Rb is not a transition metal, the concept is based on entrapping and transporting ions, Rb^+ in this case, within the porous structure of the MOF when sandwiched between Ag electrical contacts (Figure 6.23). The cyclic voltammetry (CV) of the fabricated Ag/Rb MOF/Ag device is typical of memristors. Specifically, the CV scans feature with pinched hysteresis, and exhibit high and low conductivity (Figure 6.23). The resistive effect is ionic in origin because increasing the concentration of Rb^+, or the presence of moisture, which enhances the dissociation of RbOH to Rb^+ and OH^-,

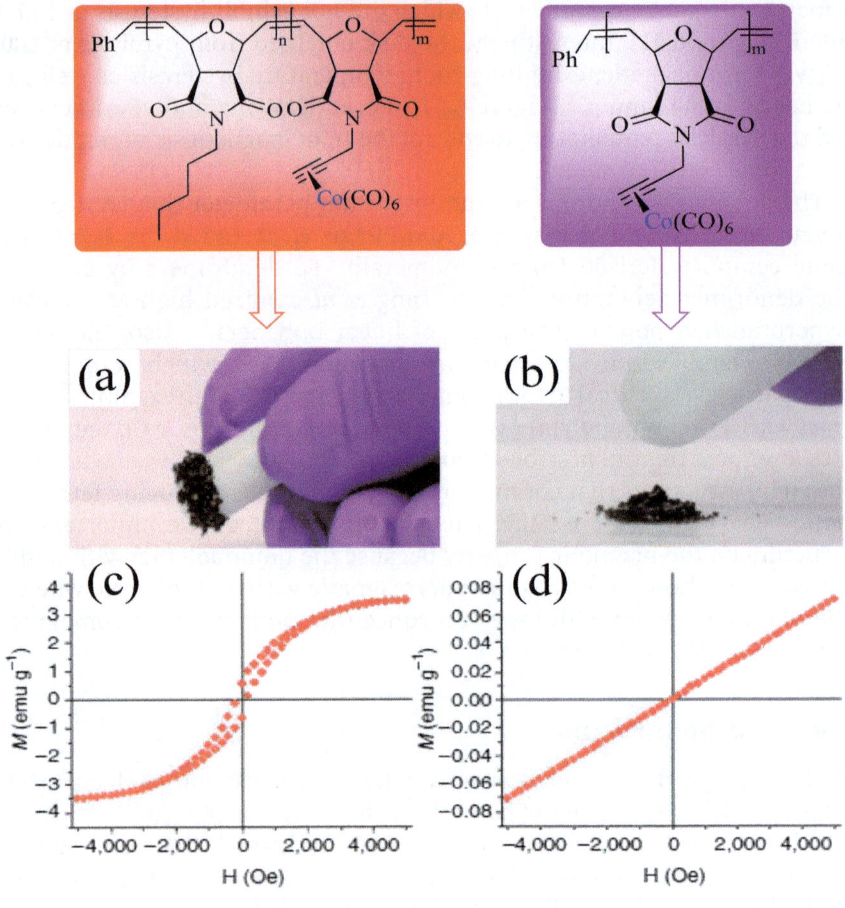

Figure 6.22 Magnetic property block copolymer and homopolymer. (a) and (b) Response of thermally annealed block copolymer and homopolymer to a magnetic stir rod. (c) and (d) Magnetization curves of thermally annealed block copolymer and homopolymer. Reprinted by permission from Macmillan Publishers Ltd: *Nature Communications* (ref. 82), copyright (2011).

increases memristance (Figure 6.23). In this case, the MOF structure is integral to the ionic resistive effect because the sub-nanometer channels of the MOF ensure the selective transport of ions. This MOF-based memristor writes, reads, and erases electrical data. To write an "on" state, Grzybowski and coworkers scanned from 0 V to 10 V, oxidizing the right electrode and reducing the left electrode. Applying a positive voltage "reads" the stored electrical data, which correspond to the flow of large Faradaic current to oxidize the left electrode. Conversely, scanning from 0 V to 10 V writes the "off" state with the reduction of the right electrode

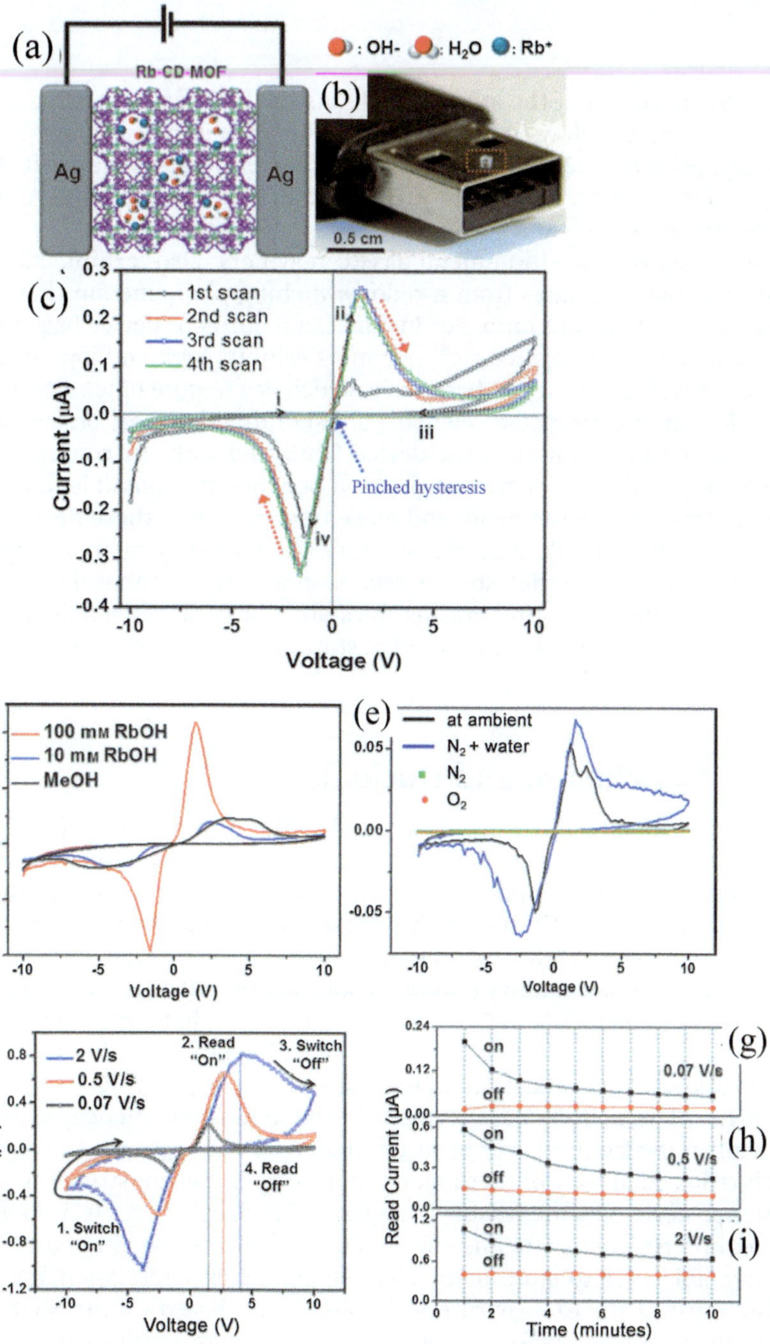

Figure 6.23 (a) Fabricated MOF-based memory device, (b) MOF crystal (red circle) on a flash memory stick, (c) CV scans showing high and low conductivities as well as pinched hysteresis, (d) effect of RbOH concentration on memristance, (e) effect of moisture on memristance, (f) read, write, and erase capabilities of the MOF memory device, (g–i) volatility of the MOF memory device. Reproduced from S. M. Yoon, S. C. Warren and B. A. Grzybowski, Storage of Electrical Information in Metal–Organic-Framework Memristors, *Angewandte Chemie, International Edition*,[89] John Wiley and Sons, © 2014 WILEY-VCH Verlag GmbH & Co. KGaA, Weinheim.

and the oxidation of the left electrode. As the left electrode is oxidized, no Faradaic current flows when the "off" state is read by applying a positive voltage (Figure 6.23). Whereas the "on" state is volatile because it changes with successive reading, the "off" state remains stable after several readings (Figure 6.23).[89]

Redox-active transition-metal-based polymers also exhibit the memristance that originates from a redox switching phenomenon that enables a large on–off current ratio. For instance, a memristor device based on conjugated or non-conjugated Co^{3+} polymers exhibits high and low conducting states as well as a pinched hysteresis, which is a feature of all memristors.[90] The low conducting state, which corresponds to the off or leakage current, is 100 times lower in the device fabricated with the conjugated polymer compared with the non-conjugated polymer. It is possible to reversibly and permanently write, read, and erase information in these devices as evidenced by their random-access memory and read-only memory capacities. For instance, the "write" state is read several times without decay, and the process is repeatable after erasing the state. The conjugated polymer gives a more robust device, which is better at storing information than the non-conjugated polymer.[90]

6.5 Conclusion and Outlook

Transition-metal-containing macromolecules exhibit some fundamental properties such as magnetism, redox activity, and chelation chemistry that are relevant in many applications. For instance, their magnetic and redox properties are investigated in the storage of magnetic and electrical data, respectively. Their ability to interact with several ligands remains the driving force in the construction of porous coordination polymers or MOFs, which are frontline materials in CO_2 capture as well as in heterogeneous catalysis. Indeed, these macromolecules hold a lot of promise in the development of heterogeneous catalysts, data storage platforms, and CO_2 capture technologies. The exciting advancement in the field is commendable, but gaps remain at the frontiers. For instance, materials scientists need to exploit the rich redox chemistry of these macromolecules in the construction of memristance devices. A memristor is an emerging data storage technology that relies on a resistive switching effect, and the ability of transition-metal-based macromolecules to store many charges may offer some landmark results. Also, most of the CO_2 capture technologies are based on the MOF lattice, with the dendritic scaffold under-explored. Through a well-thought-out synthetic strategy, the nano-cavities in dendrimers can be designed to capture CO_2, while the core, branches, and termini are constructed into catalytic centers to promote the conversion of CO_2 to value-added products. With an unrelenting effort, the unique chemistry of transition metals can be explored at the macromolecular scale to solve some of the problems of our time and those of the future.

References

1. A. S. Abd-El-Aziz, C. Agatemor and N. Etkin, in *Functional Metallosupramolecular Materials*, ed. J. G. Hardy and F. H. Schacher, Royal Society of Chemistry, 2015, pp.87–119.
2. G. R. Whittell, M. D. Hager, U. S. Schubert and I. Manners, *Nat. Mater.*, 2011, **10**, 176.
3. A. S. Abd-El-Aziz, C. Agatemor and N. Etkin, *Macromol. Rapid Commun.*, 2014, **35**, 513.
4. R. Valenzuela, *Magnetic Ceramics*, Cambridge University Press, 2005.
5. D. W. Keith, *Science*, 2009, **325**, 1654.
6. P. Williamson, *Nature*, 2016, **530**, 153.
7. R. S. Haszeldine, *Science*, 2009, **325**, 1647.
8. C. Lastoskie, *Science*, 2010, **330**, 595.
9. M. Huber and R. Knutti, *Nat. Geosci.*, 2012, **5**, 31.
10. National Research Council, *Advancing the Science of Climate Change*, The National Academies Press, Washington, DC, United States, 2010.
11. R. Wood, *Nature*, 2008, **453**(43), 45.
12. United States Environmental Protection Agency, *Climate Change Indicators in the United States*, EPA-430-R-16-004, United States Environmental Protection Agency, Washington DC, United States, 2016.
13. United States Environmental Protection Agency, *Inventory of U.S. Greenhouse Gas Emissions and Sinks 1990–2015*, EPA-430-P-17-001, United States Environmental Protection Agency, Washington DC, United States, 2017.
14. United States Environmental Protection Agency, April 2017, accessed July 2017, https://www.epa.gov/ghgemissions/overview-greenhouse-gases#carbon-dioxide.
15. R. Vaidhyanathan, S. S. Iremonger, G. K. Shimizu, P. G. Boyd, S. Alavi and T. K. Woo, *Science*, 2010, **330**, 650.
16. Z. Yao, Y. Chen, L. Liu, X. Wu, S. Xiong, Z. Zhang and S. Xiang, *ChemPlusChem*, 2016, **81**, 850.
17. X. Hu, Q. Gong, R. Zhong, X. Wang, C. Qin, H. Wang, J. Li, K. Shao and Z. Su, *Chem.–Eur. J.*, 2015, **21**, 7238.
18. R. Vaidhyanathan, S. S. Iremonger, K. W. Dawson and G. K. Shimizu, *Chem. Commun.*, 2009, 5230.
19. H. Choi and M. P. Suh, *Angew. Chem., Int. Ed.*, 2009, **48**, 6865.
20. O. Shekhah, Y. Belmabkhout, Z. Chen, V. Guillerm, A. Cairns, K. Adil and M. Eddaoudi, *Nat. Commun.*, 2014, **5**, 4228.
21. T. Li, J. E. Sullivan and N. L. Rosi, *J. Am. Chem. Soc.*, 2013, **135**, 9984.
22. A. R. Millward and O. M. Yaghi, *J. Am. Chem. Soc.*, 2005, **127**, 17998.
23. S. Nandi, P. De Luna, T. D. Daff, J. Rother, M. Liu, W. Buchanan, A. I. Hawari, T. K. Woo and R. Vaidhyanathan, *Sci. Adv.*, 2015, **1**, e1500421.
24. A. M. Fracaroli, H. Furukawa, M. Suzuki, M. Dodd, S. Okajima, F. Gándara, J. A. Reimer and O. M. Yaghi, *J. Am. Chem. Soc.*, 2014, **136**, 8863.
25. W. Qiao, H. Xu, P. Cheng and B. Zhao, *Cryst. Growth Des.*, 2017, **17**, 3128.

26. P. Nugent, Y. Belmabkhout, S. D. Burd, A. J. Cairns, R. Luebke, K. Forrest, T. Pham, S. Ma, B. Space, L. Wojtas, M. Eddaoudi and M. J. Zaworotko, *Nature*, 2013, **495**, 80.
27. M. Bosch, M. Zhang and H. Zhou, *Adv. Chem.*, 2014, **2014**, 1.
28. M. Dinca, A. F. Yu and J. R. Long, *J. Am. Chem. Soc.*, 2006, **128**, 8904.
29. J. M. Taylor, A. H. Mahmoudkhani and G. K. Shimizu, *Angew. Chem., Int. Ed.*, 2007, **46**, 795.
30. Y. Bae, D. Dubbeldam, A. Nelson, K. S. Walton, J. T. Hupp and R. Q. Snurr, *Chem. Mater.*, 2009, **21**, 4768.
31. Y. Xie, T. T. Wang, X. H. Liu, K. Zou and W. Q. Deng, *Nat. Commun.*, 2013, **4**, 1960.
32. W. Clegg, R. W. Harrington, M. North and R. Pasquale, *Chem.–Eur. J.*, 2010, **16**, 6828.
33. M. North and R. Pasquale, *Angew. Chem.*, 2009, **121**, 2990.
34. J. Melèndez, M. North and R. Pasquale, *Eur. J. Inorg. Chem.*, 2007, **2007**, 3323.
35. M. North, R. Pasquale and C. Young, *Green Chem.*, 2010, **12**, 1514.
36. C. Beattie, M. North, P. Villuendas and C. Young, *J. Org. Chem.*, 2013, **78**, 419.
37. M. North, P. Villuendas and C. Young, *Tetrahedron Lett.*, 2012, **53**, 2736.
38. J. Meléndez, M. North, P. Villuendas and C. Young, *Dalton Trans.*, 2011, **40**, 3885.
39. M. Cozzolino, K. Press, M. Mazzeo and M. Lamberti, *ChemCatChem*, 2016, **8**, 455.
40. S. Hammes-Schiffer, *Acc. Chem. Res.*, 2017, **50**, 561.
41. Y. Xia and X. Yang, *Acc. Chem. Res.*, 2017, **50**, 450.
42. C. M. Friend and B. Xu, *Acc. Chem. Res.*, 2017, **50**, 517.
43. D. Wang, D. Denux, J. Ruiz and D. Astruc, *Adv. Synth. Catal.*, 2013, **355**, 129.
44. E. Mieczyńska, T. Borkowski, M. Cypryk, P. Pospiech and A. Trzeciak, *Appl. Catal., A*, 2014, **470**, 24.
45. A. S. Roy, J. Mondal, B. Banerjee, P. Mondal, A. Bhaumik and S. M. Islam, *Appl. Catal., A*, 2014, **469**, 320.
46. Z. Zhou, C. He, L. Yang, Y. Wang, T. Liu and C. Duan, *ACS Catal.*, 2017, **7**, 2248.
47. M. Gupta, D. De, S. Pal, T. K. Pal and K. Tomar, *Dalton Trans.*, 2017, **46**, 7619.
48. P. Servin, R. Laurent, L. Gonsalvi, M. Tristany, M. Peruzzini, J. Majoral and A. Caminade, *Dalton Trans.*, 2009, 4432.
49. D. Méry and D. Astruc, *J. Mol. Catal. A: Chem.*, 2005, **227**, 1.
50. S. Gatard, S. Nlate, E. Cloutet, G. Bravic, J. Blais and D. Astruc, *Angew. Chem., Int. Ed.*, 2003, **42**, 452.
51. G. E. Oosterom, J. N. Reek, P. C. Kamer and P. W. van Leeuwen, *Angew. Chem., Int. Ed.*, 2001, **40**, 1828.
52. D. Wang, C. Deraedt, J. Ruiz and D. Astruc, *Acc. Chem. Res.*, 2015, **48**, 1871.

53. Y. Ribourdouille, G. D. Engel, M. Richard-Plouet and L. H. Gade, *Chem. Commun.*, 2003, 1228.
54. Z. Wang, G. Deng, Y. Li, Y. He, W. Tang and Q. Fan, *Org. Lett.*, 2007, **9**, 1243.
55. B. Yi, Q. Fan, G. Deng, Y. Li, L. Qiu and A. S. Chan, *Org. Lett.*, 2004, **6**, 1361.
56. A. S. Abd-El-Aziz, D. A. Armstrong, S. Bernardin and H. M. Hutton, *Can. J. Chem.*, 1996, **74**, 2073.
57. A. S. Abd-El-Aziz, A. Piorko, C. C. Lee and R. G. Sutherland, *Can. J. Chem.*, 1989, **67**, 1618.
58. A. S. Abd-El-Aziz, W. Boraie and N. T. Gavel, *Organometallics*, 1999, **18**, 1562.
59. R. G. Sutherland, C. Zhang, A. Piórko and C. C. Lee, *Can. J. Chem.*, 1989, **67**, 137.
60. P. Anbarasan, H. Neumann and M. Beller, *Chem.–Eur. J.*, 2010, **16**, 4725.
61. P. Anbarasan, H. Neumann and M. Beller, *Chem.–Eur. J.*, 2011, **17**, 4217.
62. A. V. Ushkov and V. V. Grushin, *J. Am. Chem. Soc.*, 2011, **133**, 10999.
63. J. T. Reeves, C. A. Malapit, F. G. Buono, K. P. Sidhu, M. A. Marsini, C. A. Sader, K. R. Fandrick, C. A. Busacca and C. H. Senanayake, *J. Am. Chem. Soc.*, 2015, **137**, 9481.
64. J. Magano and J. R. Dunetz, *Chem. Rev.*, 2011, **111**, 2177.
65. M. Hilbert and P. Lopez, *Science*, 2011, **332**, 60.
66. M. Castells, *End of Millennium: The Information Age: Economy, Society, and Culture*, John Wiley & Sons, 2010, vol. III.
67. G. M. Church, Y. Gao and S. Kosuri, *Science*, 2012, **337**, 1628.
68. A. S. Abd-El-Aziz, C. Agatemor, N. Etkin and R. Bissessur, *J. Mater. Chem. C*, 2017, **5**, 2268.
69. A. Berenbaum, M. Ginzburg-Margau, N. Coombs, A. J. Lough, A. Safa-Sefat, J. E. Greedan, G. A. Ozin and I. Manners, *Adv. Mater.*, 2003, **15**, 51.
70. Q. Dong, W. Qu, W. Liang, F. Tai, K. Guo, C.-W. Leung, Y. H. Lo and W.-Y. Wong, *J. Mater. Chem. C*, 2016, **4**, 5010.
71. K. Liu, C.-L. Ho, S. Aouba, Y. Zhao, Z. Lu, S. Petrov, N. Coombs, P. Dube, H. E. Ruda, W.-Y. Wong and I. Manners, *Angew. Chem., Int. Ed.*, 2008, **47**, 1255.
72. Q. Dong, G. Li, C.-L. Ho, M. Faisal, C.-W. Leung, P. W. Pong, K. Liu, B. Z. Tang, I. Manners and W.-Y. Wong, *Adv. Mater.*, 2012, **24**, 1034.
73. Q. Dong, G. Li, H. Wang, P. W. Pong, C. Leung, I. Manners, C.-L. Ho, H. Li and W.-Y. Wong, *J. Mater. Chem. C*, 2015, **3**, 734.
74. Q. Dong, W. Qu, W. Liang, K. Guo, H. Xue, Y. Guo, Z. Meng, C.-L. Ho, C.-W. Leung and W.-Y. Wong, *Nanoscale*, 2016, **8**, 7068.
75. T. Gong, Z. Gao, W. Bian, F. Tai, W. Liang, W. Liang, Q. Dong and C. Dong, *J. Organomet. Chem.*, 2016, **819**, 237.
76. M. J. MacLachlan, M. Ginzburg, N. Coombs, T. W. Coyle, N. P. Raju, J. E. Greedan, G. A. Ozin and I. Manners, *Science*, 2000, **287**, 1460.
77. M. Ginzburg, M. J. MacLachlan, S. M. Yang, N. Coombs, T. W. Coyle, N. P. Raju, J. E. Greedan, R. H. Herber, G. A. Ozin and I. Manners, *J. Am. Chem. Soc.*, 2002, **124**, 2625.

78. B. Jiang, W. L. Hom, X. Chen, P. Yu, L. C. Pavelka, K. Kisslinger, J. B. Parise, S. R. Bhatia and R. B. Grubbs, *J. Am. Chem. Soc.*, 2016, **138**, 4616.
79. B. Jiang, D. Nykypanchuk, M. K. Endoh, X. Chen, B. Qian, K. Kisslinger, T. Koga, J. B. Parise and R. B. Grubbs, *Macromolecules*, 2016, **49**, 853.
80. J. Shi, C. J. Jim, F. Mahtab, J. Liu, J. W. Lam, H. H. Sung, I. D. Williams, Y. Dong and B. Z. Tang, *Macromolecules*, 2010, **43**, 680.
81. Y. Zha, H. D. Thaker, R. R. Maddikeri, S. P. Gido, M. T. Tuominen and G. N. Tew, *J. Am. Chem. Soc.*, 2012, **134**, 14534.
82. Z. M. Al-Badri, R. R. Maddikeri, Y. Zha, H. D. Thaker, P. Dobriyal, R. Shunmugam, T. P. Russell and G. N. Tew, *Nat. Commun.*, 2011, **2**, 482.
83. J. Zhang, Y. Yan, J. Chen, W. M. Chance, J. Hayat, Z. Gai and C. Tang, *Chem. Mater.*, 2014, **26**, 3185.
84. J. M. Tour and T. He, *Nature*, 2008, **453**, 42.
85. D. B. Strukov, G. S. Snider, D. R. Stewart and R. S. Williams, *Nature*, 2008, **453**, 80.
86. L. Chua, *Appl. Phys. A*, 2011, **102**, 765.
87. R. Waser, R. Dittmann, G. Staikov and K. Szot, *Adv. Mater.*, 2009, **21**, 2632.
88. S. Vongehr and X. Meng, *Sci. Rep.*, 2015, **5**, 11657.
89. S. M. Yoon, S. C. Warren and B. A. Grzybowski, *Angew. Chem., Int. Ed.*, 2014, **53**, 4437.
90. A. Bandyopadhyay, S. Sahu and M. Higuchi, *J. Am. Chem. Soc.*, 2011, **133**, 1168.
91. Y. Chen, G. Liu, C. Wang, W. Zhang, R. Li and L. Wang, *Mater. Horiz.*, 2014, **1**, 489.
92. Z. Wang, D. Nminibapiel, P. Shrestha, J. Liu, W. Guo, P. G. Weidler, H. Baumgart, C. Wöll and E. Redel, *ChemNanoMat*, 2016, **2**, 67.
93. P. Cheng, K. Sun and Y. H. Hu, *Nano Lett.*, 2015, **16**, 572.
94. S. D. Worrall, M. A. Bissett, W. Hirunpinyopas, M. P. Attfield and R. A. Dryfe, *J. Mater. Chem. C*, 2016, **4**, 8687.
95. L. Pan, Z. Ji, X. Yi, X. Zhu, X. Chen, J. Shang, G. Liu and R. Li, *Adv. Funct. Mater.*, 2015, **25**, 2677.
96. B. Cui, Z. Mao, Y. Chen, Y. Zhong, G. Yu, C. Zhan and J. Yao, *Chem. Sci.*, 2015, **6**, 1308.

Subject Index